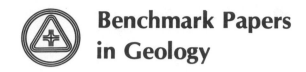

Benchmark Papers
in Geology

Series Editor: Rhodes W. Fairbridge
Columbia University

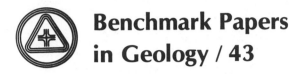

**Benchmark Papers
in Geology / 43**

A BENCHMARK® Book Series

ANCIENT CONTINENTAL
DEPOSITS

Edited by

Franklyn B. Van Houten
Princeton University

Dowden, Hutchinson
& Ross, Inc.

STROUDSBURG, PENNSYLVANIA

Copyright © 1977 by **Dowden, Hutchinson & Ross, Inc.**
Benchmark Papers in Geology, Volume 43
Library of Congress Catalog Card Number: 77-5623
ISBN: 0-87933-287-5

79   78   77   1   2   3   4   5
Manufactured in the United States of America.

LIBRARY OF CONGRESS CATALOGING IN PUBLICATION DATA
Main entry under title:
Ancient continental deposits.
   (Benchmark papers in geology ; 43)
   Includes index.
   1. Rocks, Sedimentary—Addresses, essays, lec-
tures. I. Van Houten, Franklyn Bosworth, 1914–
QE471.A59   551.3'04   77-5623
ISBN 0-87933-287-5

Exclusive Distributor: **Halsted Press**
A Division of John Wiley & Sons, Inc.
ISBN: 0-470-99283-2

SERIES EDITOR'S FOREWORD

The philosophy behind the "Benchmark Papers in Geology" is one of collection, sifting, and rediffusion. Scientific literature today is so vast, so dispersed, and, in the case of old papers, so inaccessible for readers not in the immediate neighborhood of major libraries that much valuable information has been ignored by default. It has become just so difficult, or so time consuming, to search out the key papers in any basic area of research that one can hardly blame a busy man for skimping on some of his "homework."

This series of volumes has been devised, therefore, to make a practical contribution to this critical problem. The geologist, perhaps even more than any other scientist, often suffers from twin difficulties—isolation from central library resources and immensely diffused sources of material. New colleges and industrial libraries simply cannot afford to purchase complete runs of all the world's earth science literature. Specialists simply cannot locate reprints or copies of all their principal reference materials. So it is that we are now making a concerted effort to gather into single volumes the critical material needed to reconstruct the background of any and every major topic of our discipline.

We are interpreting "geology" in its broadest sense: the fundamental science of the planet Earth, its materials, its history, and its dynamics. Because of training and experience in "earthy" materials, we also take in astrogeology, the corresponding aspect of the planetary sciences. Besides the classical core disciplines such as mineralogy, petrology, structure, geomorphology, paleontology, and stratigraphy, we embrace the newer fields of geophysics and geochemistry, applied also to oceanography, geochronology, and paleoecology. We recognize the work of the mining geologists, the petroleum geologists, the hydrologists, the engineering and environmental geologists. Each specialist needs his working library. We are endeavoring to make his task a little easier.

Each volume in the series contains in Introduction prepared by a specialist (the volume editor)—a "state of the art" opening or a summary of the object and content of the volume. The articles, usually some twenty to fifty reproduced either in their entirety or in significant extracts, are selected in an attempt to cover the field, from the key papers of the last century to fairly recent work. Where the original works are in foreign languages, we have endeavored to locate or commission translations. Geologists, because of their global subject, are often acutely aware of the oneness of our world. The selections cannot, therefore, be restricted to any one country, and whenever possible an attempt is made to scan the world literature.

To each article, or group of kindred articles, some sort of "highlight commentary" is usually supplied by the volume editor. This commentary should serve to bring that article into historical perspective and to emphasize its particular role in the growth of the field. References, or citations, wherever possible, will be reproduced in their entirety—for by this means the observant reader can assess the background material available to that particular author, or, if he wishes, he, too, can double check the earlier sources.

A "benchmark," in surveyor's terminology, is an established point on the ground, recorded on our maps. It is usually anything that is a vantage point, from a modest hill to a mountain peak. From the historical viewpoint, these benchmarks are the bricks of our scientific edifice.

RHODES W. FAIRBRIDGE

PREFACE

While considerable current interest is now trained upon sediments in ocean basins, we are here concerned with those ancient ones that never reached the open sea. How and where were they trapped? What processes operated on them in the continental realm? How does a geologist attempt to read the history recorded in them? This Benchmark volume is devoted to studies that answer these questions—papers that afford insights into fruitful methods of analysis and modes of interpretation of ancient continental (nonmarine to paralic[1]) detrital deposits. In addition, the volume focuses on the more recent directions of investigation rather than on historical development, and it focuses on analysis of rocks rather than on processes that produced them.

My principal objective in selecting the papers was to provide a reasonable representation of major nonmarine to paralic environments recorded in the more common kinds of sedimentary rocks. Yet I did not want to neglect altogether the diagnostic, though relatively minor, facies such as eolian, lacustrine, and soils. Even so, some significant aspects have necessarily been omitted. These include such topics as the inheritance and transformation of clay minerals in detrital deposits (Millot 1970, pp. 135–168, 302–322); the origin of pigment in nonmarine red beds (Van Houten 1973); modes of deposition of volcaniclastic deposits (Pettijohn 1975, pp. 299–315; Schmincke 1974); the concentration of uranium in nonmarine conglomerates and sandstones (Land et al. 1970, pp. 163–167; Fischer 1974); and diagenetic overprinting on detrital deposits (Dapples 1967; Müller 1967; Fairbridge 1975).

Inasmuch as I focused on those studies that elaborate productive techniques and methods of analysis, articles that consist mainly of paleogeographic reconstructions were excluded. Moreover, it was necessary to select short contributions in order to assemble an adequate variety of examples. Nevertheless, most of the papers included have been shortened by omitting sections dealing with introductory and general matters, regional depositional history, and other subjects less relevant to the theme of this volume. Sections concerned with

1. In an early discussion of sedimentation and tectonics, Tercier (1940) recognized the role of detrital marine-nonmarine transition assemblages of the paralic realm and stressed the development of thick sequences in this environment. Today the term *paralic* is used more generally for intertonguing shallow marine and marginal nonmarine deposits of coastal environments.

analysis and modes of reasoning were favored over those presenting descriptive detail. Some of my colleagues may well feel that a favorite article or area of interest has been slighted. I apologize for the oversight. Given the constraints, many excellent papers had to be left out, but I have called attention to some of the slighted ones in the editor's comments on the papers selected.

Pertinent suggestions and criticisms that Rhodes W. Fairbridge, Series Editor, provided have improved this volume. Peter F. Friend, Andrew D. Miall, and James R. Steidtmann helped with the selection of papers. To each of these geologists I am most grateful. I also acknowledge the generosity of those authors who provided copies of their papers for reproduction in the book.

FRANKLYN B. VAN HOUTEN

REFERENCES

Dapples, E. C. 1967. Diagenesis of sandstones. In *Diagenesis in Sediments,* eds. G. Larsen and G. V. Chilingar. Amsterdam: Elsevier Publ. Co., pp. 91–126.

Fairbridge, R. W. 1975. Epidiagenetic silicification. *Internat. Congr. Sedimentology, 9th, Theme 7,* pp. 49–54.

Fischer, R. P. 1974. Exploration guides to new uranium districts and belts. *Econ. Geology* **69:**362–376.

Lang, A. H., et al. 1970. Economic Minerals of the Canadian Shield. In *Geology and Economic Minerals of Canada,* J. R. W. Douglas, ed. Canada Geol. Survey Econ. Geology Rept. No. 1, pp. 163–167.

Millot, G. 1970. *Geology of Clays.* New York: Springer-Verlag, pp. 1–429.

Müller, G. 1967. Diagenesis in argillaceous sediments. In *Diagenesis in Sediments,* eds. G. Larsen and G. V. Chilingar. Amsterdam: Elsevier Publ. Co., pp. 127–178.

Pettijohn, F. J. 1975. *Sedimentary Rocks.* New York: Harper and Row, pp. 1–628.

Schmincke, H. V. 1974. Pyroclastic rocks. In *Sediments and Sedimentary Rocks, 1,* ed. H. Füchtbauer. New York: Halsted Press Div., pp. 160–189.

Tercier, J. 1940. Depots marins actuels et series geologique. *Eclogae Geol. Helvetiae* **32:**47–100.

Van Houten, F. B. 1973. Origin of red beds, a review—1961–1972. *Annual Rev. Earth and Planetary Sci., 1,* pp. 39–61.

CONTENTS

Contents

PART II: DELTAIC AND PARALIC COMPLEXES

CONTENTS BY AUTHOR

INTRODUCTION

Nonmarine to paralic detrital deposits are characterized by general features that direct our interests and can restrict our interpretations. These traits include a wide range of grain size and sedimentary structures as well as rapid local facies changes. The latter, together with a common paucity of fossils, makes it difficult to establish a rigorous stratigraphic framework for many of the deposits. Because these sediments accumulated relatively near their sources, provenance exerted considerable control on their mineral composition, and their tectonic setting was a basic factor in determining the sedimentary product. Each of these characteristics pervades the studies selected for this volume.

Early in the history of geology, attempts to interpret the origin of ancient continental deposits showed that detailed observations produced useful information. For example, Sorby (1859, 1880) pioneered the analysis of bedding structures in sandstone as a clue to the conditions of deposition, and he applied quantitative methods to the study of those structures (1908). Yet, "the monumental advances Sorby made went ignored until the thirties of the present century" (Allen 1963, p. 223; see also Folk 1965). Meanwhile, there was some interest in broad regional reconstructions derived mainly from stratigraphic studies like those of Walther (1893–1894) and Barrell (1909) or in general descriptions of modern environments and their products, as summarized by Twenhofel (1939). During those years identifying clasts and minerals (Mackie 1896; Gilligan 1920, Brammall 1928) and measuring grain sizes (Udden 1914) became the principal guides to reconstructing source areas and transport distance, and these interests dominated sedimentology for several decades. More recently considerable effort was spent on measuring sedimentary structures to determine dispersal patterns (Potter and Pettijohn 1963). The results led to a useful classification of paleocurrent models (Selley 1968). Some excellent recent studies have focused on sedimentary petrography, such as Füchtbauer's (1967) analysis of the Cenozoic sandstones in the Molasse north of the Alps, but this area of interest is not specifically represented among the papers reprinted in this volume. For a more complete review of the historical development of sedimentology, consult Pettijohn (1975, pp. 2–8).

As detailed knowledge of modern environments became more available (Allen 1965; Reineck and Singh 1973), increased interest developed in the ancient ones. Achievements of that concern have recently been summarized in *Recognition of Ancient Environments* (Rigby and Hamblin, eds., 1972), with which this Benchmark volume has affinities.

Productive analysis of ancient nonmarine to paralic sedimentary rocks requires at least a simple sorting out of their major depositional environments and associated facies (Crosby 1972; Pettijohn 1975, pp. 543–545). Fairbridge (1958) has discussed the relevant concept of consanguineous associations of sediments, interrelated by ancestry, environment, and evolution. A common classification of facies and environments consists of:
Piedmont-valley flat complex
    Alluvial-fan facies
        debris flow and braided-stream deposits
    Alluvial-plain facies
        meandering-stream and flood-plain deposits; swamp and lake
        and dune-field deposits
Deltaic complex
    Alluvial-plain, distributary-channel, and delta-fringe deposits
Interdeltaic paralic complex
    Tidal-flat deposits; beach and barrier-island deposits

Papers assembled here are arranged in sections progressing from proximal fans to distal paralic environments. In addition to affording a survey of sediments trapped along the system from inland source to continental margin, this range of facies permits a comparison of sand-bodies of rivers, distributary channels, beaches, and barrier bars, which commonly are difficult to differentiate.

Three facets of interest dominate much of the current research on detrital sediments and sedimentary rocks, and these facets are emphasized in the papers selected. *First,* considerable effort has been devoted to the hydrodynamics of sedimentation, to flow regimes and associated bedforms as observed in flumes and modern streams (Allen 1965, 1970; Middleton 1965; Schumm 1972, pp. 303–348; Harms et al. 1975). The lessons learned are currently being applied to ancient stream deposits.

Second, of all the major depositional environments, that of the delta has received the most attention, as recorded in the many papers assembled in useful "delta" volumes (for example, Morgan 1970; Boussard 1975). More recently tidal flats and tidal deposits, both modern and ancient, have been described in detail (Ginsburg 1975). These studies demonstrate a consistent and predictable relation among nonmarine and paralic facies. This in turn has led to the recognition of vertical successions of facies (stratification sequences) as guides to identifying ancient sedimentary environments (Visher 1965; Selley 1970) and to constructing facies models (Blatt et al. 1972, pp. 185–206; Pettijohn 1975, pp. 545– 557; Walker 1976) essentially by applying the generality Walther (1894; Middleton 1973) stated long ago that sediments accumulating beside one another in space succeed one another in vertical profile. More specifically, increased knowledge of detrital deposits has permitted a sorting out of major kinds of sand-bodies according to their environment of deposition (Potter 1967; LeBlanc 1972; MacKenzie 1972; Busch 1974). Conversely, understanding of environments and their depositional processes aids in predicting sandbody morphology—the size, shape trend, and internal structure of sandstone reservoirs.

Third, as description of ancient nonmarine to paralic deposits achieved adequate detail, the abundant data have become susceptible to statistical analysis. Much of this work in the past was based on grain-size distribution (Krumbein 1934; Pettijohn 1975, pp. 34–52) and cross-bedding measurements (Potter and Pettijohn 1963, pp. 23–113). More recently, alternating and cyclic patterns of stratification sequences as well as other facies associations have been quantified (Duff and Walton 1962; Read and Dean 1967; Demirmen 1972).

Each of these three themes has given a new importance to detailed reporting and analysis of primary sedimentary structures such as those cataloged in useful atlases by Pettijohn and Potter (1964), Gubler et al. (1966), and Conybeare and Crook (1968), and it has reemphasized the role of field geology. Coincidently, machine-readable mark-sense forms for recording field data systematically and rapidly have been devised (Alexander-Marrack et al. 1970; Friend et al. 1976, pp. 13–38). Of course, all such studies must be set in an established stratigraphic framework and depend on fossils for dating the rocks and indicating

3

biological aspects of paleoenvironments. Although these matters are not the principal concerns of the papers selected, they do receive increasing attention toward the paralic realm.

Regardless of changing interests and new questions, studies of non-marine to paralic deposits will necessarily focus on particular episodes of earth history such as the Devonian and Carboniferous Periods when appropriate conditions of deposition prevailed and on those geologic provinces where relevant rocks are well displayed. In spite of a real effort to make this collection of papers as cosmopolitan as possible, both geologically and geographically, it has an unavoidable national and stratigraphic bias.

REFERENCES

Alexander-Marrack, P. D.; Friend, P. F.; and Yeats, A. K. 1970. Mark sensing for recording and analysis of sedimentological data. In *Data Processing in Biology and Geology,* ed. J. L. Cutbill, *Systematics Assoc. Pub., Spec. Vol* **3.** London: Acad. Press, pp. 1–16.

Allen, J. R. L. 1963. Henry Clifton Sorby and the sedimentary structures of sands and sandstones in relation to flow conditions. *Geol. en Mijnbouw* **42:** 223–228.

——— 1965. A review of the origin and characteristics of Recent alluvial sediments. *Sedimentology* **5:**89–191.

——— 1970. *Physical Processes in Sedimentation.* New York: Am. Elsevier Publ. Co., pp. 1–248.

Barrell, J. 1906. Relative geologic importance of continental, littoral, and marine sedimentation. *Jour. Geology* **14:**316–356.

Blatt, H.; Middleton, G.; and Murray, R. 1972. *Origin of Sedimentary Rocks.* Englewood Cliffs, New Jersey: Prentice-Hall, pp. 1–634.

Brammall, A. 1928. Dartmoor detritals; a study in provenance. *Geologists' Assoc. Proc.* **39:**1–27.

Broussard, M. L. S., ed. 1975. *Deltas, Models for Exploration.* Houston, Texas: Houston Geol. Soc., pp. 1–555.

Busch, D. A. 1974. Stratigraphic traps in sandstones—exploration techniques. *Am. Assoc. Petroleum Geologists Mem. 21,* pp. 1–174.

Conybeare, C. E. B., and Crook, K. A. W. 1968. *Manual of Sedimentary Structures.* Bur. Min. Res., Geol., Geophys., (Canberra, Australia) Bull. 102, pp. 1–327.

Crosby, E. J. 1972. Classification of sedimentary environments. *Soc. Econ. Paleontologists and Mineralogists Spec. Publ. 16,* pp. 4–11.

Demirmen, F. 1972. Mathematical search procedures in facies modeling in sedimentary rocks. In *Mathematical Models of Sedimentary Processes,* ed. D. F. Merriam. New York: Plenum Press, pp. 81–114.

Duff, P. McL. D., and Walton, E. K. 1962. Statistical basis for cyclothems: A quantitative study of the sedimentary successions in the East Pennine coalfield. *Sedimentology* **1:**235–255.

Fairbridge, R. W. 1958. What is a consangineous association? *Jour. Geology* **66:**319–324.

Folk, R. L. 1965. Henry Clifton Sorby (1826–1908), the founder of petrography. *Jour. Geol. Education* **13:**43–47, 93.

Friend, P. F.; Alexander-Marrack, P. D.; Nicholson, J.; and Yeats, A. K. 1976. Devonian sediments of east Greenland, 1., *Medd. Grönland* **206:**1–56.

Füchtbauer, H. 1967. Die sandsteine in der Molasse nördlich der Alpen. *Geol. Rundschau* **56:**266–300.

Gilligan, A. 1919. The petrography of the Millstone Grit of Yorkshire. *Geol. Soc. London Quart. Jour.* **75:**251–294.

Ginsburg, R. N., ed. 1975. *Tidal Deposits.* New York: Springer-Verlag, pp. 1–428.

Gubler, Y., ed. 1975. *Essai de nomenclature et caracterization des principales structures sedimentaires.* Paris: Edit. Technip, pp. 1–291.

Harms, J. C.; Southard, J. B.; Spearing, D. R.; and Walker, R. G. 1975. Depositional environments as interpreted from primary sedimentary structures and stratification sequences. *Soc. Econ. Paleontologists and Mineralogists,* Short Course 2, Dallas, pp. 1–161.

Krumbein, W. C. 1934. Size frequency distribution of sediments. *Jour. Sed. Petrology* **4:**65–77.

LeBlanc, R. J. 1972. Geometry of sandstone reservoir bodies. *Am. Assoc. Petroleum Geologists, Mem. 18,* pp. 133–190.

MacKenzie, D. B. 1972. Primary stratigraphic traps in sandstone. *Am. Assoc. Petroleum Geologists, Mem. 16,* pp. 47–63.

Mackie, W. 1896. The sands and sandstones of eastern Moray. *Edinburgh Geol. Soc. Trans.* **7:**148–172.

Middleton, G. V., ed. 1965. Primary sedimentary structures and their hydrodynamic interpretation. *Soc. Econ. Paleontologists and Mineralogists Spec. Pub. 12,* pp. 1–265.

——— 1973. Johannes Walther's law of the correlation of facies. *Geol. Soc. America Bull.* **84:**979–987.

Morgan, J. P., ed. 1975. Deltaic sedimentation, modern and ancient. *Soc. Econ. Paleontologists and Mineralogists Spec. Pub. 15,* pp. 1–312.

Pettijohn, F. J. 1975. *Sedimentary Rocks.* New York: Harper and Row, pp. 1–628.

———, and Potter, P. E. 1964. *Atlas and Glossary of Primary Sedimentary Structures.* New York: Springer-Verlag, pp. 1–370.

Potter, P. E. 1967. Sandbodies and sedimentary environments: A review. *Am. Assoc. Petroleum Geologists Bull.* **51:**337–365.

———, and Pettijohn, F. J. 1963. *Paleocurrents and Basin Analysis.* New York: Acad. Press, pp. 1–296.

Read, W. A., and Dean, J. M. 1967. A quantitative study of a sequence of coal-bearing cycles in the Namurian of central Scotland, 1. *Sedimentology* **10:**137–159.

Reineck, H. E., and Singh, I. B. 1973. *Depositional Sedimentary Environments.* New York: Springer-Verlag, pp. 1–439.

Rigby, J. K., and Hamblin, W. K., eds. 1972. Recognition of ancient sedimentary environments. *Soc. Econ. Paleontologists and Mineralogists Spec. Pub. 16,* pp. 1–340.

Schumm, S. A. 1972. *River Morphology.* Stroudsburg, Pennsylvania: Dowden, Hutchinson & Ross, pp. 1–421.

Selley, R. C. 1968. A classification of paleocurrent models. *Jour. Geology* **76:**99–110.

———— 1970. *Ancient Sedimentary Environments*. New York: Cornell Univ. Press, pp. 1–237.

Sorby, H. C. 1859. On the structures produced by the current present during the deposition of stratified rocks. *The Geologist* **12:**137–147.

———— 1908. On the application of quantitative methods to the study of the structure and history of rocks. *Geol. Soc. London Quart. Jour.* **64:**171–232.

Twenhofel, W. H. 1939. *Principles of Sedimentation*. New York: McGraw-Hill Book Co., pp. 1–610.

Udden, J. A. 1914. The mechanical composition of clastic sediments. *Geol. Soc. America Bull.* **25:**655–744.

Visher, G. S. 1965. Use of the vertical profile in environmental reconstruction. *Am. Assoc. Petroleum Geologists Bull.* **49:**41–61.

Walker, R. G. 1976. Facies models. 1. General introduction. *Geosci. Canada* **3:** 21–24.

Walther, J. 1893–1894. *Einleitung. in die geologie als historische Wissenschaft*. Jena: Fischer Verlag, 3 vols; pp. 1–1055.

Part I

PIEDMONT-VALLEY FLAT COMPLEX

Editor's Comments
on Papers 1, 2, and 3

ALLUVIAL-FAN AND PIEDMONT-PLAIN FACIES

Although the general morphology of alluvial fans in deserts had long been known and debris flow was early recognized as an important transport agent, detailed information about their deposits became available only recently through papers like that of Blissenbach (1954). These recent reports emphasized the distinctive patterns of deposits in western United States as well as the transport role of both debris flows and braided streams. With this background Bluck (1965, 1967) was able to demonstrate that careful study of conglomerates in Devonian and Triassic sequences of Great Britain discriminates between debris flow and braided-stream facies of fanglomerates. Colin and Walker (1973) have published the first detailed description of an Archean alluvial fan based on a rigorous analysis of a conglomeratic sequence in southern Canada. Williams (1969) identified a pediment beneath a late Precambrian alluvial fan in northwestern Scotland. Bull (1972) has summarized criteria for recognizing ancient alluvial-fan deposits.

Detailed description of proximal fan facies, based mainly on clast composition and distribution in the Cenozoic Swiss Molasse (Matter 1964; Gasser 1968), has led to reliable reconstruction of fans at the foot of the rising Alps. Van Houten (1974) has reviewed the relation between major stages of fanglomerate accumulation in a foredeep and tectonic development of the associated mountain belt. Alluvial fans also formed in fault basins, as in the early Mesozoic rift valleys in eastern North

America (Wessel 1969) and in grabens along transform fracture zones (Crowell 1974).

Most of the ancient alluvial fans, including those described in Papers 1 to 3, were built largely of nonvolcanic debris. Nevertheless, analysis of fanglomerates derived from coeval volcanic piles (Swanson 1966; Van Houten 1976) provides basic information about dispersal mechanisms. In proximal facies, dispersal mechanisms involve debris flows transported by turbulent and laminar viscous flowage (Fisher 1971; Schmincke and Swanson 1967) on very gentle slopes (Rodine and Johnson 1976) where the load accumulates in part as base-surge deposits (Schmincke et al. 1973; Sheridan and Updike 1975) composed of pyroclastic detritus carried at high velocity in ground-hugging, gas-rich clouds.

Many of the more distal fan and piedmont-plain deposits were distributed by braided streams. Knowledge of this dispersal mechanism in the past has been amplified by comparison of modern and ancient examples (Smith 1970), as well as by study of braided-stream sediments in the late Precambrian Torridonian Formation of northwestern Scotland (Selley 1965; Banks 1973), in Carboniferous coal measures in eastern United States (Mrakovich and Coogan 1974) and Wales (Kelling 1968), in a late Jurassic succession in northwestern New Mexico (Campbell 1976), and the Cretaceous deposits of Banks Island, Arctic Canada (Miall 1976), and of braid-plain deposts in Early Cretaceous graben fill in southern England (Allen 1975). The common contorted and disrupted bedding in sandy deposits of this sort are quicksand structures produced by expulsion of water and generated in part by current turbulence (Selley 1969). Interbedded deposits of debris flows and braided streams are well displayed in the Devonian deposits of western Norway (Nilson 1968, 1969) and the late Paleozoic New Red Sandstone of Scotland (Steel 1974).

The three papers selected discuss both the process and product associated with the ancient debris flows and braided streams. Application of current understanding of alluvial-fan facies to an ancient example is particularly well illustrated by Miall's (1970) discussion of Devonian alluvial fans in Arctic Canada (Paper 1). He analyzed dispersal structures, together with the grain size, shape, and composition of clasts, to reconstruct a bajada composed of 11 confluent fans formed largely by subaerial debris flow.

Cant and Walker (1975) recognized eight distinct facies in a Devonian sandstone in Québec (Paper 2). Relations among the facies and their sedimentary structures suggest that channel development of a braided-stream system began by scouring and deposition of intraclast lag. After this, unidirectional dunes migrated down channel, followed by lateral movement of transverse bars, and ending with vertical accretion of rippled silt and mud.

Although fanglomerates commonly record the plane-bed phase of the upper flow regime, they rarely preserve the antidune (standing-wave) phase. Hand et al. (1969), however, found nearly symmetrical antidunes on the surface of fine-grained conglomerate in a Triassic fan in Massachusetts (Paper 3). Their data suggest that this bedform, with upstream-dipping cross-laminae, was produced by very shallow, rapidly flowing flash-flood sheetwash across the fan surface.

With insight and guidance of the sort provided by these papers, coarse deposits of piedmont provinces can now be interpreted and reconstructed with considerable accuracy.

REFERENCES

Allen, P. 1975. Wealden of the Weald: A new model. *Geologists' Assoc. Proc.* **86:**389–436.

Banks, N. L. 1973. Falling-stage feature of a Precambrian braided stream: Criteria for sub-aerial erosion. *Sedimentology* **10:**147–154.

Blissenbach, E. 1954. Geology of alluvial fans in semiarid regions. *Geol. Soc. America Bull.* **65:**175–189.

Bluck, B. J. 1965. The sedimentary history of some Triassic conglomerates in the Vale of Glamorgan, South Wales. *Sedimentology* **4:**225–245.

———— 1967. Deposition of some Upper Old Red Sandstone conglomerates in the Clyde area: A study in the significance of bedding. *Scottish Jour. Geol.* **3:**139–167.

Bull, W. G. 1972. Recognition of alluvial-fan deposits in the stratigraphic record. *Soc. Econ. Paleontologists and Mineralogists Spec. Pub. 16,* pp. 63–83.

Campbell, C. V. 1976. Reservoir geometry of a fluvial sheet sand. *Am. Assoc. Petroleum Geologists Bull.* **60:**1009–1020.

Crowell, J. C. 1974. Sedimentation along the San Andreas Fault, California. *Soc. Econ. Paleontologists and Mineralogists Spec. Pub. 19,* pp. 292–303.

Colin, C. T., and Walker, R. G. 1973. Sedimentology, stratigraphy, and crustal evolution of the Archean greenstone belt near Sioux Lookout, Ontario. *Canadian Jour. Earth Sci.* **10:**817–845.

Fisher, R. V. 1971. Features of coarse-grained, high-concentration fluids and their deposits. *Jour. Sed. Petrology* **41:**916–927.

Gasser, U. 1968. Die innere zone der subalpinen Molasse des Entlebuchs (kt. Luzern), Geologie und Sedimentologie. *Eclogae Geol. Helvetiae* **61:**229–319.

Kelling, G. 1968. Patterns of sedimentation in Rhondda Beds of South Wales. *Am. Assoc. Petroleum Geologists Bull.* **52:**2369–2386.

Matter, A. 1964. Sedimentologische untersuchungen in östlichen Napfgebiet. *Eclogae Geol. Helvetiae* **57:**315–428.

Miall, A. D. 1976. Paleocurrent and paleohydraulic analysis of some vertical profiles through a Cretaceous braided stream deposit, Banks Island, Arctic Canada. *Sedimentology* **23:**459–483.

Mrakovich, J. V., and Coogan, A. H. 1974. Depositional environment of the Sharon Conglomerate Member of the Pottsville Formation in northeastern Ohio. *Jour. Sed. Petrology* **44:**1186–1199.

Nilsen, T. H. 1968. The relationship of sedimentation to tectonics in the Solund Devonian district of southwestern Norway. *Norges Geol. Undersökelse Nr. 259,* pp. 5–108.

—— 1969. Old Red sedimentation in the Buelandet-Vaerlandet Devonian district, western Norway. *Sed. Geology* **3:**35–57.

Rodine, J. D., and Johnson, A. M. 1976. The ability of debris, heavily freighted with coarse clastic materials, to flow on gentle slopes. *Sedimentology* **23:**213–234.

Schmincke, H-U., and Swanson, D. A. 1967. Laminar viscous flowage structures in ash-flow tuffs from Gran Canaria, Canary Islands. *Jour. Geol.* **75:**641–664.

Schmincke, H-U.; Fisher, R. V.; and Waters, A. C. 1973. Antidune and chute and pool structures in the base-surge deposits of the Laacher See area, Germany. *Sedimentology* **20:**553–574.

Selley, R. C. 1965. Diagnostic characteristics of fluviatile sediments in the Torridonian Formation (Precambrian) of northwest Scotland. *Jour. Sed. Petrology* **35:**366–380.

—— 1969. Torridonian alluvium and quicksands. *Scottish Jour. Geol.* **5:**328–346.

Sheridan, M. F., and Updike, R. G. 1975. Sugarloaf Mountain tephra—a Pleistocene rhyolitic deposit of base-surge origin in northern Arizona. *Geol. Soc. America Bull.* **86:**571–581.

Smith, N. D. 1970. The braided stream depositional environment—comparison of the Platte River with some Silurian clastic rocks, north central Appalachians. *Geol. Soc. America Bull* **81:**2993–3014.

Steel, R. J. 1974. New Red Sandstone floodplain and piedmont sedimentation in the Hebridean province, Scotland. *Jour. Sed. Petrology* **44:**336–357.

Swanson, D. A. 1966. Tieton volcano, a Miocene eruptive center in the southern Cascade Mountains, Washington. *Geol. Soc. America Bull.* **77:**1293–1314.

Van Houten, F. B. 1974. Northern Alpine Molasse and similar Cenozoic sequences in southern Europe. *Soc. Econ. Paleontologists and Mineralogists Spec. Pub. 19,* pp. 260–273.

—— 1976. Late Cenozoic volcaniclastic deposits, Andean foredeep, Colombia. *Geol. Soc. America Bull.* **87:**481–495.

Wessel, J. M. 1969. A paleocurrent and petrographic study of the Triassic Mount Toby Conglomerate and Turners Falls Sandstone of north central Massachusetts. M.S. thesis, University of Massachusetts, pp. 1–157.

Williams, G. E. 1969. Characteristics and origin of a Precambrian pediment. *Jour. Geol.* **77:**183–207.

11

1

Reprinted from pp. 556, 560–571 of *Jour. Sed. Petrology* **40**(2):556–571 (1970)

DEVONIAN ALLUVIAL FANS, PRINCE OF WALES ISLAND, ARCTIC CANADA[1]

ANDREW D. MIALL[2]

Geology Department, University of Ottawa, Ottawa, Canada

ABSTRACT

Lower-Middle Devonian continental conglomerates, with minor sandstones, form a belt 150 miles long and up to 10 miles wide in eastern Prince of Wales Island, bordering an uplifted area of Lower Paleozoic and Precambrian rocks (Boothia Uplift). The succession forms part of the Peel Sound Formation; it reaches 1000 feet in thickness and towards the west the coarse clastics intertongue into sandstones, which in turn grade into shales and carbonates. Conglomerate clast types and directional sedimentary structures indicate derivation from the Boothia Uplift.

The conglomerates are poorly sorted and mean grain size lies in the pebble range, comparing closely with modern alluvial fan deposits. Areal variations in clast lithologies are interpreted in terms of dispersal on eleven alluvial fans. Areal variations in grain size and clast roundness are interpreted as inter-fan in origin (gross differences between adjacent fans). Measurements of maximum clast size revealed intra-fan changes, namely decrease with distance of transport.

Clast changes in vertical section are attributed to varying rates of uplift and erosion in the source area. Comparison of grain size and textural characteristics of the conglomerates with those of modern deposits suggests a debris flood mode of origin.

[Editor's Note: Material has been omitted at this point.]

RESULTS
General description

The bulk of the Peel Sound conglomerate succession consists of cobble and boulder conglomerates. Clasts reach a diameter of 5 feet (150 cm) but boulders larger than 20 inches (50 cm) are rare. Bedding is normally faint to absent except where there are interbedded sandstone layers. The latter occur at vertical intervals varying from about 5 to 100 feet (1.5 to 30 m); they average about 2 feet (60 cm) in thickness and are generally lenticular in form.

The conglomerate matrix is sandstone with a similar petrography to that of the interbedded sandstone lenses. Thin section point counts give the following average percentage composition: quartz 25, feldspar 2, weathered grains (limonite stained, probably originally ferromagnesian minerals) 22, carbonate detrital grains and cement 45; plus minor variable amounts of chert, quartzite and shale, totalling approximately 6 percent. According to the classification of Pettijohn (1957) this composition falls in the range of a lithic greywacke. There are no marked changes in sandstone mineralogy from north to south along the Conglomerate Facies belt, whereas the conglomerate clast lithologies vary widely.

Five of the seven principle clast lithologies can be identified as fragments from the Boothia Uplift area (table 1). The westward decrease in grain size away from the uplift and the evidence of paleocurrents supports the deduction that a landmass in the area of the Boothia axis was the source of all the Peel Sound clastic sediment.

The conglomerates have been analyzed for vertical and areal variations in grain size and clast characteristics. Most of the field measurements discussed below were made in stream gorge sections between 200 and 800 feet (60–240 m) above sea level, but since there are no marker horizons in the Conglomerate Facies there is rarely any accurate stratigraphic control of the outcrops. The bedding is, however, virtually horizontal throughout the area studied and it therefore seems likely that most measurements have been made on the same 6–800 feet (180–240 m) of beds.

Directional structures

In figs. 6a to 6c paleocurrent readings have been grouped into small sub-areas and local vector means calculated for each. Although there are few readings a westerly transport direction is indicated, and for the other facies of the Peel Sound Formation the paleocurrent pattern is very similar.

The reasons for the scarcity of sedimentary structures are not clear. As will be discussed below, the lack of a well developed fabric in the conglomerates may be readily explained, but the paucity of structures in the sandstones is less easy to understand. Arctic weathering conditions are characterised by very low rainfall (as distinct from snow fall) and the washing and etching effects associated with running water must, therefore, be of very limited importance. Without this weathering process sedimentary structures may remain hard to detect. Lamination is often obscure or absent in the Peel Sound sandstones, reflecting a lack of grain size sorting and colour banding. The latter may have been removed by post-depositional alteration.

Whole-rock grain size analysis

Mean grain size M_z and standard deviation σ_I have been calculated for seventeen conglomerate samples (table 2). The average mean grain size is -5.7ϕ (5.0 cm) which is in the coarse pebble range. The average standard deviation is 2.64 phi units (very poorly sorted). These values indicate torrential deposition, and together with the high degree of roundness and sphericity (discussed below) and the typically fluvial aspect of the interbedded sandstone lenticles the

evidence suggests a sub-aerial, alluvial fan origin for the Peel Sound conglomerates.

At Bellot Cliff (loc. 31, see fig. 2) vertical variations in grain size were measured but the results reveal no statistically significant trends. The Prescott Island localities in table 2 are arranged in descending order to correspond to increasing distance from the upturned base of the Peel Sound Formation, and this arrangement reveals a marked westward decrease in mean grain size and an unexpected increase in standard deviation. These results will be discussed below in connection with the roundness and sphericity measurements.

Pettijohn (1957, p. 50) and Spencer (1963) have noted that in most clastic sediments certain grain size ranges, particularly silt ($+4$ to $+7\phi$) and coarse sand (-2 to $+1.5\phi$), are scarce, and it was suggested that this may be a reflection of the way in which most rocks break down under the influence of normal subaerial weathering processes. Spencer suggested that all clastic sediments were composed of three 'fundamental populations' which were combined in different proportions according to the mode of transport and deposition. These three populations are gravel (-2 to -3.5ϕ), sand ($+1.5$ to $+4.0\phi$), and clay ($+7$ to $+9\phi$).

TABLE 2.—A) *Conglomerate data, Prescott Island*

loc. no.	cong. M_z	cong. σ_I	Aston R	Aston S	Gneiss R	Gneiss S	Aston R: σ_I	Aston %	Gneiss %
101A	−6.1	2.4	0.70	0.69	0.67	0.70	0.098	31	64
101B	−6.3	2.5	0.74	0.68	—	—	0.085	32	51
100A	−6.6	2.4	0.72	0.64	0.69	0.70	0.101	24	71
100B	−6.8	2.6	0.72	0.68	0.69	0.70	0.069	31	65
105	−6.1	2.3	0.72	0.70	0.67	0.69	0.074	29	70
106	−5.9	2.2	0.72	0.69	0.67	0.69	0.089	26	96
107	−4.8	3.3	0.71	0.66	0.68	0.70	0.064	23	71
110	—	—	0.70	0.68	0.67	0.65	0.079	19	74
108A	−4.8	3.8	0.68	0.73	0.65	0.67	0.093	30	62
108B	−4.8	3.0	0.69	0.70	0.65	0.71	0.069	14	83
109	−4.8	3.4	0.67	0.70	0.66	0.66	0.085	14	82
111	—	—	0.68	0.67	0.67	0.69	0.079	18	80

TABLE 2.—B) *Conglomerate data, Bellot Cliff*

loc.	cong. M_z	cong. σ_I	Aston R	Aston S	Aston R: σ_I	Aston %	Gneiss %
31A	−6.7	2.2	0.62	0.69	0.100	33	13
31B	−6.3	2.2	0.67	0.68	0.113	36	22
31C	−5.3	3.0	0.69	0.69	0.096	27	30
31D	—	—	0.69	0.72	0.116	44	23
31E	−6.6	3.0	0.71	0.71	0.150	36	19
31F	—	—	—	—	—	45	24

Localities in (A) arranged in descending order from east to west (see fig. 2) Localities in (B) arranged in descending order from bottom to top of section M_z and σ_I: mean grain size and standard deviation (Folk and Ward, 1957) R and S: roundness and sphericity (Krumbein, 1941) %: number of clasts percent in conglomerate

13

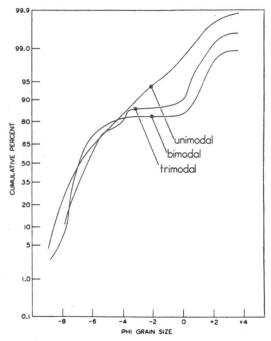

FIG. 3.—Typical grain size cumulative curves for the Peel Sound conglomerates.

Inspection of the sorting curves for the Peel Sound conglomerates shows that one sample is unimodal, twelve are bimodal and four are trimodal (fig. 3). The most prominent mode in all cases lies in the cobble range, varying from -6.7 to -7.7ϕ (10.2 to 20.5 cm). All samples except that from loc. 31B have a secondary mode in the medium to coarse sand range: $+2.0\phi$ to -0.8ϕ (0.25 to 1.75 mm). A very weak third mode in the pebble range is present in the samples from locs. 31C, 31E, 101B and 106. The size range is from -4.0ϕ to -5.4ϕ (1.6 to 4.2 cm). These figures suggest several departures from Spencer's three fundamental populations. The cobble mode was not present in any of Spencer's samples, whereas clay is scarce in the Peel Sound conglomerates. The subsidiary pebble and sand modes in the conglomerates compare only approximately with the corresponding two populations of Spencer (1963). In both cases the range of modal size for the Peel Sound samples is two phi units coarser than the range of median size suggested by Spencer. (Spencer states that although many clastic sediments have polymodal or skewed size distributions, the two or three fundamental populations of which they are composed each show log-normal distribution. If this is the case then the mode and median of each fraction will be the same, and the two measures can be directly compared). Simi-

lar disparities are indicated by the analyses of thirty-six Triassic continental sandstones by Cadigan (1961). Most of these have mean grain sizes within Spencer's sand population range, but a 'granite wash alluvium' and a 'coarse arkose' have mean grain sizes of $+0.18$ and $+0.85\phi$ respectively which compare with the Peel Sound sand fraction and fall in one of Spencer's scarce ranges.

Most of the sandstones that support Spencer's (1963) hypothesis of fundamental populations are sediments that were deposited in environments where a high degree of sorting and reworking is common, eg. the aeolian and marine environments. In immature sediments, formed rapidly with little time for sorting and abrasion, the modal size ranges may be much coarser. Certain grain size ranges are scarce, and this may well be due to weathering characteristics, but their size range is probably not fundamental in the sense intended by Spencer.

Clast lithotype analysis

This method of study has proved to be the most useful in the investigation of the Peel Sound conglomerates. The marked vertical and areal variations in clast proportions that have been ennumerated by these counts may be readily interpreted in terms of processes active at the present day on modern alluvial fans.

Vertical clast variations: One of the most marked differences between the Lower and the Upper Peel Sound Formation is the difference in the clast composition of the conglomerates. In fact the two parts of the formation may be subdivided solely on this basis, for over a limited vertical interval of the succession several changes take place together: a) Lithologies other than conglomerate die out rapidly upwards, decreasing from more than 50% to less than 10%. b) The clast lithologies in the conglomerates change in relative proportions (discussed below). c) The clast size increases; the maximum intermediate diameter changes from approximately 25 cm to more than 50 cm, and commonly in excess of 100cm. d) Changes take place in the conglomerate matrix which correspond to changes in the major clasts.

These variations take place over a vertical interval of less than 50 feet (15 m). A typical pattern of clast variation is illustrated in fig. 4 which is generalised from a series of sections ten miles north of Cape Brodie. The change in the relative clast proportions demonstrates the 'reversed stratigraphy' effect. Generally speaking, downcutting in a source area will expose successively older rocks, unless the source area geology is particularly complex or the strata are all vertical. In the derived sediments the proportion of older material will thus be expected to increase upwards in a vertical succession. Such is the case in fig. 4: the gneisses of the Precambrian basement increase up the section at the expense of the younger Hunting dolomites. A very similar pattern is present in the Lower Peel Sound sections near Transition Bay. At southern Pandora Island, Mount Matthias and Bellot Cliff, the Lower Peel Sound conglomerates are rich in Read Bay limestone, which likewise diminishes in importance up the section, from 90-100% of the total clasts, to less than 1%, plus an occasional derived invertebrate fossil. In the Upper Peel Sound, the conglomerates are coarse and polymict, containing abundant gneiss, gabbro, quartz sandstone and dolomite clasts.

The rapid change between the Lower and the Upper Peel Sound Formation suggests that a marked rejuvenation of the source area was taking place at the time. This would account for the coarsening of the clasts, the increase in the proportion of conglomerate, and the increasing importance of older material, notably basement metamorphic rocks, in the derived sediments. There is no direct evidence that the rejuvenation took place everywhere at the same time, but the magnitude of the changes effected would suggest that this was the case. Most modern alluvial fans are to be found in areas of recent or present day tectonic activity (Blissenbach, 1954, p. 180; Bull, 1964, p. 5). Their abundance in the deserts of the southwestern United States reflects the youthful nature of the landscape as

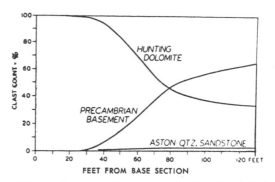

FIG. 4.—Variations in clast composition in vertical sections of the Lower-Upper Peel Sound boundary, generalised from exposures ten miles north of Cape Brodie.

well as the aridity of the climate, for it is only in areas of considerable relief that such thick wedges of coarse and immature sediment are able to accumulate.

There is further evidence to suggest that many of the conglomerates are syntectonic in the form of a fold structure exposed ten miles north of Cape Brodie. Figure 5 illustrates the two limbs of the syncline which have completely different clast compositions. No evidence of major faulting, shearing or brecciation was discovered in the outcrop, although the hinge area of the fold is incompletely exposed. The syncline is part of the upturned cover rocks along the margins of the Boothia Uplift. To the east vertical and overturned older Paleozoic rocks stand against the Precambrian core of the Uplift; to

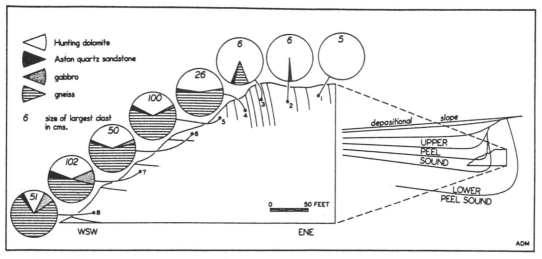

FIG. 5.—Clast count variations within the synclinal fold bordering the Boothia Uplift near Cape Brodie (left), and a possible explanation for the different clast compositions on the two limbs: a syn-depositional fold (right).

the west the rocks grade into the finer grained facies of the Peel Sound Formation.

The cape Brodie structure is interpreted as a syndepositional fold, and the suggested mode of origin is illustrated in figure 5. The nearly oligomict conglomerates comprising the base of the Lower Peel Sound represent the deposits formed during or following an early, gentle elevation of the Boothia Uplift. Later movement was more marked. Uplift may have taken place along steep reversed faults, as suggested by Kerr and Christie (1965, fig. 3), but if so, faulting is confined to the basement rocks. In the cover rocks movement is expressed by folding. The accelerated movement caused much coarser debris than before to be transported out of the uplifted area, while the early oligomict conglomerates were themselves tilted up and eroded. Erosion cut down into the basement rocks, and debris rapidly became enriched in metamorphic rock material. Coarse, polymict conglomerates accumulated in progressively on-lapping wedges along the edges of the uplift while the latter was still active. They are, therefore, syntectonic conglomerates. Continued movement must have allowed the removal of much of the newly formed conglomerate wedge, but the earlier parts of it are themselves folded, as demonstrated by the beds from which clast counts 5 and 6 were taken. The conglomerates at these two points are of almost identical composition, showing that they probably represent one horizon.

It is not known whether similar structures are developed elsewhere along the margin of the Boothia Uplift. Similar superimposition of fan wedges has been described by Blissenbach (1954) and has been attributed to tectonic activity in areas of modern fan formation.

Within the polymict cobble and boulder conglomerates of the Upper Peel Sound Formation there is less variation in clast types. At Bellot Cliff, loc. 31, eight clast counts made over approximately 350 feet of vertical section suggest an upward decrease in Aston quartzite and an increase in basement gneiss material (table 2). However, the correlation coefficients r of the two trends (clast percent:vertical height) are low (−0.352 and +0.52, respectively; p = > 0.10 for both these correlations), and so they may be considered to be statistically not proven, although the trends do make sense geologically for they demonstrate the same reversed stratigraphy effect.

Areal clast variations: In four areas of the Conglomerate and Conglomerate-Sandstone Facies sufficient field counts were made to reveal clast dispersal patterns. Results are illustrated in figures 6 and 9. In each of figures 6a, b, c two or three clast types have been chosen to illustrate the most distinctive dispersal patterns and isopleth lines have been drawn to connect points of equal clast percentage.

In the Bellot Cliff area (fig. 6a) dispersal patterns may be grouped into three distinct types: (a) a pattern which shows the highest percentage in the Russell Island area, typified by the grey-green quartz sandstone tentatively identified as a variety of the Aston Formation; (b) Patterns which show highest clast percentages immediately to the south and southeast of Bellot Cliff, typified by the Aston quartz sandstone (also shown by the gneiss, gabbro and basic volcanic patterns); (c) Patterns showing apparently random distribution of the clast lithology, typified by the Hunting dolomite. The meager paleocurrent evidence obtained from clast imbrication and structures in the interbedded sandstones is included.

At Mount Matthias there are two distinct clast dispersal patterns. The gneiss clasts comprise up to 18 percent of the congomerate towards the northwest corner of the area, but are entirely absent in the south and east. Gabbro clasts, on the other hand, are abundant in the east and steadily decrease towards the west to a minimum of 7 percent.

In southern Pandora Island the conglomerate consists largely of Precambrian gneiss and Hunting dolomite, but the relative porportions of these lithologies changes from north to south. On nearby Prince of Wales Island dolomite is absent, but there is a small percentage of Read Bay limestone, which disappears to the west.

These patterns may be readily explained on the basis of a sedimentary model analogous to the modern bajada, which is defined as the detrital slope immediately below the zone of erosion in an arid landscape, e.g. 'a series of confluent alluvial fans along the base of a mountain range' (Howell, 1960). Modern bajadas abound in the deserts of the southwestern United States. The Peel Sound Formation may be usefully compared with the example described by Bull (1964) in Western Fresno County, California, which consists of a belt of alluvial fans fifty miles long and up to eighteen miles wide (fig. 7). In such a fan system each of the measureable parameters such as mean and maximum clast size, lithotype percentage, roundness and sphericity etc., may be controlled by either intra-fan or inter-fan variations or both acting together.

Intra-fan changes are those relating to (a) distance of transport, which is more or less proportional to the amount of abrasion of the

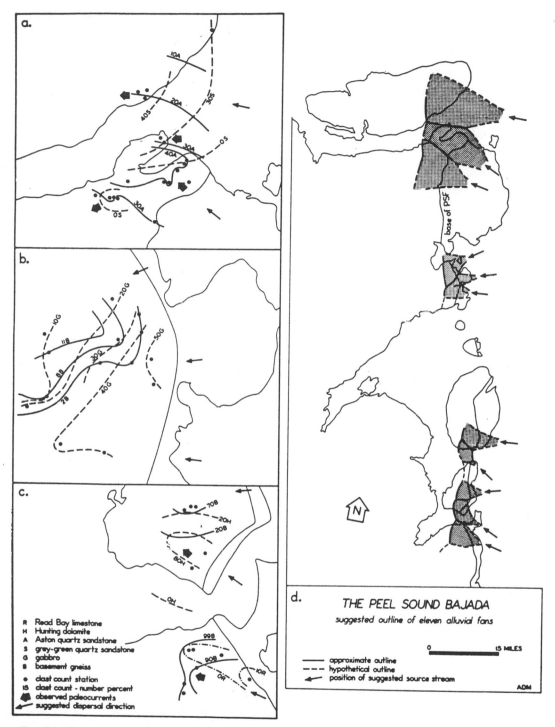

Fig. 6.—Clast count isopleth lines in three areas of eastern Prince of Wales Island: a. Bellot Cliff, Aston and grey-green quartz sandstone patterns; b. Mount Matthias, basement gneiss and gabbro patterns; c. Pandora Island and nearby Prince of Wales Island, basement gneiss, Hunting dolomite and Read Bay limestone patterns. 6d: reconstructed bajada showing the pattern of alluvial fans bordering the Boothia Uplift.

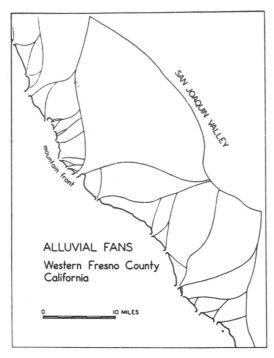

FIG. 7.—A modern bajada. From Bull (1964).

clasts, and (b) changes in slope caused by rejuvenation or pediplanation of the source area. The vertical clast variations described above probably reflect the second cause: rejuvenation in the source area. On a fan surface roundness and sphericity may be expected to increase with distance of transport (Blissenbach, 1954; Lustig, 1965; Bluck, 1965) although it has been shown that sphericity values do not always vary in such a simple way (discussed below). Mean and maximum clast size will tend to decrease with increasing distance of transport, as the power of the transporting medium wanes and the material is abraded. Clast lithology counts will show little variation on a given fan surface if all the clasts have the same density and resistance to erosion, and tend to produce similar shapes when fractured. Transport will not then be selective, and abrasion will reduce all clasts to sand size at a similar rate. It is thought that few of the areal clast type variations described in this section are related to causes of intra-fan origin—an exception is discussed below.

Figure 7 illustrates the importance of inter-fan variations in clast characteristics. Each alluvial fan is derived from a different catchment area and if the geology of the source area is variable then clast lithotypes may be very different on adjacent fans. Similarly, the great variation in fan sizes corresponds to a wide range in the size of the catchment area and the length of the feeder streams. Detrital material arriving at the mountain front in one of the larger stream valleys will have undergone considerably more abrasion than the deposits on the small fans, which may be little more than extended scree slopes. Varying fan size is thus likely to be the cause of large differences in roundness, sphericity and grain size characteristics in the resulting sediments.

Inter-fan variability in clast parameters may thus be of much greater magnitude than the variability to be found on single alluvial fans, and it is on this basis than most of the areal variations in the Peel Sound conglomerates are interpreted.

In the Bellot Cliff area (fig. 6a) it is suggested that the grey-green quartz sandstones were derived from the east, implying that an area northwest of Aston Bay, Somerset Island, now occupied by the sea, was the only important source of this lithology. The paleocurrent evidence, such as it is, supports this interpretation. The four clast types showing the second pattern (as typified by the Aston quartzites in fig. 6a) were probably derived in greater quantity from the southeast, and again the paleocurrent evidence is consistent with the suggestion. The isopleth lines shown in fig. 6a can be attributed to deposition on two fans overlapping and mixing in the vicinity of Bellot Cliff. The fans are not well defined because of the tendency of distributaries to wander over the fan surface, so that in time parts of the depositional slope will have been affected by streams from at least two sources, with resultant mixing of detrital materials (Denny, 1967).

Given the size of the present outcrop, the fans must have extended at least twelve miles from apex to toe, although an unknown amount at the proximal end has been removed by erosion. The distinctive dispersal patterns thus appear to have been created by mixing of material from two source areas of markedly different surface geology. The random distribution of the Hunting dolomite clasts can be explained on the supposition that this lithology was derived in equal quantities from both areas.

The distribution pattern of gneiss clasts at Mount Matthias (fig. 6b) may be explained in much the same way. An alluvial fan probably entered the area from the northeast, carrying approximately 20% gneiss material, and the detritus was mixed with other sediment entering from the east, where the Precambrian gneisses were not exposed. The gradual westerly decrease in gabbro material across the area may be due to its much lower resistance to weathering and erosion. In present day outcrops of the conglomerate the gabbro clasts are often weathered to a friable powder, whereas the gneiss and quartz sandstone show virtually no evidence of weathering, except for oxidation crusts.

The evidence from the Pandora Island area (fig. 6c) suggests the former existence of three alluvial fans. In central Pandora Island gneiss and Hunting dolomite make up the bulk of the conglomerate in the approximate ratio of 5:1, whereas in the southern part of the island the ratio is approximately 1:8. Paleocurrent evidence is very meager, but the clast counts suggest two distinct sources of detritus. This is confirmed by the existence of a single three-foot seam of highly dolomitic conglomerate interbedded with the gneiss-rich deposits at one exposure in central Pandora Island. The thin dolomite-rich layer probably represents a tongue from the southern fan that retained its identity in the zone of fan merging.

On the mainland south of Pandora Island dolomite is absent, but near the eastern margin of the Peel Sound outcrop the conglomerate contains 14% Read Bay limestone. The mixture suggests a third discrete fan source, for the presence of this limestone in the higher, coarser Peel Sound conglomerates is unusual. The Read Bay Formation is the youngest in the succession of source rocks and might thus be expected to have been removed first when erosion commenced. It is much commoner in the basal parts of the Peel Sound Formation. At the time of accumulation of the conglomerates the Reed Bay was still a young formation and was probably not very resistant to erosion. This might explain its rapid disappearance to the west on the fan under discussion.

The reconstructed outlines of the alluvial fans are shown in figure 6d. There is little doubt that others were also present but the complete pattern cannot be reconstructed because of lack of exposure. It should be emphasized that the outlines drawn in fig. 6d can only be approximated because boundaries shift from time to time due to fan migration. The 'hypothetical' boundaries are included for the sake of clarity, but in many cases there is little or no evidence for their actual position. Source streams are placed so as to be consistent with paleocurrent data, wherever present.

Maximum clast size

In general measurements of this parameter show that the largest clasts are found in the areas where the lithology measured is present in the greatest proportion, i.e. close to the source of the material. The maximum size diminishes towards the west and towards areas of low clast

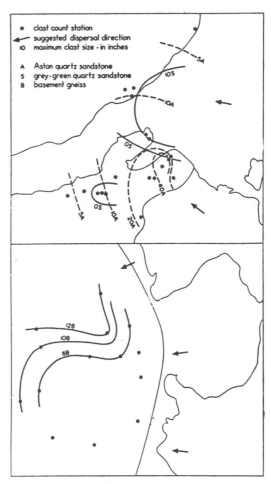

FIG. 8.—Isopleth lines showing variations in maximum clast size. Areas shown as for figs. 6a, b: Bellot (top) and Mount Matthias (bottom).

percentage (fig. 8). Maximum clast size values give a concentric pattern of isopleth lines and the direction of transport is interpreted to be more or less perpendicular to these lines, the diminution in size being due to abrasion during transport across the fan surface.

Roundness and sphericity

Vertical variations in these parameters were measured at Bellot Cliff, loc. 31. Five samples of Aston quartz sandstone spaced out over approximately 350 feet (105m) of section show increasing roundness with height. ($r = 0.986$, $p = < 0.01$), whereas sphericity shows no statistically significant variations ($r = 0.324$, $p = > 0.10$). The roundness changes may be attributed to reduction of relief in the source area, with the effect of reducing stream slope and increas-

TABLE 3.—*Conglomerate clast parameter trends, south coast of Prescott Island*

parameter	trend with dist. to W	n	r	p
M$_z$	decrease	10	−0.88	<0.001
σI	increase	10	+0.73	<0.02
Aston R	decrease	12	−0.61	<0.05
gneiss R	?decrease	11	−0.27	>0.10
Aston S	none	12		
gneiss S	none	11		
Aston %	decrease	12	−0.87	<0.001
gneiss %	increase	12	+0.76	<0.01

data from table 2.

ing stream length by headward erosion, thus allowing more time for abrasion.

This explanation is consistent with the small changes in clast lithologies detected in the same section (discussed above) for it suggests that in contrast to early Peel Sound times the period during which the polymict Upper Peel Sound conglomerates were formed was one of comparative tectonic stability. The abrupt changes in clast composition that mark the Lower-Upper Peel Sound boundary suggest rejuvenation, with erosion cutting down into progressively deeper levels of the source area. During Upper Peel Sound times down-cutting probably gave way to valley-widening as tectonic activity ceased; clast types hardly changed, and material was transported at a progressively slower rate, as reflected in the higher roundness values.

Lateral changes in roundness and sphericity of Aston quartz sandstone and Precambrian gneiss were investigated along the south coast of Prescott Island (table 2). Sphericity appears to vary randomly, but roundness figures show an unexpected decrease towards the west (the supposed direction of transport). Several other unusual results obtained on this traverse are summarized in table 3. Taken as a whole the various trends cannot be explained by assuming that sampling was carried out along the axis of a single fan, although the sample sites were chosen with this aim in view, and it was intended to show how the different clast parameters vary with distance of transport.

The results may be explained by postulating that the conglomerate was deposited by two alluvial fans converging in the area that is now southwestern Prescott Island (fig. 9). One fan, richer in Aston quartz sandstone material entered the area from the east, and another, with a high proportion of gneiss clasts entered from the Pandora Island area to the southeast. If the latter was more proximal it would be expected to contain material that was more poorly sorted

with clasts of lower mean roundness. The resulting mixture of material in the southwest corner of Prescott Island would produce the apparently anomalous trends summarized in table 3. In support of the possibility that the Pandora fan was more proximal is the fact that on Pandora Island the base of the Peel Sound Formation is displaced at least a mile to the west relative to Prescott Island, possibly reflecting a bend in the original mountain front from which the sediments were derived. Unfortunately paleocurrent indicators are almost absent in the area.

The virtually random pattern of clast sphericity values probably reflects at least in part an incorrect sampling procedure. Krumbein (1942), Blissenbach (1954) and Bluck (1964) found no significant relationship between sphericity and distance of transport. Bluck claims that measurements on a Triassic fanglomerate (Bluck, 1965) and a Devonian fanglomerate (1967) showed an increase of sphericity with distance, but his graphs are based on very few readings. He argues that many factors may act to obscure the relationship, such as jointing tendencies controlling the initial particle shape and selective transport moving certain shapes faster than others. Sneed and Folk (1958) show that sphericity varies markedly with small changes in size and lithology. The variability of the present results may be due to any or all of these causes, but the principal factor is probably that samples were taken from too wide a size range —one whole phi class instead of one third of a phi class, as recommended by Sneed and Folk (1958).

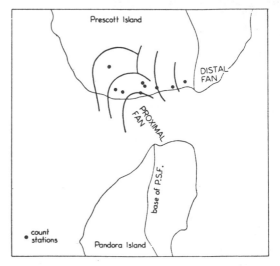

FIG. 9.—Suggested origin of the conglomerates at the south end of Prescott Island: two converging alluvial fans.

20

MODE OF DEPOSITION

The Peel Sound conglomerates are coarse, poorly sorted, and they lack a well developed clast framework. Clast orientation is generally random—cross bedding and imbrication are rarely seen, and the conglomerates are interbedded with minor sandstones of typical fluvial aspect. These are the chief characteristics of 'mudflow' deposits as summarized by Blissenbach (1954, p. 185-187), who states that 5-40% of all the alluvial fan deposits studied by him in Arizona were laid down by this mechanism. The sandstones represent deposits formed during the waning stages of mudflows.

It is thought that the bulk of the Peel Sound conglomerates were formed by deposition from successive sub-aerial mudflows, although it is important to add that this term is somewhat misleading. Mudflows, as described by Blackwelder (1928) and various other authors from eye-witness accounts of modern floods, are occasional catastrophic events in semi-arid regions, and they consist of a mobilised mass of coarse, often bouldery material, moving something like a lava flow. However, the flows may actually contain only a small proportion of mud. Published figures indicate a range from more than 30% clay-silt (Bull, 1964) to less than 10% (Lustig, 1965) and Lustig's measurements were made on only the granule-clay fraction of the conglomerates. In the case of the Peel Sound rocks the more general term 'debris flood' deposits is more appropriate, for the conglomerates contain less than 1% of the clay-silt fraction (whole-rock grain size measurements).

The Peel Sound conglomerates may be compared with a recent debris flood deposit described by Sharp and Nobles (1953). These flows originated in a mountainous area, on a slope of up to 32°, but after the first three miles the angle became 3°, and near the end of the flow, which extended for fifteen miles, the angle was less than 1°. Rare boulders up to six feet (2 m) in diameter were seen in the deposits, those of 2–3 feet (60–120 cm) were common, and some of 18 inches (45 cm) were carried as far as twelve miles (19 km) from the source.

The total combined width of the Peel Sound Conglomerate and Conglomerate-Sandstone Facies reaches ten miles (16 km) and although the largest boulders present are 5 feet (1.5 m) in diameter, the deposits of the Conglomerate-Sandstone Facies (the outermost three miles of the alluvial fans) rarely contain clasts larger than 10 inches (25 cm) in diameter. An unknown amount of the Peel Sound conglomerates may have been removed by erosion at the proximal end of the fans, and the debris floods may have originated anywhere within the watershed of the Boothia Uplift which averages 40 miles (64km) in width. Nevertheless, the scale of the Peel Sound deposits and the scale of the recent flood of Sharp and Nobles (1953) appear to be comparable.

Sorting values of the Peel Sound conglomerates compare with other recent debris flood deposits. Standard deviation σ_I ranges from 1.82 to 3.36 phi units with an average of 2.52. Lustig (1965, modern fans) and Bluck (1967, Devonian fans) obtained much lower values at 1.3 to 2.7 and 0.72 to 1.15 phi units respectively. Bluck's figures are surprisingly low; they fall in the 'moderately sorted' range of Folk (1968, p. 46). Neither of these workers included the whole rock in their measurements. Krumbein (1942) obtained values of 2.37 to 2.86 phi units, measuring canyon flood material, and other workers studying modern alluvial fan deposits obtained values ranging from 4.1 to 6.2 phi units (Bull, 1964) and 1.0 to 3.8 phi units (Landim and Frakes, 1968).

Debris floods are commonest at the present day in areas with a rainfall of 10-20 inches per annum (Blissenbach, 1954, p. 177), but the reason for this is that only under semi-arid conditions is vegetation scarce, allowing surface run-off to build up rapidly to flood proportions. Chawner (1935) has shown how removal of vegetation by a forest fire may allow catastrophic floods to occur in areas that normally experience unspectacular surface run-off. During Peel Sound times (Lower to Middle Devonian) land vegetation had a very restricted distribution (Seward, 1969); the evidence suggests that most land plants were restricted to swampy or low lying country and evolution had probably not proceeded so far as to produce a vegetation cover in high, mountainous areas. There is, therefore, no reason for equating the presence of debris flood deposits in the Peel Sound Formation with an arid or semi-arid climate.

CONCLUSIONS

1) Directional sedimentary structures: these are rare except in the interbedded sandstone lenticles. An upstream clast imbrication is occasionally present in the conglomerates, but its scarcity is consistent with the suggested debris flood origin of these deposits.

2) Clast lithotype analysis: counts of the proportions of the various clast types present enable alluvial fan dispersal patterns to be drawn. The presence of soft or easily weathered lithologies (eg. Read Bay limestone, gabbro, in the present study) enables direction of transport to be deduced, for these clasts decrease in numbers

21

570 ANDREW D. MIALL

rapidly with distance of transport.

3) Maximum clast size analysis: variability may be attributed to abrasion and to varying size of source material on different alluvial fans. The present study has shown that in general dispersal patterns indicated by measurements of maximum clast size for each lithology are very similar to those indicated by clast lithotype analyses.

4) Whole-rock grain size analysis :mean grain size and standard deviation show statistically significant variations from one alluvial fan to another. The standard deviation of the conglomerates varies between approximately 2.0 and 4.0 ϕ, which compares well with published values for modern debris flood deposits. Analysis of cumulative curves has shown that the conglomerates have polymodal size distributions, and that the most prominent grain size modes are approximately two phi size classes coarser than the common modes of well sorted clastic sediments such as aeolian and marine deposits. The principle modes are coarse sand (-0.8 to $+2.0\phi$), pebbles (-4.0 to $+5.4\phi$), and cobbles (-6.7 to -7.7ϕ), and they probably indicate primary deficiencies in certain grain size ranges, relating to rock weathering and erosion characteristics.

5) Clast roundness: measurements of this parameter on clasts of one phi class interval (-6 to -7ϕ) have shown a high average roundness (up to 0.74 for quartzite clasts). Statistically significant variations detected in vertical and lateral directions have been related to distance and rate of transport and to inter-fan changes.

6) Clast sphericity: measurements of this parameter revealed a high average sphericity (up to 0.73 for quartzite clasts in the -6 to -7ϕ size class). However, sampling in vertical and lateral directions revealed apparently random variations, probably because the size range measured was too large. Other measurements of

clast shape might prove more profitable.

7) Grain size and textural characteristics of the conglomerates suggest a debris flood origin. Interbedded sandstones were probably formed during the waning stages of the floods.

8) The results of the present study have indicated several methods of investigation that can profitably be used in studies of ancient continental conglomerate deposits. In general terms, vertical variations in conglomerate characteristics may be related to evolutionary processes on single alluvial fans and changes in their source areas, whereas lateral variations, if measured over a sufficiently large area, will reflect differences between adjacent fans. Down fan and across fan changes are likely to be small in comparison to the variability displayed between fans of markedly different size and source geology.

ACKNOWLEDGEMENTS

The present study forms part of a doctoral dissertation presented to the University of Ottawa. Field work was supported by grants from the National Research Council of Canada, the Geological Survey of Canada and the Department of Indian Affairs and Northern Development. Thanks are due to J. C. Sproule and Associates Ltd. for considerable expediting assistance at Resolute Bay, and to Outboard Marine (Canada) Ltd. for the loan of a 'Snowcruiser' for two field seasons.

Thanks are extended to D. L. Dineley (leader of the University of Ottawa group) for advice and assistance, to B. R. Rust for much assistance at all stages of the work and for critically reading the manuscript, and to J. A. Donaldson and G. deV. Klein who also read the manuscript. Field assistance was given by D. S. Broad, D. Langley, P. Dodson and J. Thorpe.

REFERENCES

BEATY, C. B., 1963, Origin of alluvial fans, White Mountains, California and Nevada: Assoc. American Geographers Annals, v. 53, p. 516–535.
BLACKADAR, R. G., AND CHRISTIE, R. L., 1963, Geological Reconnaissance, Boothia Peninsula, and Somerset, King William and Prince of Wales Islands, District of Franklin: Canada Geol. Survey, paper 63-19, 15 p.
BLACKWELDER, ELIOT, 1928, Mudflow as a geological agent in semi-arid mountains: Geol. Soc. America Bull., v. 39, p. 465–484.
BLISSENBACH, ERICH, 1954, Geology of alluvial fans in semiarid regions: Geol. Soc. America Bull., v. 65, p. 175–190.
BLUCK, B. J., 1964, Sedimentation of an alluvial fan in southern Nevada: Jour. Sedimentary Petrology, v. 34, p. 395–400.
——— 1965, The sedimentary history of some Triassic conglomerates in the Vale of Glamorgan, South Wales: Sedimentology, v. 4, p. 225–245.
——— 1967, Deposition of some Upper Old Red Sandstone conglomerates in the Clyde area: a study in the significance of bedding: Scot. Jour. Geology, v. 3, p. 139–167.
BROAD, D. S., 1968, Lower Devonian Heterostraci from the Peel Sound Formation, Prince of Wales Island, Northwest Territories: M.Sc. thesis, Univ. of Ottawa, 138 p.
——— DINELEY, D. L., AND MIALL, A. D., 1968, The Peel Sound Formation of Prince of Wales and adjacent island: A preliminary report: Arctic, v. 21, p. 84–91.

22

BROWN, R. L., DALZIEL, I. W. D., AND RUST, B. R., 1969, The structure, metamorphism and development of the Boothia Arch., Arctic Canada: Canadian Jour. Earth Sci., v. 6, p. 525–543.

BULL, W. B., 1964, Alluvial fans and near surface subsidence in Western Fresno County, California: U.S. Geol. Survey Prof. Paper 437-A, 71 p.

CADIGAN, R. A., 1961, Geological interpretation of grain size distribution measurements of Colorado Plateau sedimentary rocks: Jour. Geology, v. 69, p. 121–144.

CHAWNER, W. D., 1935, Alluvial fan flooding, the Montrose, California, flood of 1934: Geog. Rev., v. 25, p. 225–263.

CURRAY, J. R., 1960, Sediments and history of Holocene transgression, Continental Shelf, Northwest Gulf of Mexico, p. 221–266, in (Shepard, F. P., Phleger, F. B., and Van Andel, T. H., eds.), Recent sediments, Northwest Gulf of Mexico. Amer. Assoc. Petrol. Geol., Tulsa, Oklahoma.

DENNY, C. S., 1967, Fans and pediments: Am. Jour. Sci., v. 265, p. 81–105.

DINELEY, D. L., 1965, Notes of the scientific results of the University of Ottawa expedition to Somerset Island, 1964: Arctic, Jour. Arctic Inst. North Amer., v. 18, p. 55–57.

——— 1966, Geological studies in Somerset Island, University of Ottawa expedition, 1965: Arctic, Jour. Arctic Inst. North Amer., v. 19, p. 270–277.

DONALDSON, J. A., 1966, Marion Lake map-area, Quebec-Newfoundland, Canada Geol. Survey, Mem. 388, 85 p.

FOLK, R. L., 1968, Petrology of Sedimentary rocks. Hemphill's, Austin, Texas, 170 p.

FOLK, R. L., AND WARD, W. C., 1957, Brazos River bar: a study in the significance of grain size parameters: Jour. Sed. Petrology, v. 27, p. 3–26.

FORTIER, Y. O., and others, 1963, Geology of the north-central part of the Arctic archipelago, Northwest Territories (Operation Franklin): Canada Geol. Survey, Mem. 320, 671 p.

FRIEDMAN, G. M., 1958, Determination of sieve-size distribution from thin section data for sedimentary petrological studies: Jour. Geology, v. 66, p. 394–416.

HOOKE, R. LeB., 1967, Processes on arid-region alluvial fans: Jour. Geology, v. 75, p. 438–460.

——— 1968, Steady state relationships on arid-region alluvial fans in closed basins: Am. Jour. Sci., v. 266, p. 609–629.

HOWELL, J. V., 1960, coordinating chairman: Glossary of Geology and related sciences, 2nd ed. Am. Geol. Inst., Washington, D.C., 325 p.

KERR, J. W., AND CHRISTIE, R. L., 1965, Tectonic history of Boothia Uplift and Cornwallis Fold Belt, Arctic Canada: Amer. Assoc. Petrol. Geol., Bull., v. 49, p. 905–926.

KRUMBEIN, W. C., 1941, Measurement and geologic significance of shape and roundness of sedimentary particles: Jour. Sedimentary Petrology, v. 11, p. 64–72.

——— 1942, Flood deposits of Arroyo Seco, Los Angeles County, California: Geol. Soc. America Bull., v. 53, p. 1355–1402.

LANDIM, P. M. B., AND FRAKES, L. A., 1968, Distinction between tills and other diamictons based on textural characteristics: Jour. Sedimentary Petrology, v. 38, p. 1213–1223.

LUSTIG, L. K., 1965, Clastic sedimentation in Deep Springs Valley, California: U.S. Geol. Survey Prof. Paper 325-F, p. 131–192.

MIALL, A. D., 1969, The sedimentary history of the Peel Sound Formation, Prince of Wales Island, Northwest Territories: Ph.D. dissertation, Univ. of Ottawa, 279 p.

MIALL, A. D., 1970, Continental marine transition in the Devonian of Prince of Wales Island, Northwest territories; Canadian Jour. Earth Sci., v. 7, p. 125–144.

MIALL, A. D., in press, A new Lower Devonian rock unit in the Canadian Arctic Islands: A discussion: Canadian Jour. Earth Sci.

ORMISTON, A. R., 1967, Lower and Middle Devonian trilobites of the Canadian Arctic Islands: Canada Geol. Survey Bull., 153, 148 p.

PACK, F. J., 1923, Torrential potential of desert waters: Pan Am. Geol., v. 40, p. 349–356.

PETTIJOHN, F. J., 1957, Sedimentary Rocks. Harper and Brothers, New York, 2nd edition, 718 p.

SEWARD, A. C., 1959, Plant life through the ages: Hafner, New York, 603 p.

SHARP, R. P., AND NOBLES, L. H., 1953, Mudflow at Wrightwood, southern California: Geol. Soc. Amer. Bull., v. 46, p. 547–560.

SNEED, E. D., AND FOLK, R. L., 1958, Pebbles in the Lower Colorado River, Texas. A study in particle morphogenesis: Jour. Geol., v. 66, p. 114–150.

SPENCER, D. W., 1963, The interpretation of grain size distribution curves of clastic sediments: Jour. Sedimentary Petrology, v. 33, p. 180–190.

TUKE, M. F., DINELEY, D. L., AND RUST, B. R., 1966, The basal sedimentary rocks in Somerset Island, Northwest Territories: Canadian Jour. Earth Sci., v. 3, p. 697–711.

23

2

Reprinted from pp. 102, 104–119 of *Canadian Jour. Earth Sci.* **13**(1):102–119 (1976)

Development of a braided-fluvial facies model for the Devonian Battery Point Sandstone, Québec

DOUGLAS J. CANT AND ROGER G. WALKER

Department of Geology, McMaster University, Hamilton, Ontario L8S 4M1

Received 29 May 1975

Revision accepted for publication 15 September 1975

Eight distinct facies have been defined in a 110 m thick section of the Lower Devonian Battery Point Sandstone near Gaspé, Québec. The first is a scoured surface overlain by massive sandstone with mudstone intraclasts. Facies A and B are trough cross-bedded sandstones, with poorly- and well-defined stratification, respectively. Facies C and D consist of large isolated, and smaller multiple, sets of planar cross-stratified sandstones, respectively. Facies E comprises large sandstone-filled scours, facies F comprises ripple cross stratified fine sandstones with interbedded mudstones, and facies G comprises sets of very low angle cross-stratified sandstones.

The overall context of the Battery Point Sandstone, the presence of rootlets, and the abundance of trough and planar–tabular cross bedding, all suggest a generally fluvial environment of deposition. Analysis of the facies sequence and interpretation of the primary sedimentary structures suggest that channel development began by scouring, and deposition of an intraclast lag. Above this, the two trough cross bedded facies indicate unidirectional dune migration downchannel (vector mean direction 291°). The large planar tabular sets are associated with the trough cross bedded facies, but always show a large (almost 90°) paleoflow divergence, suggesting lateral movement of in-channel transverse bars. The smaller planar tabular sets occur higher topographically in the fluvial system, and the rippled silts and muds indicate vertical accretion. Because of the very high ratio of in-channel sandy facies to fine-grained vertical accretion facies, and because of the evidence of lateral migration of large in-channel bars, the Battery Point River appears to resemble modern braided systems more than meandering ones.

Huit faciès distincts ont été définis dans une section de 110 m d'épaisseur de grès du Dévonien inférieur de Battery Point, près de Gaspé, Québec. Le premier est une surface d'érosion sumontée de grès massif à fragments de mudstone. Les faciès A et B sont des grès à stratifications entrecroisées du type 'fosse'; ces stratifications sont peu évidentes dans le premier de ces faciès, très nettement marquées dans le second. Les faciès C et D consistent en séries de grès à stratifications entrecroisées du type 'plan', épaisses et isolées pour C, minces mais multiples pour D. Le faciès E comprend des chenaux à remplissage de grès, le faciès F des grès fins à ripple-marks, et le faciès G des ensembles gréseux à stratifications entrecroisées sous un angle très faible.

Le contexte général du grès de Battery Point, la présence de radicelles et l'abondance de stratifications entrecroisées de type 'fosse' et de type 'plan', tout suggère un milieu de sédimentation fluviatile. L'analyse séquentielle des faciès et des structures sédimentaires syngénétiques montrent que la formation du chenal débuta par une forte érosion, laissant sur place des fragments intraclastiques. Au-dessus, les deux faciès à stratification entrecroisée du type 'fosse' indiquent la migration vers l'aval d'une dune unidirectionnelle (direction moyenne: 291°). Les grands ensembles à stratification entrecroisée de type 'plan' sont associés avec ceux du type 'fosse', mais montrent toujours une grande divergence de direction d'écoulement (près de 90°), ce qui suggère un mouvement latéral de barres transversales.

Les ensembles de lits entrecroisés plans se présentent plus haut dans la topographie du système fluviatile, et les silts et boues à ripple-marks indiquent une accrétion verticale.

La proportion très élevée des faciès sableux par rapport aux faciès fins d'accrétion verticale et l'évidence d'une migration latérale de grandes barres intérieures font que la Rivière de Battery Point évoque plus un système anastomosé actuel qu'une rivière à méandres.

[Traduit par le journal]

[Editor's Note: Material has been omitted at this point.]

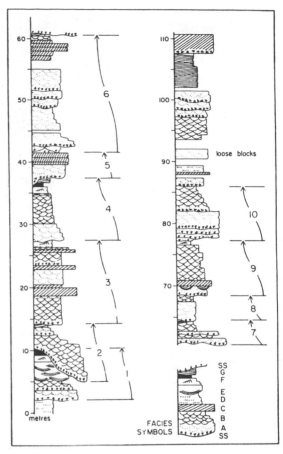

Facies Descriptions

To attempt the description and interpretation of the 110 m thick sequence (Fig. 2) in one piece is clearly crude and imprecise. It is necessary to subdivide the sequence, if possible, into smaller, homogeneous 'pieces' that might occur only once, or may be repeated many times within the 110 m sequence. Subdivision into 'pieces', or facies, is based upon careful measurement of the section and documentation of all physical and biological features. Facies were only defined formally after we were very familiar with the section, and hence with the variability associated with each of our proposed facies—that is, after we had some appreciation for how different, or how intergradational the facies were. Following de Raaf *et al*. 1965, we use the term 'facies' for a unit of rock distinguished in the field by lithology, sedimentary structures, and organic features from adjacent facies, both vertically and laterally in the section. The facies will be interpreted *after* the facies sequence has been established, so that the interpretation can be based upon their specific characteristics *and* their overall context.

FIG. 2. Measured section through part of the Battery Point Formation. Zero datum is about 50 to 100 m above the base of the formation. Facies lettering scheme is explained in the text. Blank portions of the column represent small faults or covered intervals; we do not believe that there are any major dislocations in this part of the area, but the indicated thicknesses where faults or cover are present may be inaccurate. Ten generally fining-upward sequences, defined by scoured surfaces, are indicated; above about 86 m the outcrop becomes too discontinuous to define more sequences. The sketched projection of the beds to the right is an approximate measure of mean grain size; maximum projection indicates coarse sandstone (about 1 mm).

Scoured Surfaces (SS)

The scoured surface facies (Fig. 3) consists of erosion surfaces overlain by layers of massive, coarse sandstone with abundant mudstone clasts. The maximum depth of erosion observed was about 5 m. Overlying the erosion surface, the mixture of mudstone clasts and coarse massive sandstone reaches a maximum thickness of about 25 cm. As well as the major scoured surfaces, there are other intraclast-strewn layers (commonly within facies A or B) that represent minor breaks in sedimentation as the clasts are deposited. The scoured surfaces are not facies in quite the same sense as the facies described below, but they are recognizable 'states' within the sedimentation sequence, and will be analyzed as such below.

Poorly Defined Trough Cross-Bedded Facies (A)

This facies (Fig. 4) rests upon scoured, intraclast-strewn surfaces (SS, Fig. 3), and is composed of poorly defined sets of trough cross

25

FIG. 3.   Scoured surface (SS), cutting down about 5 m stratigraphically from top left to person's hand, lower right. Backpack rests on trough cross bedding (facies B); above and to the left is a prominent isolated scour (facies E) overlain by rippled sandstone and siltstone, with interbedded mudstone (facies F). Photo from 5 to 10 m on Fig. 2.

FIG. 4.   Poorly defined trough cross bedding, facies A. Crude troughs can be seen, particularly below the 30 cm scale. This is an 'end-member' example of facies A, and with improving definition of the troughs, and with better sorting, facies A grades into facies B. Photo at 62 m, Fig. 2.

FIG. 5.   Well defined trough cross bedding, facies B. Scale is 16 cm long. Photo at 96 m, Fig. 2.

bedding, with trough depths averaging about 30 cm (range 15 to 60 cm). In most occurrences of the facies, the troughs are regularly stacked on one another, with uniform depth of scour. Trough width is variable (up to 3 m), and individual troughs have been traced downcurrent for up to 10 m. In plan view, the very wide troughs have straight internal strata with curvature only at trough margins—the smaller troughs have more arcuate internal strata.

The internal strata are poorly defined because of the poor sorting and lack of fine material necessary to show up the cross strata. Pebbles up to 5 cm in diameter are scattered through the troughs, rather than concentrated at their bases, or else form discontinuous inclined strata, which rarely persist through the total thickness of the set. Mudstone intraclasts up to 5 cm diameter occur at the bases of some troughs, rendering the internal organization of the troughs even poorer. The mean grain size of the sand is about 0.75 mm (coarse), but the mean ranges from about 0.5 to 1.0 mm in different occurrences of the facies. Individual occurrences of the facies vary in thickness from about 60 cm to 4.3 m.

Well Defined Trough Cross-Bedded Facies (B)

This facies is composed of well defined sets of trough cross bedding (Fig. 5), with trough depths averaging 15 to 20 cm (range 10 to 45 cm). The troughs are regularly stacked on top of each other, but in some individual occurrences of the facies, trough depths decrease upward. Trough widths vary from 30 cm to 1.5 m. Individual troughs have been traced downcurrent for up to 6.2 m, and in plan view, the internal cross strata are almost semi-circular. In vertical section parallel to flow, the cross strata curve only slightly at the toe, giving an appearance similar to planar–tabular cross bedding. In at least one occurrence, the sets climb downcurrent at about 5° to regional bedding.

The sets are composed of well sorted medium sand, and the well defined cross strata are shown up by fine, darker colored layers. Also, many individual cross strata are graded, further emphasizing the layering. The mean grain size is about 0.4 mm (medium sand), ranging from about 0.25 mm to 0.8 mm (coarse sand) in different occurrences of the facies. A few of the coarser sets have granules and pebbles concentrated at their bases.

We believe there is a gradation from the poorly to well defined trough cross bedded facies, and some occurrences were difficult to assign positively to either one.

Large Scale, Planar–Tabular Cross-Bedded Facies (C)

The planar–tabular sets comprising this facies

27

FIG. 6.   Isolated set of planar–tabular cross bedding, facies C. Note straight foresets and angular bottom contacts. Scale is 16 cm long, and set is about 1 m thick. Photo at 19 m, Fig. 2.

occur singly, or in cosets of up to 4 sets (Fig. 6). Sets average about 60 cm in thickness (range 30 cm to 3 m). The sets are extensive, with planar foresets having been observed up to 18 m wide, and sets persisting in the downcurrent direction for at least 20 m. Every occurrence of this facies (except one) is developed on a very flat surface that truncates one of the two troughed facies (A or B) (Fig. 7). The exception rests on an intraclast-strewn scour surface (SS). The tops of the planar–tabular sets are also flat truncation surfaces, commonly with large scours cut into them both parallel and perpendicular to the paleocurrent direction of the planar–tabular set (paleocurrents are discussed later). In vertical section parallel to flow, the sets have straight foresets with sharply angular bottom contacts. The foreset geometry remains constant in the downstream direction in any one set, except where reactivation surfaces truncate the foresets. In plan view, the foresets are straight across the entire width of outcrop.

The inclined foresets are shown up either by thin partings of finer sand, or because of normal grading from bottom to top of individual inclined layers. The layers vary from 1 to 8 cm in thickness. Grain size is generally in the coarse sand range (average 0.75 mm, range 0.5 to 1.0

mm), and in the rare sets where pebbles are present, they are scattered within the set rather than being concentrated at the base.

Small Scale Planar–Tabular Cross-Bedded Facies (D)

This facies is composed of small scale, planar–tabular sets of cross-bedding (Fig. 8), with individual sets ranging in thickness from 12 to 30 cm. Cosets are composed of between 2 and 10 individual sets, averaging 5 or 6 sets. Each set is very persistent in a downcurrent direction, and most can be traced for at least 10 m. Within cosets, paleocurrent directions tend to be fairly constant, with a maximum range of about 20° on either side of the mean.

The small scale cross-bed cosets are developed on flat truncation surfaces on top of facies B or C. They are normally truncated above by flat surfaces, and more rarely by irregular scours. Cross-bedding within sets is normally sharply angular without reactivation surfaces, and the inclined strata are shown up by thin, darker, slightly finer layers.

The sets are composed of well sorted, fine to medium sand, with mean sizes ranging from about 0.25–0.5 mm in different cosets. As individual sets become thicker, there is a transition

FIG. 7. Isolated set of planar–tabular cross bedding (facies C, top of photo) overlying broad troughs (facies B). Paleoflow for trough cross bedding is toward top left, almost exactly perpendicular to paleoflow of the planar tabular set (toward right). Scale is foreground 25 cm long. Photo at 25.5 m, Fig. 2.

FIG. 8. Multiple sets of small scale planar–tabular cross bedding, facies D. Scale is 15 cm long. Photo at 26.5 m, Fig. 2.

29

FIG. 9.   Large isolated scour, facies E, with stratification filling the scour conforming to the shape of the lower surface. Scour cuts into well defined trough cross bedding (facies B). Dark bars on notebook each 5 cm, numbers are tenths of feet. Photo at 6 m, Fig. 2.

into facies C, the large scale planar–tabular cross-bedding, and sets about 30 cm in thickness could probably be assigned to either facies.

Asymmetrical Scour Facies (E)

This facies consists of large, asymmetrical scours and scour fillings, up to 45 cm deep and 3 m wide (Fig. 9). The scours cut into each other, and into underlying troughed facies (A and B), and occurrences of the asymmetrical scour facies have a flat, erosionally truncated top. The shape of the scours ranges from spoon shaped to flat bottomed, with some very shallow, almost flat scours.

The main difference between the scour fillings and the two troughed facies (A and B) lies in the geometry of the infilling strata. In the asymmetrical scours, the layers are not at the angle of repose, but are parallel to the lower bounding surface. The layers show up by darker, finer partings, and within some scours there are minor erosion surfaces and 2–3 cm sets of cross-lamination. In some places where the layers can be seen in plan view, a moderately well developed parting lineation is oriented parallel to the inferred axes of spoon-shaped scours.

Most scours are filled with well sorted sand (mean diameters range from 0.25 to 0.75 mm), but some have pebbly sand fills. One scour fill is coarse and pebbly at the upstream end, fining to well sorted medium sand at the downstream end of the scour.

Rippled Sandstone and Mudstone Facies (F)

This facies includes cross-laminated sandstones (Fig. 10), and alternating cross-laminated sandstones and mudstones (Fig. 11).

The two occurrences of ripple cross-laminated sandstones without mudstones are 45 and 90 cm thick, respectively. One consists of tabluar sets of cross-lamination, and the other consists of trough cross-lamination with abundant organic material.

There are two occurrences of the interbedded fine sandstones (mean size < 0.25 mm) and mudstones. The first is 1.5 m thick and consists of three coarsening-upward sequences, which grade from basal mudstones into trough cross-laminated fine sandstone, and finally into granule sandstone. The sandstones capping each coarsening-upward sequence have sharp, bioturbated tops. The second occurrence is also 1.5 m thick, but consists of at least five sandstone-to-mudstone fining-upward sequences. The sandstones

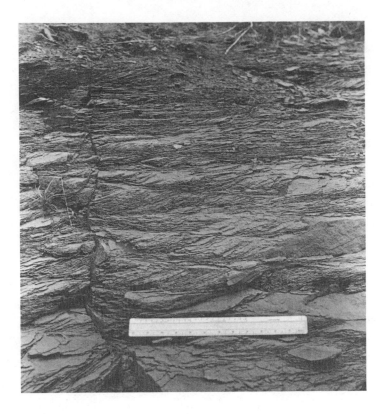

Fig. 10.   Small scale cross laminated sandstones, facies F. Scale is 30 cm long. Photo at 27 m, Fig. 2.

have sharp bases and commonly contain climbing ripples with eroded stoss sides. Other sandstones are structureless, probably due to bioturbation.

Low Angle Stratified Sandstone Facies (G)

This facies consists of sets (1 to 2, rarely up to 8) of low angle (< 10°) stratified sandstones ranging in thickness from 30 to 90 cm (Fig. 12). The sets are commonly continuous across the width of outcrop (minimum 10 m), but in places truncate each other vertically and laterally. In one occurrence the sets change laterally to more trough-like shapes, and in a different occurrence the single set resembles a very shallow, very wide, trough cross-bed. Each occurrence is developed on a surface that may be flat or eroded into shallow scours, and all occurrences

are truncated by erosional intraclast-strewn surfaces.

Within each set, the sweeping slightly curved layers are very continuous laterally, and in almost all cases have an original dip of less than 10°, flattening to horizontal at the base of each set. Individual layers are about 1 cm thick, and are well defined by alternating finer layers.

In places, the top surfaces of the layers show a well developed parting lineation, which is parallel to the depositional *strike* of the inclined layers. The grain size is fine sand, with mean grain sizes in the range 0.10 to 0.25 mm. Comminuted plant material is abundant.

Petrology of the Sandstones

The objectives of our work did not include a detailed study of petrology. The average com-

FIG. 11. Alternating layers of fine siltstone and ripple cross laminated fine sandstone (facies F), truncated by a massive sandstone with mudstone intraclasts (scoured surface, SS). Dark bars on notebook each 5 cm, numbers are tenths of feet. Photo at 9 m, Fig. 2.

position of five point-counted thin sections from different levels in the 120 m sequence is shown in Table 1; this composition falls into the subarkose class of Folk (1968).

The cement is dominantly clay, probably kaolinite, but there is also some quartz cementation that is younger than the clay. Hematite is dispersed throughout the cement. The analyses suggest a 'granite'-rich source area, with some volcanic contribution: this is consistent with an Acadian–Caledonian mountain source.

Facies Sequence: Analysis and Distillation

Having defined eight facies, we are faced with the problem of analyzing their sequence and, if possible, distilling their sequence into a generalized or idealized form that can be used as a basis for a facies model. Many models have been presented in the literature without any real discussion of their formulation. We feel that it is important to discuss how our model is formulated, and to show where we are making subjective

TABLE 1. Average composition of five sandstone specimens

Constituent	Percentage
Quartz	66
Orthoclase	10
Volcanic rock fragments	7
Sedimentary rock fragments	5
Chert	3
Hematite	3
Perthite	2
Plagioclase	2
Cement	1
Others*	1

*Includes metamorphic rock fragments, zircon, muscovite, biotite, chlorite, pyrite, and epidote.

judgements, and where our conclusions can be substantiated statistically.

There are two main ways that the raw facies sequence (Fig. 13) can be visualized. The upward transitions can either be shown with arrows in a 'Facies Relationship Diagram' (Fig. 14), or they can be tallied and converted into probabili-

FIG. 12. Below notebook, interbedded fine siltstones and sandstones of facies F. Immediately above notebook, a fine sandstone (60 cm) shows low angle inclined stratification (facies G), and is truncated irregularly by coarser sandstones showing poorly and well defined trough cross stratification (facies A and B). Photo at 37 m, Fig. 2.

ties, shown in Table 2. This transition probability matrix is concise, and has been used by many workers; the most recent application to fluvial sediments was by Miall (1973), who cites most of the earlier literature. The main objection to the use of such a matrix is that it does not show the nature of the transitions (gradational, sharp, channelled), but in the Battery Point Formation where sharp and channelled contacts predominate, this is not a serious objection (see de Raaf *et al.* 1965, fig. 22).

In order to comprehend the facies relationship diagram (Fig. 14), some simplification is necessary. There are at least three possible approaches: (1) arbitrary rejection from Figure 14 of those transitions that occur only once; (2) arbitrary rejection from Table 2 of those transitions that have a low probability of occurring; and (3) examination of the data to see which transitions occur more, and less, frequently than if the facies were in a random sequence. The third approach

D. A,B,E,F.B.A,G.B,C,A,C,B,C,B,D,F/
A,B,D,F,G. A,C,A / A.A /C,B,G. A,B,
A,B,G. A,, B. A,E,C,B,F. A,B. A,C,D/
A / B.A. A / G.C

. SS, SCOURED SURFACE
, SHARP CONTACT
,, GRADATIONAL CONTACT
/ FAULT OR COVERED INTERVAL

FIG. 13. Sequence of facies observed in the study area. From stratigraphic base, upward, the sequence reads D, A, B, E, *etc.*

was first suggested by Selley (1970; see also Harms *et al.* 1975), and has been refined by Miall (1973). From the raw data, a transition probability matrix was calculated for the observed transitions (Table 2). Then a second matrix was calculated assuming the same abundance of facies but in a random sequence; thus, the probability of transition from facies A to any

TABLE 2. Observed transition probabilities

	SS	A	B	C	D	E	F	G
SS		.800	.133	.067				
A	.154		.462	.231		.077		.077
B	.308	.077		.154	.154	.077	.077	.154
C		.286	.571		.143			
D	.333						.667	
E				.500			.500	
F	.667							.333
G	1.000							

TABLE 3. Transition probabilities for random sequence

	SS	A	B	C	D	E	F	G
SS		0.320	0.245	0.151	0.075	0.038	0.075	0.094
A	0.280		0.260	0.160	0.080	0.040	0.080	0.100
B	0.259	0.315		0.148	0.074	0.037	0.074	0.093
C	0.237	0.288	0.220		0.068	0.034	0.068	0.085
D	0.222	0.270	0.206	0.127		0.032	0.063	0.079
E	0.215	0.262	0.200	0.123	0.062		0.062	0.077
F	0.222	0.270	0.206	0.127	0.063	0.032		0.079
G	0.226	0.274	0.210	0.129	0.065	0.032	0.065	

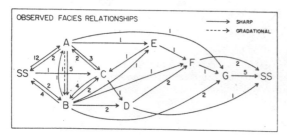

FIG. 14. Facies relationship diagram showing all observed transitions. Numbers indicate how many times a particular upward transition occurred.

other facies depends only upon the observed relative abundances of the other facies (Table 3). A chi-squared test showed that the observed facies transitions were statistically significantly different from random. A *difference* matrix was then calculated (observed minus random), which highlights those transitions that have a higher or lower probability of occurring than if the sequence were random (Table 4—computational details in Miall 1973).

We can now reconstruct a facies relationship diagram showing the transitions that occur more frequently than random. The nature of the transitions can be shown by reference to the raw data, and different weights of arrows can be used to show subjectively how much more com-

monly certain transitions occur (Fig. 15). By trial and error, it appears for the Battery Point data that a reasonable amount of simplification occurs if only the entries in the difference matrix (Table 4) greater than +0.05 are shown in Fig. 15. Lower difference values indicate transitions little different from random. Because our purpose is simplification, we have not shown such transitions in Fig. 15, although we are

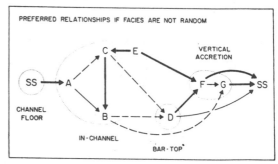

FIG. 15. Facies relationship diagram, showing only those transitions that occur more commonly than random. Diagram is derived from data in Table 4, and only shows transitions whose observed-minus-predicted transition probabilities exceed 0.05. Dotted arrows show probabilities in the range 0.05 to 0.10; light solid arrows, 0.10 to 0.30; heavy solid arrows, probabilities > 0.30. Four facies associations have been interpreted and circled on the diagram; the asterisk against Bar Top is explained in the text.

TABLE 4. Observed minus random transition probabilities

	SS	A	B	C	D	E	F	G
SS		+0.480	−0.112	−0.084	−0.075	−0.038	−0.075	−0.094
A	−0.126		+0.202	+0.071	−0.080	+0.037	−0.080	−0.023
B	+0.049	−0.238		+0.006	+0.080	+0.040	+0.003	+0.061
C	−0.237	−0.002	+0.351		+0.075	−0.034	−0.068	−0.085
D	+0.111	−0.270	−0.206	−0.127		−0.032	+0.604	−0.079
E	−0.215	−0.262	−0.200	+0.377	−0.062		+0.438	−0.077
F	+0.445	−0.270	−0.206	−0.127	−0.063	−0.032		+0.254
G	+0.774	−0.274	−0.210	−0.129	−0.063	−0.032	−0.065	

aware (and will continue to be aware in our interpretation) that we have arbitrarily rejected some of the data.

Facies and Facies Sequence: Interpretation and Model Building

We consider here the interpretation of Fig. 15, and the way in which this interpretation, along with interpretations of individual facies, can be presented as a *local model*.

First, it is clear in Fig. 15 that, above the scoured surface, facies A, B, and C (trough cross-bedded and planar–tabular cross-bedded sandstones) tend to occur in close mutual association. The trough cross-bedded sandstones (facies A and B) represent unidirectional migration of dunes, and the planar–tabular sets probably represent movement of larger bed forms. The occurrence of all three facies above intraclast-strewn scoured surfaces suggests deposition within a channel. This interpretation will be elaborated below.

The smaller scale of the cross bedding in facies D, together with its tendency to occur above facies B and C (Fig. 15), suggests deposition in topographically higher parts of the fluvial system, perhaps as small bed forms migrating on top of larger sand flats or bars in fairly shallow water, or in the uppermost parts of channels after most of the channel has been filled. For brevity, both of these topographically higher environments have been combined as bar top* in Figs. 15 and 16. The asterisk is intended to remind readers that more than just bar top is implied.

The particular usefulness of this type of facies analysis shows up in the interpretation of facies E, the large asymmetrical scours. In Fig. 15, facies E is related to facies C and F; a re-examination of the difference matrix (Table 4) also shows that the A → E and B → E transitions occur more

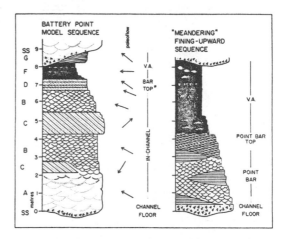

FIG. 16. The Battery Point model sequence is a sketch of the data in Fig. 15, showing average facies thicknesses as well as facies sequence. Facies E is omitted because it does not occur a sufficient number of times to allow its typical position to be sketched. Paleocurrent arrows are averages of all the data applicable to each facies: the grand vector mean for facies A and B together is 291° (standard deviation 40°). The 'meandering' fining-upward sequence is drawn to the same scale, using the data on thicknesses and sequence of structures given by Allen (1970). V.A. indicates vertical accretion, and within the V.A. deposits, calcareous concretions, desiccation cracks, and rootlets are shown symbolically.

frequently than random. This relationship of E with A, B, and C suggests that the formation of the asymmetrical scours takes place *in-channel* rather than in topographically higher parts of the system (scouring on the tops of point bars, for example, has been described by Harms *et al.* 1963).

Finally, Fig. 15 shows that facies F and G consistently occur stratigraphically above all other facies, and are consistently followed by scoured surfaces. The alternation of cross-laminated fine sandstones and siltstones with mudstones (facies F) suggests *vertical accretion* in disused parts of the floodplain, with subsequent erosion and re-establishment of an active channel. Facies G has enigmatic sedimentological characteristics, but because of its consistent position in the facies sequence and its upper flow regime features, it must represent vertical flood stage accretion of sand in an overbank or non-channelled area.

The transitions shown in Fig. 15 can be expressed as a stratigraphic column (Fig. 16). This figure attempts to show the characteristic thicknesses of the various facies, and the number of times they can be interbedded in a typical Battery Point facies sequence. Not all sequences are so complete (see Fig. 2); for example, the sequence SS → A → B → SS implies channel cutting, dune migration, and then renewed cutting before any of the topographically higher facies could be formed or preserved. Facies E has been omitted from the stratigraphic column in Fig. 16—the two occurrences of the facies are in different contexts in the facies sequence, and hence, we do not have enough consistent data to idealize its position in Fig. 16.

Paleocurrent Directions, and the Use of $\Delta\Theta$

In the field, paleocurrent directions were measured directly on such features as plan views of trough axes, and direction of dip of foresets of planar–tabular cross-bedding. For any one occurrence of an individual facies, individual readings were averaged to give a single direction associated with the deposition of that facies.

The bulk of the data (Fig. 16) is from the trough and planar cross-bedded facies. Combining facies A and B, 32 occurrences of the facies show a grand vector mean flow direction of 291° (standard deviation, 40°). Because these are the basal in-channel facies, their flow directions are probably close to that of the regional paleoslope, which therefore had a west-north-west dip.

Facies C (planar–tabular cross-bedded sandstones) characteristically has a paleoflow direction that is at a high angle to the *immediately* underlying and overlying facies (commonly facies B). This angular difference (Fig. 16) can best be demonstrated by using a new way of presenting paleocurrent data—that is, the $\overline{\Delta\Theta}$ matrix. If an occurrence of facies i (flow direction Θ_i) is directly followed above by facies j (flow direction Θ_j), then $\Delta\Theta_{ij}$ is the absolute value of the difference in directions:

$$[1] \qquad \Delta\Theta_{ij} = |\Theta_i - \Theta_j|$$

Care must be taken with azimuths close to north: with hypothetical readings of 350° and 010°, $\Delta\Theta$ is 20°, not 340°!

The average value of $\Delta\Theta_{ij}$ (written $\overline{\Delta\Theta}_{ij}$) for all facies pairs (i directly followed by j) can be calculated from

$$[2] \qquad \overline{\Delta\Theta}_{ij} = \frac{1}{n}\sum |\Theta_i - \Theta_j|$$

where n is the number of times facies i is followed directly by facies j, where paleocurrent data is available for both facies. The $\overline{\Delta\Theta}$ matrix for the Battery Point facies is shown in Table 5. The highest angular divergences (excluding single observations) are B → C, C → A, and C → B. This indicates a very strong tendency for the single planar–tabular sets of facies C to occur above B at a high angle, and to be followed in turn by facies A or B at a high angle.

These relationships can be interpreted to imply that the planar–tabular cross-bedding represents the deposits of some form of bar or 'microdelta' that grows laterally in an active channel (B → C). After construction of the planar–tabular set, small dunes resume migration down the channel in their original direction (C → B). The planar–tabular sets probably represent lateral growth of an existing mid-channel transverse bar. This type of bar growth has been documented in detail for the Platte River by Smith (1971, 1972). These paleocurrent data are interesting, because they again suggest that braided streams have relatively high paleocurrent variances. This point was emphasized by Smith (1971, 1972), but was criticized on theoretical grounds by Banks and Collinson (1974), who discussed the preservability of many of the features observed by Smith. However, deposits formed by lateral bar growth appear to be preserved in the Battery Point formation, and hence we agree with Smith (1972, 1974) that lateral bar growth is important, and that it gives rise to relatively high paleocurrent variances.

Nature of the Battery Point River: Comparison With Other Fluvial Systems

The combination of the overall stratigraphic setting, the presence of rare rootlets, and the scoured surfaces overlain by trough cross bedding (with some preservation of vertical accretion deposits) indicate very strongly that the

TABLE 5. $\overline{\Delta\Theta}$ matrix, showing average paleocurrent divergences between different facies (see text for details)

	SS	A	B	C	D	E	F	G
SS		45*	23					
A		18	5			7*		20*
B		30*	45	59	36	10*		55*
C		75	67	28				
D					23			
E				43*		27*	22*	
F								
G								

NOTE: 50 facies transitions are represented in this table; readings marked * represent single observations, all others are averages.

section of Battery Point studied represents a fluvial environment. Fortunately, there exists a well developed generalized facies model for shallow *meandering* rivers that can be used as a basis for comparison. The latest and most general statement of this meandering sequence has been given by Allen (1970), and is reproduced to scale in Fig. 16. Deposition takes place above a channel floor lag by lateral accretion of a point bar. The upper part of the bar is rippled, and is overlain by vertical accretion deposits.

It can be seen that the Battery Point local model differs considerably from the meandering fluvial model. The major differences are:

(1) The Battery Point is characterized by preservation of a thick sandstone sequence, the bulk of which represents the deposits of sinuous-crested dunes moving on the channel floor.

(2) Along with the abundant trough cross-bedding, the Battery Point contains thick, isolated, planar–tabular cross-bed sets. They always have high paleocurrent divergences from the channel floor trough cross-bedding. Thick planar tabular sets do not appear to be an important part of the meandering river model.

(3) There is relatively little preservation of vertical accretion deposits in the Battery Point river system.

We believe that these differences are sufficiently great that the Battery Point local model cannot be considered as a mere variant of the general meandering model. The abundance of trough cross-bedding, the evidence of mid-channel transverse bars with lateral growth of planar–tabular sets of cross-bedding, and the meagre preservation of vertical accretion deposits all suggest a *braided*, or low sinuosity pattern for the Battery Point River.

Unfortunately, there is no existing model for

sandy braided deposits with which the Battery Point can be compared. The most important studies of modern sandy braided streams are those of Doeglas (1962), Collinson (1970), and Smith (1970, 1971, 1972). All give excellent descriptions of the various morphological elements in the river, and discuss their formation, but the three authors do not emphasize the stratigraphic sequence that might form if the deposits were preserved. In ancient rocks, braided (or low sinuosity) channel interpretations have been suggested for Devonian rocks in Spitsbergen (Moody-Stuart 1966) and part of the Late Pennsylvanian Coal Measures of South Wales (Kelling 1968). Kelling emphasizes fining-upward sequences, with a typical one beginning with a lag gravel, and passing up into cross-stratified sandstones, ripple-laminated sandstones, flat laminated siltstones, mudstone, underclay, and coal (Kelling 1968, fig. 4). The paleocurrent vectors for the cross-stratified sandstones agree fairly closely with the axial trend of the channels, but higher in the cycle, there is considerable paleocurrent variance. This is comparable with the Battery Point (Fig. 16), especially with respect to the alignment of the trough cross bedding with the assumed channel trend, but the up-to-180° divergences in ripple orientation are greater in the Coal Measures than in the Battery Point.

Because of the limited number of studies, we are unable to say how the Battery Point compares with other sandy braided deposits. We are also unable to locate the Battery Point in the spectrum between highly sinuous meandering river types, and highly anastomosing braided types characterized by many active channels and rapidly migrating bars.

In suggesting the existence of bars within the

Battery Point channel, we have relied on the evidence of the single, thick, planar–tabular sets of cross-bedding, with their anomalous paleocurrent directions. These sets overlie trough cross-bedding, implying aggradation of the channel floor before bar formation. The planar–tabular sets are themselves overlain by trough cross-bedding that shows the original paleocurrent direction (Fig. 16), implying yet more channel filling to, and above, the level of the bars. We therefore suggest that aggradation of the channel floors was an important process leading to preservation of sand bodies in the geological record. This was earlier suggested by Moody-Stuart (1966) for low sinuosity (braided) sandy streams, but was emphatically denied by Allen (1970) in his discussion of Moody-Stuart. Allen (1970, pp. 311–314) contends that all preservation of coarse members in alluvial fining-upward sequences (presumably of any type) is due to lateral accretion, as in the well understood classical meandering model. We have shown above (and in Fig. 16) that the Battery Point model is quite unlike the classical meandering model, and our present understanding of the evidence seems to suggest that vertical aggradation was more important than Allen would imply ("vertical deposition of the coarse members as envisaged by Moody-Stuart (1966) can therefore be ruled out"; Allen 1970, p. 313). Vertical aggradation has been implied for some modern braided rivers, for example, the Lower Yellow River of China (Chien 1961, p. 745), and is presently being studied by Cant (1975) in the South Saskatchewan River.

We envisage the initial formation of a new scour, and the lining of the scour with mud clasts and massive sandstone (facies SS). Rapid migration of sinuous crested dunes, with coarse bed load, gave rise to the poorly defined trough cross bedding (A), but as some aggradation took place, the coarser material ceased to move, the dunes advanced more slowly, and the well defined trough cross bedding (B) formed. At places within the stream, large transverse bars with foreset slopes up to 3 m high were formed; they migrated both downstream and laterally, to form planar–tabular sets of cross bedding (C). When these bars became inactive, further aggradation led to their burial by more trough cross bedding (B). At places within the channels, unusually deep isolated scours were cut and filled (facies E), and after extensive aggradation, smaller

straight crested dunes or sand waves up to about 30 cm high migrated across the sand flats or shallow channels now occupying this part of the river. As the main flow became diverted elsewhere, the area was in places covered by rippled silts; occasional overbank flooding contributed vertical accretion muds (facies F). More severe flooding probably gave rise to the low angle, cross stratified sandstones (facies G); this particular interpretation is suggested by the position of facies G in the overall facies sequence, and by the presence of upper flow regime sedimentary structures. With overall subsidence of the basin, the area just described was later reoccupied, and another overall fining-upward sequence was deposited.

Discharge and Width of the Battery Point River

The deepest scoured channel that we observed was about 5 m; it was filled entirely with trough cross bedding. If we make the reasonable assumption that at some time (bankfull discharge) dunes were moving on the bed in a flow depth of 5 m, then we can estimate the discharge of the Battery Point River.

The mean flow velocities required for dunes (composed of medium sand) to migrate in shallow flows (up to 40 cm) lies in the range 50–100 cm/s (Harms *et al.* 1975, fig. 2–3). For flows a few metres deep, the velocities are upward of 100 cm/s (Harms *et al.* 1975, figs. 2–10, and 2–11); we shall use a value of 120 cm/s in the calculations below.

Also, for dunes we can estimate an average resistance to flow. The dimensionless Chezy discharge coefficient, $C/\sqrt{g}$, ranges from 8 to 12 for dunes; we will use a value of 10.

The Chezy equation is

$$[3] \qquad \overline{U} = C(DS)^{1/2}$$

where $\overline{U}$ = mean velocity (cm/s), C is the Chezy discharge coefficient, D is mean flow depth (cm), and S is slope.

Substituting in [3], we have

$$[4] \qquad \begin{array}{c} 120 = 10 \cdot g^{1/2} (500 \cdot S)^{1/2} \\ \text{or} \\ S \approx 2.9 \times 10^{-4} \end{array}$$

It has been shown that in a general way, braided and meandering streams can be separated empirically by their slope (S) and discharge (Q) relationships (Leopold and Wolman 1957), the equation being

[5] $S = 0.06 \, Q^{-0.44}$

Thus, for a given slope the discharge is generally greater than that given by [5] for braiding, and less for meandering. If the Battery Point River were braided, on a slope of 2.9×10^{-4}, the bankfull discharge may have exceeded 100 000 ft^3/s (2840 m^3/s). This is about one sixth of the mean annual discharge of the Mississippi, and implies a fairly large river draining the Acadian mountains.

At a mean velocity of about 1 m/s, the cross sectional area of the river must have been 2840 sq. m to accommodate bankfull flow. Assuming an average depth at bankfull to be 5 m, the channel width must have been about 567 m. Adding in the various bars and islands within the river, the original width may have been of the order of 1–2 km.

Status of Sandy Fluvial Facies Models

A well-established facies model should serve four important functions (Walker, *in* Harms *et al.* 1975): (1) because it is a distillation of many examples, the model is a *norm* with which other examples can be compared and contrasted; (2) because the model embodies the descriptive essence of many examples, it acts as a *framework* and *guide* for future observations; (3) because the model emphasizes the important genetic features of many examples, it can act as a generalized basis for *hydrodynamic interpretation*; and (4) because the model emphasizes the genetic relationships between its components, it can act as a *predictor* in new situations where only parts of the system may be exposed.

The generalized meandering model of Allen (1970; our Fig. 16) fulfills all of the above functions about as well as any existing facies model. A possible objection to the model is that in its statistical formulation, Allen (1970) may have blurred together the deposits of (1) active lateral accretion of point bars, (2) chute cutoffs, and (3) neck cutoffs. These three types of alluvial deposition give rise to three recognizably distinct fining-upward sequences. Nevertheless, with the existence of the *norm*, departures from it, in the form of neck or chute cutoff deposits, are more easily identified. Thus Allen's model is a norm; it has also guided other observations, and has acted as a basis for hydrodynamic interpretation (Allen 1970, pp. 314–317). It can also be used successfully as a predictor in new situations.

Extrapolation from the meandering model to braided situations is clearly not possible, and this emphasizes the importance of establishing some form of preliminary braided model. The model ideally will be descriptive, as is our local model, but the more examples that can be incorporated and distilled into a general model, the better it will serve as a basis for interpretation. The braided model is presently not sufficiently well formulated to resolve the interpretive disagreement between Allen and Moody-Stuart. Input into the model must come from detailed studies of ancient rocks believed to be braided in origin, and from recent sediment studies of the type done on sandy braided streams by Collinson (1970), Smith (1970), and Cant (1975). As more studies are made, the 'generalized' braided model will gain in its predictive power, and we will have a better geological understanding of the spectrum of braided river types.

Acknowledgments

Many of the ideas in this paper have been stimulated by discussions with G. V. Middleton, J. C. Harms, J. B. Southard, and D. B. Spearing; some of the ideas were presented at an SEPM Short Course in Dallas in April, 1975. We are indebted to the Society of Economic Paleontologists and Mineralogists for permission to reproduce Figs. 13 and 14. The text has been improved by the comments of P. McCabe. The work was supported by the National Research Council of Canada, and the field expenses of the first author were covered by a grant from the Geological Society of America. We thank these colleagues and agencies for their interest in, and support of, our work; we are solely responsible for the facts and interpretations presented herein.

ALLEN, J. R. L. 1965. A review of the origin and characteristics of recent alluvial sediments. Sedimentology, **5**, pp. 89–191.
———— 1970. Studies in fluviatile sedimentation: a comparison of fining-upwards cyclothems, with special reference to coarse-member composition and interpretation. J. Sediment. Petrol. **40**, pp. 298–323.
ALLEN, J. R. L. and FRIEND, P. F. 1968. Deposition of the Catskill facies, Appalachian region: with notes on some other Old Red Sandstone basins. *In*: Late Paleozoic and Mesozoic continental sedimentation, northeastern North America. (G. deV. Klein, *Ed*.) Geol. Soc. Am. Spec. Pap. 106, pp. 21–74.
BANKS, N. L. and COLLINSON, J. D. 1974. Discussion of "some sedimentological aspects of planar cross-strati-

fication in a sandy braided river". J. Sediment. Petrol. 44, pp. 265–267.

BOUCOT, A. J., CUMMING, L. M. and JAEGER, H. 1967. Contributions to the age of the Gaspé Sandstone and Gaspé Limestone. Geol. Surv. Can. Pap. 67-25, 27 p.

BULLARD, E. C. EVERETT, J. E. and SMITH, A. G. 1965. The fit of the continents around the Atlantic. In: A symposium on continental drift. (P. M. S. Blackett, E. C. Bullard, and S. K. Runcorn, Eds.) R. Soc. London Phil. Trans. 258A, pp. 41–51.

CANT, D. J. 1975. Sandy braided stream deposits in the South Saskatchewan River. Geol. Soc. Am. Abst. with Prog. 7 (no. 6), p. 731.

CHIEN, NING. 1961. The braided stream of the lower Yellow River. Sci. Sin. 10, pp. 734–754.

COLLINSON, J. D. 1970. Bed forms of the Tana River, Norway. Geog. Ann. 52A, pp. 31–56.

DE RAAF, J. F. M., READING, H. G., and WALKER, R. G. 1965. Cyclic sedimentation in the lower Westphalian of North Devon, England. Sedimentology. 4, pp. 1–52.

DOEGLAS, D. J. 1962. The structure of sedimentary deposits of braided rivers. Sedimentology 1. pp. 167–190.

EYNON, G. and WALKER, R. G. 1974. Facies relationships in Pleistocene outwash gravels, southern Ontario: a model for bar growth in braided rivers. Sedimentology, 21, pp. 43–70.

FOLK, R. L. 1968. Petrology of sedimentary rocks. Hemphills, Austin, Texas. 170 p.

HARMS, J. C., MacKENZIE, D. B., and McCUBBIN, D. G. 1963. Stratification in modern sands of the Red River, Louisiana. J. Geol. 71, pp. 566–580.

HARMS, J. C., SOUTHARD, J. B. SPEARING, D. R., and WALKER, R. G. 1975. Depositional environments as interpreted from primary sedimentary structures and stratification sequences. Soc. Econ. Paleontol. Mineral, Short Course 2, 161 p.

KELLING, G. 1968. Patterns of sedimentation in Rhondda Beds of South Wales. Am. Assoc. Pet. Geol., Bull. 52, pp. 2369–2386.

LEOPOLD, L. B. and WOLMAN, M. G. 1957. River channel patterns: braided meandering and straight. U.S. Geol. Surv. Prof. Pap. 282-B.

McGERRIGLE, H. W. 1950. The geology of eastern Gaspé. Quebec Dep. Mines, Geol. Rep. 35, 168 p.

MIALL, A. D. 1973. Markov chain analysis applied to an ancient alluvial plain succession. Sedimentology, 20, pp. 347–364.

MOODY-STUART, M. 1966. High- and low-sinuosity stream deposits with examples from the Devonian of Spitsbergen. J. Sediment. Petrol. 36, pp. 1102–1117.

SELLEY, R. C. 1970. Studies of sequence in sediments using a simple mathematical device. Geol. Soc. London Quart. J. 125, pp. 557–581.

SMITH, N. D. 1970. The braided stream depositional environment: comparison of the Platte River with some Silurian clastic rocks, North-Central Appalachians. Geol. Soc. Am. Bull. 81, pp. 2993–3014.

——— 1971. Transverse bars and braiding in the lower Platte River, Nebraska. Geol. Soc. Am. Bull. 82, pp. 3407–3420.

——— 1972. Some sedimentological aspects of planar cross-stratification in a sandy braided river. J. Sediment. Petrol. 42, pp. 624–634.

——— 1974. Some sedimentological aspects of planar cross-stratification in a sandy braided river: a reply to N. L. Banks and J. D. Collinson. J. Sediment. Petrol. 44, pp. 267–269.

ST. JULIEN, P., HUBERT, C., SKIDMORE, W. B., and BELAND, J. 1972. Appalachian structure and stratigraphy, Quebec. Int. Geol. Cong., Montreal, Guidebook for field excursion A56-C56, 99 p.

3

Copyright © 1969 by the Society of Economic Paleontologists and Mineralogists

Reprinted from *Jour. Sed. Petrology* **39**(4):1310–1316 (1969)

ANTIDUNES IN THE MOUNT TOBY CONGLOMERATE (TRIASSIC), MASSACHUSETTS[1]

BRYCE M. HAND[2]
Amherst College, Amherst, Massachusetts
JAMES M. WESSEL AND MILES O. HAYES
University of Massachusetts, Amherst, Massachusetts

ABSTRACT

Nineteen consecutive sand waves, interpreted as antidunes, occur on an exposed bedding surface of granule conglomerate within the Mount Toby Conglomerate (Triassic) near Sunderland, Massachusetts. The waves are nearly symmetrical, with smoothly rounded crests somewhat sharper than the intervening troughs. Slip faces are not developed. Average wave length is 63.3 cm, average height 6.5 cm. Cross laminae within the waves are parallel to the upstream wave fronts (maximum steepness about 22°), indicating that the antidunes were migrating in the upstream direction. Current direction is known independently from cobble imbrication in adjacent beds, regional grain size trends, oriented structures, and known source areas.

Hydrodynamic calculations indicate that the preserved antidunes formed in water about 1.6 cm deep, flowing at 100 cm/sec down a paleoslope of at least 2.7°. These conditions would be compatible with sheetwash on the surface of an alluvial fan.

Upstream-dipping cross-laminae are not limited to the unit displaying sand waves, but occur abundantly in many other beds at the Sunderland exposure and elsewhere within the Mount Toby Conglomerate. Thus, antidune structures may be far more abundant in the geologic record than has been assumed.

INTRODUCTION

Among the significant advances in sedimentology in recent years has been an increased understanding of bedforms in alluvial materials. The simple description of bedding characteristics that once sufficed is giving way to attempts to relate sedimentary structures (through the bedforms that produced them) to hydraulic conditions at the time of deposition.

As outlined by a number of workers (for example, Gilbert, 1914; Simons and Richardson, 1963; Simons, Richardson, and Nordin, 1965; Harms and Fahnestock, 1965), it is usual for ripples to develop in sand when current velocity is low, and for these to be replaced by dunes, plane bed, and antidunes at progressively greater velocities if water depth remains constant. Details of this sequence are given in the references cited.

Stratification types attributable to ripples and dunes (lower flow regime) and to the plane bed phase of the upper flow regime have been widely recognized in the sedimentary record (Visher, 1965; Klein, 1965; Gwinn, 1964). Much less common, however, are reports of structures produced by antidunes. Power (1961) reported upstream-dipping cross lamination in coarse sands of the Coso Formation, a Pleistocene alluvial fan deposit in California, and suggested its relation to antidune sedimentation. Harms and Fahnestock (1965) mentioned the occurrence of similar crossbedding in Recent deposits of Medano Creek, Colorado. Walker (1967) described wave-like forms within sandstone beds of the Hatch Formation (Devonian) of New York, which he interpreted as short-crested antidunes. This limited published record of antidune structures in ancient rocks suggests any of three possibilities: (1) antidune deposits may be insignificant in the stratigraphic record; (2) antidunes may not produce distinctive structures that are capable of preservation; or (3) such structures for the most part have been overlooked or misinterpreted. Our own observations cause us to favor the last possibility.

By far the largest number of descriptions of antidunes have come from the laboratory, beginning with the pioneer work of Gilbert (1914), who conducted experiments on antidunes and named the bedform. More recent studies have been carried out by Kennedy (1961), Simons, Richardson, and Albertson (1961), Middleton (1965), and many others.

Active antidunes can also be seen in many natural environments. They are commonplace in the swash zone of many beaches (Hayes). Rahn (1967) observed breaking antidunes associated

[1] Manuscript received December 9, 1968; revised March 10, 1969.
[2] Present Address: Department of Geology, Syracuse University, Syracuse, New York.

41

FIG. 1.—Antidunes in Mount Toby Conglomerate, 1.8 miles northeast of Sunderland, Massachusetts.

with flash flood runoff on alluvial fans. Although not labeled as such, antidune waves in the San Juan River near Mexican Hat, Utah, are shown in figure 3–2 of *Principles of Geology* (Gilluly, Waters, and Woodford, 1968). Similar waves appear in great profusion in the dust jacket photograph of *Fluvial Processes in Geomorphology* (Leopold, Wolman, and Miller, 1964), which shows a section of Muddy River, Mt. McKinley, Alaska. They can also be seen in figures 1–3 (Missouri River floodplain) and 9–16 (Quinebaug River, Connecticut) in *Physical Geology* (Longwell, Flint, and Sanders, 1969). The Quinebaug photograph is particularly interesting, documenting the fact that large quantities of sand and gravel can be deposited by a stream in the antidune phase of the upper flow regime.

GEOLOGIC SETTING

Triassic sediments of northern Massachusetts comprise a wholly continental sequence of fanglomerates, fluvial sandstones, mudstones, and lake sediments occupying a fault valley flanked by higher areas of metamorphosed Paleozoic rocks. The Mount Toby Conglomerate, which contains the antidunes described in this paper, is stratigraphically highest of the Triassic formations preserved in the area. It consists of fine to very coarse sandstone with tongues of poorly sorted pebble to boulder conglomerate, and represents a coarse border facies whose fine-grained lateral equivalents have been removed by erosion. The conglomerates become coarser and volumetrically more important toward the

eastern source, grading locally into talus breccia along the margin of the basin.

The regional paleogeography is one of coalescing alluvial fans spreading westward from a major border fault. Wessel (1969) has shown that radial paleocurrent patterns and systematic changes in grain size and clast lithology within the Mount Toby Conglomerate can be used to identify the principal centers of sediment dispersal. The structures described in this paper are preserved in sediments deposited on a fan whose apex lay about 2 miles to the northeast. The coarseness of the sediment argues for swift, probably intermittent streams flowing down steep slopes in what may have been a semi-arid climate. Further evidence for swift stream velocities is found in the dominance of planar bedding and low-angle crossbeds and virtual absence of angle-of-repose crossbeds and ripples or ripple lamination in the fan deposits. The bedding types that are present would seem to require transitional and upper flow regime conditions. If antidunes can be preserved, it is in just such an environment that one might expect to find them.

DESCRIPTION OF ANTIDUNES

The sediment waves here interpreted as antidunes occur in a nearly vertical exposure 1.8 miles northeast of Sunderland, Massachusetts (fig. 1). Nineteen consecutive wave forms are preserved on the upper surface of a bed of granule conglomerate. The bed itself can be traced continuously from one end of the expo-

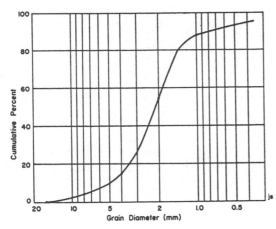

Fig. 2.—Grain size distribution for bed containing antidunes. Data from point count on polished section.

Table 1.—*Dimensions of Triassic antidunes near Sunderland, Massachusetts.*

Antidune	Height (H) in cm	Length (L) in cm	H/L
1	7		
		61	
2	7		.115
		48	
3	4		.083
		69	
4	7.5		.109
		64	
5	5.5		.086
		71	
6	7		.099
		69	
7	8		.116
		61	
8	7		.115
		69	
9	7		.102
		64	
10	6		.094
		61	
11	7		.115
		66	
12	8		.121
		64	
13	6.5		.102
		58	
14	4		.069
		64	
15	7		.110
		66	
16	7		.106
		66	
17	5		.076
		61	
18	6		.098
		58	
19	7		.120
Average	6.5	63.3	.102

← Downstream direction

sure to the other, a distance of about 14 m, with little change in character. Its average thickness is about 10 cm. Median grain size is 2.1 mm (fig. 2). Immediately above the granule conglomerate bed is a thin layer of relatively non-resistant, fine-grained sandstone. This sandstone has been stripped from the upper surface of the conglomerate, exposing the wave forms in three dimensions for about 20 cm into the cliff face.

The waves are approximately trochoidal in form, with smoothly rounded crests and troughs, the crests somewhat sharper than troughs. No avalanche faces are developed. Although individual waves are nearly symmetrical, they tend to be slightly steeper on their upstream (northeastern) flanks. Measurements of upstream and downstream faces, averaged for the six steepest waves, show inclinations of 22° and 19.5°, respectively. These measurements are made with respect to the main bedding, no allowance being made for inclination of the original fan surface.

Average wavelength is 63.3 cm, with all but two crest-to-crest measurements falling within 10 percent of this value (table 1). The regularity of the wave train is shown by a composite profile constructed for the whole length of the exposure (fig. 3). Wave height is more variable than wavelength, ranging from 4 to 8 cm and averaging 6.5 cm (table 1).

Across the width of the exposed upper surface of the bed (maximum 25 cm), wave crests appear to be parallel to one another and show no evidence of rapid change along their strike. Indeed, because the trend of the outcrop makes

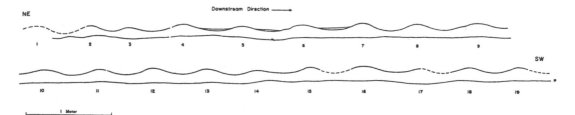

Fig. 3.—Profile of antidune set, constructed from tracings of individual waves in serial photographs.

a horizontal angle of about 60° with the strike of the wave crests (S 29 E), an observer moving 14 m from one end of the outcrop to the other also travels 7 m "across current." This suggests that individual crests must have been continuous for 7 m or more along their strike. The preserved waves may therefore be regarded as long-crested (or "two-dimensional") antidunes.

Crossbedding within waves is crude and difficult to photograph, but easily distinguished in the field. The crossbeds dip northeastward, indicating that the bedform waves were moving northeastward (the upstream direction as indicated by regional paleogeography and imbrication in associated conglomerates). Individual cross-laminae are inclined approximately parallel to the wave fronts. Because the front of an individual wave is sigmoidal, cross-laminae would be expected to display a range of dip values; field observations confirm this to be the case. Maximum steepness of laminae approaches 22°, but near the base of the bed the dips are somewhat more gentle. Presumably, above the point of inflection cross-laminae should be convex upward, but this was not definitely observed in the bed displaying antidunes.

HYDRODYNAMICS OF ANTIDUNE FORMATION

Preservation of entire sediment waves at the Sunderland exposure has provided an unusual opportunity for evaluating hydrodynamic conditions at the time of deposition.

Experiments have shown antidune wavelength (L) to be related to mean flow velocity ($\bar{v}$) in the following manner (Kennedy, 1961, p. 18):

$$\bar{v}^2 = gL/2\pi \qquad (1)$$

Accordingly, the average wavelength observed at Sunderland (63.3 cm) implies a mean current velocity of 100 cm/sec (approximately 2.2 mph).

Velocity, in turn, can be related to average depth ($\bar{d}$), slope (S) and bed resistance (f):

$$S = f\frac{\bar{v}^2}{8g\bar{d}} \qquad (2)$$

Equation (2) is a modification of the Darcy-Weisbach equation describing flow in rough pipes, in which $4\bar{d}$ has been substituted for pipe diameter and S replaces head loss per unit pipe length (Briggs and Middleton, 1965). Using equation (2) it is possible to determine combinations of $\bar{d}$, S, and f that would be compatible with the observed velocity of 100 cm/sec.

The friction factor, f, represents the combind effect of grain resistance and bedform resistance (Taylor and Brooks, 1962). However, experiments by Kennedy (1961, p. 112) show

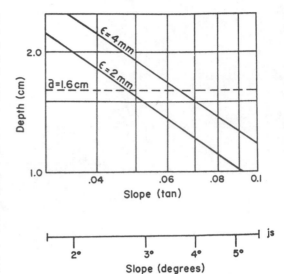

FIG. 4.—Depth-slope diagram based upon Darcy-Weisbach equation for average flow velocity 100 cm/sec. ϵ = effective size of roughness elements on bed.

that with coarse grained sediment, flow resistance over antidunes is essentially the same as that offered by a plane bed of the same material. To a first approximation, then, bedform resistance can be neglected and grain resistance alone need be considered in estimating f. Under these circumstances, the value of f depends upon Reynolds number and relative roughness, and can be read from a Moody diagram (Henke, 1966, p. 152). On the standard Moody diagram relative roughness is expressed as:

$$\frac{\text{Height of roughness elements}}{\text{Pipe diameter}}$$

In the context of open channel flow, 4 times the average depth must be used in place of pipe diameter. Average grain size can be used to represent the height of roughness elements.

Because a suitable value for f can be found, equation (2) can be used together with a Moody diagram to calculate combinations of depth and slope able to produce a known velocity over a bed of known grain size. Figure 4 shows these relations for a current velocity of 100 cm/sec and sediment of granule size.

The minimum slope necessary to produce the Sunderland antidunes can be determined by assuming that the average water depth at the time of deposition was the maximum consistent with antidune flow. Antidunes interact strongly with the free water surface, inducing stationary waves whose crests and troughs are in phase with those of the antidunes themselves. For any

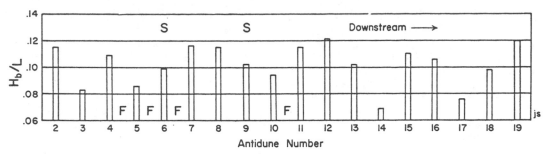

FIG. 5.—Antidune steepnesses (H_b/L) computed for individual waves. F. denotes partial filling of trough by horizontally laminated granule conglomerate. S indicates shoulder on upstream flank of antidune.

particular set of antidunes, greater water depths require steeper water waves. A limiting (maximum) water depth can therefore be found by assuming the steepest possible water waves, where $H/L = 0.142$ (Michell, reviewed in Kennedy, 1961, p. 21). Maximum steepness of antidunes in the Sunderland exposure is 0.12 (fig. 5). Following the method described by Hand (this issue) the observed value of $H_b/L = 0.12$ indicates that average water depth could not have exceeded $0.025 \times L$, or about 1.6 cm.

If the effective size of the elements of roughness on the bed was 2.0 mm (average grain size), a water depth of 1.6 cm would have required a slope of 2.7° to produce a current velocity of 100 cm/sec (fig. 4). It is probable, however, that the slope would be estimated more accurately by assuming a larger size for the elements of roughness. To represent roughness, Einstein and Barbarossa (1952) used the 65th

percentile rather than the median; Engelund (1966) used twice this value. Part of the uncertainty arises from the fact that considerable flow resistance may be offered by sediment in transport (Bagnold, 1966), an effect that is not considered in direct application of the Darcy-Weisbach equation. It must also be realized that the foregoing calculation of slope assumes that resistance offered by bedforms can be neglected. Kennedy's results (1961, p. 113) suggest that actual bed resistance may have been 10 to 15 percent greater than would have been offered by a plane bed of the same material. For all these reasons, the slope estimate of 2.7° must represent an absolute minimum.

For purposes of calculating limiting values of depth and slope, it has been assumed that the water waves accompanying the steepest antidunes were on the verge of breaking. Under these conditions, other waves within the same

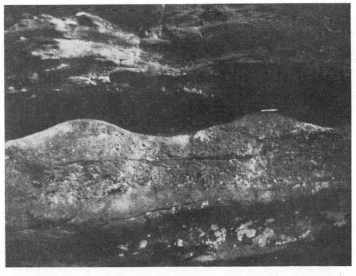

FIG. 6.—Antidunes 8 (left) and 9, with shoulder on upstream flank of 9. Wavelength is 69 cm. In granule conglomerate beds and underlying sandstone, cross-beds dip upstream (to left).

wave train would likely have exceeded critical steepness and been breaking. The profiles of antidunes 6 and 9 (figs. 3 and 6) suggest that this may very well have been the case. These particular waves have weakly developed secondary crests or shoulders on their upstream flanks. From our own flume observations, we conclude that these shoulders were built when breaking of the accompanying water waves induced flow separation near the bed.

As reported by Middleton (1965), breaking of the water waves causes antidunes to be reduced in size or obliterated altogether. In this connection, the antidune steepnesses plotted in figure 5 are of interest. Groups of well-developed (steep) antidunes are separated by others that are less well developed. Perhaps the less-steep antidunes mark places where water waves were breaking or had recently broken. Support for this interpretation may be found in the partial filling of the troughs preceding antidunes 5, 6, 7, and 11 (fig. 5). The sediment in the fill is compositionally indistinguishable from that which makes up the rest of the granule conglomerate bed, but is horizontally laminated. It may represent trough deposition that occurred when the flow was disturbed by breaking waves. The association of fill of this kind with groups of unusually low antidunes and antidunes having shoulders is at least suggestive. An alternative explanation might ascribe the trough filling to drainage moving along certain of the troughs after the main antidune flow had ceased. The coincidence with modified antidune profiles and low crests would then be fortuitous.

To summarize, we suggest that the sediment waves near Sunderland are antidunes that developed beneath a sheet of water only 1 or 2 centimeters deep, moving at about 100 cm/sec down a paleoslope of 2.7° or more. There were at least 19 antidunes in this particular train and their individual wave crests were continuous for more than 7 m transverse to the current. Breaking occurred sporadically in groups of waves separated by others that were not breaking. Scour associated with the breaking may have reduced, but did not eliminate the affected antidunes. The conditions as reconstructed here seem entirely consistent with flash flood (sheetwash) runoff on the surface of an alluvial fan.

ANTIDUNE STRUCTURES IN ADJACENT BEDS

Although only one set of preserved wave forms has been found at the Sunderland exposure, features suggesting antidune sedimentation are common to most of the beds in this exposure. The ledge with preserved antidunes ac-

tually consists of two beds of granule conglomerate, nearly identical to one another except that wave forms have not been preserved on top of the lower member (fig. 6). Cross-laminae in both units dip northeastward (upstream) at from 20° to 22°. Both beds are continuous for the length of the outcrop (14 m), indicating that deposition in the antidune phase can result in laterally continuous crossbedded units of fairly uniform thickness (here about 10 cm).

Upstream-dipping cross-lamination is also prominently displayed in several fine to coarse sandstone beds above and below the granule horizon (fig. 6). In general, inclination of cross-laminae appears somewhat more gentle in the finer grained sediments, but dips as steep as 24° occur in a bed of coarse sandstone immediately below the granule beds. Cross-bed sets in most of the sandstones are highly lenticular, closely resembling the antidune deposits produced experimentally by Middleton (1965).

In several instances, low-angle cross-beds within a single sandstone bed dip both northeastward and southwestward, indicating that the bedforms responsible for them were capable of moving either upstream or downstream. It is known that antidunes can behave in this manner (Gilbert, 1914; Kennedy, 1963), but the same cannot be said for bedforms outside the upper flow regime. Bimodality of cross-bed orientations, the modes dipping toward and away from the known source, was cited by Power (1961) as evidence for antidunes in the Coso Formation.

The features described here for the Sunderland antidune exposure are repeated again and again throughout the Mount Toby Conglomerate. Crossbedding, almost exclusively of the low-angle type, usually occurs in lenticular sets and commonly dips in opposite directions within a single exposure. Associated beds display horizontal lamination representing the plane bed phase of the upper flow regime, but dune and ripple structures (lower flow regime) are extremely rare. Conglomerates throughout the formation include cobbles and boulders that are incapable of transport except in the upper flow regime (Harms and Fahnestock, 1965).

The combined evidence indicates not only that antidune structures are capable of preservation, but that they may, in fact, be quite common in certain nonmarine deposits.

ACKNOWLEDGEMENT

The writers are grateful to John E. Sanders for reading the manuscript and offering helpful suggestions.

REFERENCES

BAGNOLD, R. A., 1966, An approach to the sediment transport problem from general physics: U. S. Geological Survey Prof. Paper 422-I, 37 p.

BRIGGS, L. I., AND MIDDLETON, G. V., 1965, Hydromechanical principles of sediment structure formation, p. 5–16 *in* Middleton, G. V., ed., Primary sedimentary structures and their hydrodynamic interpretation. Soc. Economic Paleontologists and Mineralogists, Spec. Publ. 12, Tulsa, Okla., 265 p.

EINSTEIN, H. A., AND BARBAROSSA, N. L., 1952, River channel roughness: Amer. Soc. Civil Eng. Trans. v. 117, p. 1121–1132.

ENGELUND, F, 1966, Hydraulic resistance in alluvial streams: Amer. Soc. Civil Eng., Jour Hydraulics Div., v. 92, HY2, Proc. Paper 4739, p. 315–326.

GILBERT, G. K., 1914, The transportation of debris by running water: U. S. Geol. Survey, Prof. Paper 86, 263 p.

GILLULY, JAMES, WATERS, A. C., AND WOODFORD, A. O., 1968, Principles of geology. 3rd ed., W. H. Freeman and Co., San Francisco, 687 p.

GWINN, V. E., 1964, Deduction of flow regime from bedding characteristics in conglomerates, and sandstones: Jour. Sedimentary Petrology, v. 34, p. 656–658.

HAND, B. M., 1969, Antidunes as trochoidal waves: Jour. Sedimentary Petrology, v. 39, p. 1302–1309.

HARMS, J. C., AND FAHNESTOCK, R. K., 1965, Stratification, bed forms, and flow phenomena (with an example from the Rio Grande), p. 84–115, *in* Middleton, G. V., ed., Primary sedimentary structures and their hydrodynamic interpretation. Soc. Economic Paleontologists and Mineralogists, Spec. Publ. 12, Tulsa, Okla., 265 p.

HENKE, R. W., 1966, Introduction to fluid mechanics. Addison-Wesley Publ. Co., Reading Mass., 232 p.

KENNEDY, J. F., 1961, Stationary waves and antidunes in alluvial channels. W. M. Keck Lab. Hydraulics and Water Resources, Calif. Inst. Technol., Rept. KH-R-2, 146 p.

——— 1963, The mechanics of dunes and antidunes in erodible-bed channels: Jour. Fluid Mechanics, v. 16, p. 521–544.

KLEIN, G. DE V., 1965, Dynamic significance of primary structures in the Middle Jurassic Great Oolite Series, southern England, p. 173–191 *in* Middleton, G. V., ed., Primary sedimentary structures and their hydrodynamic interpretation. Soc. Economic Paleontologists and Mineralogists, Spec. Publ. 12, Tulsa, Okla., 265 p.

LEOPOLD, L. B., WOLMAN, M. G., AND MILLER, J. P., 1964, Fluvial processes in geomorphology. W. H. Freeman and Co., San Francisco, 522 p.

LONGWELL, C. R., FLINT, R. F., AND SANDERS, J. E., 1969, Physical geology. John Wiley and Sons, Inc., New York, 685 p.

MIDDLETON, G. V., 1965, Antidune crossbedding in a large flume: Jour. Sedimentary Petrology, v. 35, p. 922–927.

POWER, W. R., 1961, Backset beds in the Coso Formation, Inyo County, California: Jour. Sedimentary Petrology, v. 31, p. 603–607.

RAHN, P. H., 1967, Sheetfloods, streamfloods, and the formation of pediments: Assoc. Amer. Geographers Ann., v. 57, p. 593–604.

SIMONS, D. B., AND RICHARDSON, E. V., 1963, Forms of bed roughness in alluvial channels: Am. Assoc. Civ. Engineers Trans., v. 128, p. 284–302.

———, AND ALBERTSON, M. L., 1961, Flume studies using medium sand (0.45mm): U. S. Geol. Survey, Water-Supply Paper 1498-E, 26 p.

SIMONS, D. B., RICHARDSON, E. V., AND NORDIN, C. F., 1965, Sedimentary structures generated by flow in alluvial channels, p. 34–52 *in* Middleton, G. V., ed., Primary sedimentary structures and their hydrodynamic interpretation. Soc. Economic Paleontologists and Mineralogists, Spec. Publ. 12, Tulsa, Okla., 265 p.

TAYLOR, R. H., AND BROOKS, N. H., 1962, Discussion of "Resistance to flow in alluvial channels," by D. B. Simons and E. V. Richardson: Amer. Soc. Civil Engineers Trans., v. 127, p. 982–992.

VISHER, G. S., 1965, Fluvial processes as interpreted from ancient and recent fluvial deposits, p. 116–132 *in* Middleton, G. V., ed., Primary sedimentary structures and their hydrodynamic interpretation. Soc. Economic Paleontologists and Mineralogists, Spec. Publ. 12, Tulsa, Okla., p. 116–132.

WALKER, R. G., 1967. Upper flow regime bed forms in turbidites of the Hatch Formation, Devonian of New York State: Jour. Sedimentary Petrology, v. 37, p. 1052–1058.

WESSEL, J. M., 1969, A paleocurrent and petrographic study of the Triassic Mount Toby Conglomerate and Turners Falls Sandstone of North Central Massachusetts: M.S. thesis, Univ. Massachusetts, 157 p.

Editor's Comments
on Papers 4, 5, and 6

ALLUVIAL-PLAIN FACIES: FLUVIAL FINING-UPWARD CYCLES

Alluvial-plain deposits constitute essentially the central realm along the spectrum of nonmarine facies. Much of this fluvial sandstone and mudstone occurs in a cyclic pattern that early geologists commonly attributed to tectonic pulses. More careful study of fluvial successions, especially in the Cenozoic molasse of southern France (Crouzel 1957) and Switzerland (Bersier 1959), revealed that they are characteristically arranged in fining-upward sequences like point-bar deposits of meandering rivers. Allen (1962, 1964, 1965) recognized their significance as a fundamental framework of Devonian alluvial-plain deposits in southern Great Britain and applied the term *cyclothem* to them. At a more general level Beerbower (1964, 1969) has reviewed the rationale for calling these sequences "cyclothems," and he discussed mechanisms that can produce them. Fluvial cyclothems differ in detail from deltaic ones, and different stratigraphic boundaries have been proposed to define them (Belt 1975). Visher (1972) has summarized the physical characteristics of fluvial deposits of this sort, and Walker (1976) has integrated them into a model.

Moody-Stuart (1966) has shown that deposits of braided and meandering streams can be differentiated by the geometry of their channel deposits, presence or absence of large-scale sigmoidal cross-stratification, presence and extent of fine-grained channel fill, position and extent of levee deposits, and paleocurrent pattern. Detailed analysis of late Cenozoic fluvial deposits in Cook Inlet, Alaska, reveals significant correlation between stream type and sandstone reservoir geometry (Hayes et al. 1976). Thin, lenticular braid-bar sandbodies of short, high-gradient braided streams on alluvial fans form a heterogeneous, discontinuous reservoir, whereas thicker, more continuous point-bar deposits of meandering streams on an alluvial plain form a thicker, more extensive reservoir.

Numerous recent studies of fining-upward alluvial-plain sequences show convincingly that they are products of meandering streams in which the coarser sediments transported by tractive currents formed point bars and channel bars whereas the finer fractions accumulated from suspension on the interchannel floodplains. These reports describe such formations as the late Devonian rocks of Scotland (Read and Johnson 1967) and Eire (Graham 1975), Carboniferous strata of southeastern United States (Briggs 1974), late Carboniferous strata of England (Jones 1973), Permo-Carboniferous rocks of northeastern Spain (Nagtegaal 1969), late Paleozoic Gondwana deposits of India (Casshyap 1970; Sengupta 1970), early Cenozoic deposits of Wyoming (Steidtmann 1971; Neasham and Vondra 1972), Cenozoic molasse of southern Europe (Van Houten 1974), and late Cenozoic molasse of Iraq (Kukal and Al-Jassim 1971).

Recognition of the basic role of fining-upward cycles induced more detailed description of alluvial-plain sequences, and this in turn supplied enough data to support meaningful statistical studies. After pioneering efforts like that of Duff and Walton (1962) on cyclic Carboniferous coal measures in England, data from Devonian Upper Old Red Sandstone and associated Carboniferous deposits in Scotland were analyzed statistically in a series of papers by Read and Dean (1967, 1976). Duff (1967) made a similar analysis of Permian coal measures in New South Wales. Among the statistical methods applied, Markov chain analysis (Vistelius 1949), which detects repetitive processes in space and time, proved effective on cyclic patterns in coal-bearing mid-Carboniferous sediments in central Scotland (Read 1969) and in Paleocene alluvial-plain deposits in Wyoming (Gingerich 1969), and it has since been used successfully in other studies. Allen (1974) has also analyzed fluvial cyclothems using the nonparametric Mann-Whitney U-test of whether two independent groups of numerical data come from the same population, and by means of an upwards facies transition probability matrix that denotes both the type and strength of the pattern and variability of a sequence and lists the probability that a given facies state will be succeeded by any other.

Three investigations of the alluvial-plain facies have been selected to represent a range from more general description to specialized statistical analysis. Friend (1965) had demonstrated that detailed observations of sedimentary structures provide basic data for reconstructing the conditions of deposition of Devonian late orogenic Wood Bay cyclothems in Spitsbergen. In 1972 Friend and Moody-Stuart published a more comprehensive report on these deposits (Paper 4). Their analysis of paleocurrents, sandstone composition, and grain-size variation portrayed three river systems. The authors also attempted to estimate the dependent variables of the systems, including alluvial grain-size, channel cross-section and plan, friction factor, strength of river flow, and channel slope.

From the experience of his several studies of fining-upward sequences, Allen (1970) produced comparative analysis of six major facies in fluvial cycles in the Devonian strata of eastern United States and Great Britain. He showed that the more variable, lower coarse member was formed by lateral accretion in sinuous streams and can be interpreted in terms of stream power and channel sinuosity. Allen (1974) pursued this kind of detailed comparison in a study of lateral variation among well-exposed Devovian cyclothems in Wales (Paper 5). The results suggest that variation *within* cyclothems was produced by local, autocyclic, changes inside the deposits, whereas variation *between* them was produced by allocyclic changes outside the sedimentary system.

In his study of Devonian rocks in Arctic Canada, Miall (1973) applied Markov chain analysis to four kinds of detrital successions (Paper 6). The results are especially illuminating because they show that for some facies such as alluvial-fan deposits Markov analysis is of little assistance in forecasting a predictable sequence, whereas for deposits with fining-upward cycles, a predictable sequence is strongly indicated.

REFERENCES

Allen, J. R. L. 1962. Petrology, origin and deposition of the highest Lower Old Red Sandstone of Shropshire, England. *Jour. Sed. Petrology* **32**:657–697.
———— 1964. Studies in fluviatile sedimentation: Six cyclothems from the Lower Old Red Sandstone, Anglo-Welsh Basin. *Sedimentology* **3**:173–198.
———— 1965. Fining-upwards cycles in alluvial successions. *Geol. Jour.* **4**:(Pt. 2):229–246.
———— 1970. Studies in fluviatile sedimentation: A comparison of fining-upwards cyclothems, with special reference to coarse-member composition and interpretation. *Jour. Sed. Petrology* **40**:298–323.
Beerbower, J. R. 1964. Cyclothems and cyclic depositional mechanisms in alluvial plain sedimentation. *Kansas Geol. Survey Bull.* **169**:31–42.

—— 1969. Interpretation of cyclic Permo-Carboniferous deposition in alluvial plain sediments in West Virginia. *Geol. Soc. America Bull.* **80:**1843–1848.

Belt, E. S. 1975. Scottish Carboniferous cyclothem patterns and their paleoenvironmental significance. In *Deltas, Models for Exploration,* ed. M. L. Broussard. Houston, Texas: Houston Geol. Soc., pp. 429–449.

Bersier, A. 1959. Sequences detritique et divagations fluviales. *Eclogae Geol. Helvetiae* **51:**854–893.

Briggs, G., ed. 1974. Carboniferous of the southeastern United States. *Geol. Soc. America Spec. Paper 148,* pp. 1–361.

Casshyap, S. M. 1970. Sedimentary cycles and environments of deposition of the Barakar Coal Measures of Lower Gondwana, India. *Jour. Sed. Petrology* **40:**1302–1317.

Crouzel, F. 1957. Le Miocene continental du bassin d' Aquitaine. *France Service Carte Géol.* **54**(1956):1–264.

Duff, P. McL. D. 1967. Cyclic sedimentation in the Permian coal measures of New South Wales. *Geol. Soc. Australia Jour.* **14**(Pt. 2):293–307.

Duff, P. McL. D., and Walton, E. K. 1962. Statistical basis for cyclothems: A quantitative study of the sedimentary succession in the East Pennine coalfield. *Sedimentology* **1:**235–256.

Friend, P. F. 1965. Fluviatile sedimentary structures in the Wood Bay Series (Devonian) of Spitsbergen. *Sedimentology* **5:**39–68.

Gingerich, P. D. 1969. Markov analysis of cyclic alluvial sediments. *Jour. Sed. Petrology* **39:**330–332.

Graham, J. R. 1975. Deposits of a near-coastal fluvial plain—The Toe Head Formation (Upper Devonian) of Southwest Cork, Eire. *Sed. Geology* **14:**45–61.

Hayes, J. B.; Harms, J. C.; and Wilson, T. 1976. Contrasts between braided and meandering stream deposits, Beluga and Sterling Formations (Tertiary), Cook Inlet, Alaska. In *Recent and Ancient Sedimentary Environments in Alaska,* ed. T. P. Miller. Anchorage, Alaska: Alaska Geol. Soc., pp. J1–J27.

Jones, P. C. 1972. Quartz arenite—lithic arenite facies in the fluvial foreland deposits of the Trenchard Group (Westphalian), Forest of Dean, England. *Sed. Geology* **8:**177–198.

Kukal, Z., and Al-Jassim, J. 1971. Sedimentology of Pliocene molasse sediments of the Mesopotamian geosyncline. *Sed. Geology* **5:**57–81.

Moody-Stuart, M. 1966. High- and low-sinuosity stream deposits, with examples from the Devonian of Spitsbergen. *Jour. Sed. Petrology* **36:**1102–1117.

Nagtegaal, P. J. C. 1969. Sedimentology, paleoclimatology, and diagenesis of post-Hercynian continental deposits in the south-central Pyrenees, Spain. *Leidse Geol. Meded.* **42:**143–238.

Neasham, J. W., and Vondra, C. F. 1972. Stratigraphy and petrology of the lower Eocene Willwood Formation, Bighorn Basin, Wyoming. *Geol. Soc. America Bull.* **83:**2167–2180.

Read, W. A., and Dean, J. M. 1967. A quantitative study of a sequence of coal-bearing cycles in the Namurian of central Scotland, 1. *Sedimentology* **9:**137–156.

—— 1976. Cycles and subsidence: Their relationship in different sedimentary and tectonic environments in the Scottish Carboniferous. *Sedimentology* **23:**107–120.

Read, W. A., and Johnson, S. R. H. 1967. The sedimentology of sandstone formations within the Upper Old Red Sandstone and Lowest Calciferous limestone measures west of Stirling, Scotland. *Scottish Jour. Geol.* **3**(Pt. 2):242–267.

Sengupta, S. 1970. Gondwana sedimentation around Bheemaram (Bhimaram), Pranhita-Godavari Valley, India. *Jour. Sed. Petrology* **40:**140–170.

Steidtmann, J. R. 1971. Origin of the Pass Peak Formation and equivalent Early Eocene strata, central western Wyoming. *Geol. Soc. America Bull.* **82:**159–176.

Van Houten, F. B. 1974. Northern Alpine Molasse and similar Cenozoic sequences in southern Europe. *Soc. Econ. Paleontologists and Mineralogists Spec. Pub. 19,* pp. 260–273.

Visher, G. S. 1972. Physical characteristics of fluvial deposits. *Soc. Econ. Paleontologists and Mineralogists Spec. Pub. 16,* pp. 84–97.

Vistelius, A. B. 1949. The mechanism of formation of sedimentary beds. *Akad. Nauk SSSR Doklady* **65:**191–194.

Walker, R. G. 1976. Facies models. 3. Sandy fluvial systems. *Geosci. Canada* **3:**101–109.

4

Reprinted from pp. 1, 7, 22–32, 34–47, 68–71 of *Norsk Polarinst. Skr.* **157**:1–77 (1972)

Sedimentation of the
Wood Bay Formation (Devonian) of Spitsbergen:
Regional analysis of a late orogenic basin

P. F. FRIEND AND M. MOODY-STUART

Abstract

Our object has been the analysis of the fresh-water, late orogenic, Wood Bay Formation, which outcrops over an area 150 km by 75 km, and is up to 3 km thick.

Expeditions from Cambridge University have collected systematic sedimentological data, mainly measuring "vertical" sections. Vertebrate fossils were used to define three informal time-rock sub-divisions of the Formation.

Analysis of palaeocurrents, sandstone composition and grain-size variation have allowed us to distinguish three river systems which flowed towards a northern area of clay flats. Rivers of the eastern system were large, north and north-north-west flowing, bed-load rivers of braided type. They were relatively heavily laden with sand-grade sediment, which was rich in feldspar. In contrast, the rivers of the western system were small, eastward flowing, mixed-or-suspended load rivers of high sinuosity. They were relatively less loaded with sand-grade detritus, which was poor in feldspar, particularly in "Lower" Wood Bay time. The central system flowed north, but was similar to the western system in other respects.

We suggest that the eastern system drained an area to the south-east that was very large compared with the small drainage areas of the western system. Calculation of the relative size of these areas depends on estimation of the relative effective precipitation of rain. If this relationship is estimated, and an actual area worked out for a western river, areas, denudation rates and relief can be suggested for all the systems. Suggestions for the relief are 150 m in the west, and 6000 m in the south-east. Estimates of vertical movement can be made using these figures and a simplified model of tectonism followed by isostatic adjustment, erosion and sedimentation.

Fold trends provide evidence that the Upper Devonian eastern boundary fault (Balliolbreen Fault) is a left-lateral, strike-slip fault. The isolated Devonian succession round Hornsund can most easily be fitted to the regional pattern in the north, by supposing that the Hornsund rocks have been moved by major left-lateral strike-slip faulting too.

[*Editor's Note:* Material has been omitted at this point.]

53

Grain-size variation

LOCALITY GRAIN-SIZE

As a first stage in considering variation of grain-size in the Wood Bay Formation, we have calculated a mean grain-size for each sedimentological section. Our method has been to note the grain-size at quarter metre intervals up the section, and determine the average of these grain-sizes. Plots of these data are presented in Fig. 11.

The most striking feature of these plots of average grain-size is the way in which eastern areas are consistently coarser than the central and western areas.

SEMI-CYCLES

Definition

Analysis in terms of fining-upwards cycles (or cyclothems) has become a valuable technique in the investigation of those fluvial formations that consist of successions of alternating siltstone and sandstone units. Examples of this technique have been common in recent papers on the Old Red Sandstone of Southern Britain (ALLEN 1964, 1965a, 1970), Eastern U.S.A. (ALLEN and FRIEND 1968; ALLEN 1970) and Spitsbergen (FRIEND 1961, 1965; MOODY-STUART 1966).

It was our object to analyse these cycles quantitatively on a regional scale. We therefore required a definition of the cycle which could be objectively applied. We have not attempted to define a cycle based on the simultaneous multivariate analysis of many characters (grain-size, structures, scale), although this would be an interesting study in itself. We have rather adopted a univariate approach by defining a cycle on the basis of one variable alone, grain-size. This approach is simple to apply and easy to interpret.

To distinguish these cycles from others, we have called them semi-cycles (ALEXANDER-MARRACK, FRIEND, and YEATS 1971). Our definition of a fining-up semi-cycle (F) is illustrated in Fig. 12. We apply the name "coarsening-up semi-cycle" (C) where a coarser set at the top of a fining-up semi-cycle is succeeded by an even coarser set. This is also illustrated in Fig. 12, where it can be seen that a coarsening-up semi-cycle may consist of one set only.

Maximum grain-size and sandstone percentage

The next feature we examined was the maximum grain-size in each semi-cycle. For this purpose we ignored the presence of pebbles, because, in the Wood Bay Formation, these rarely form more than thin and scattered zones in sandstones.

Fig. 13 shows the variation in maximum grain-size. It is apparent that the western areas generally contain the same maximum grain-sizes as the eastern areas, even though their average grain-sizes (Fig. 11) are consistently less.

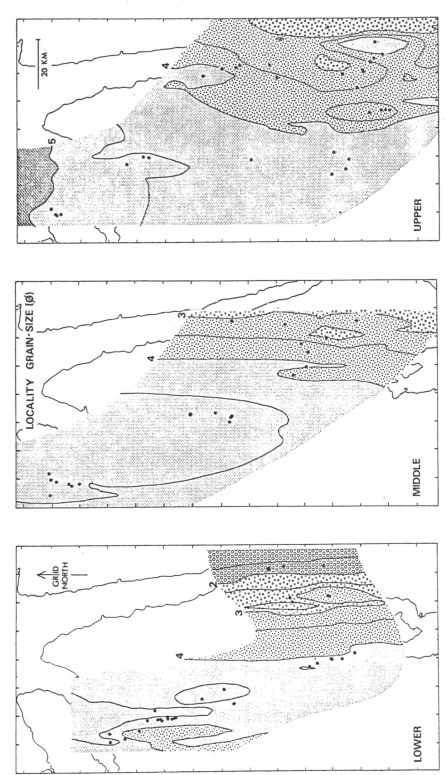

Fig. 11. *Standard map area (outlined on Fig. 1), for the three informal time-rock units, showing locality grain-size in Ø units. (> 4Ø is siltstone, 4Ø–3Ø very fine sandstone, 3Ø–2Ø fine sandstone, <2Ø medium and coarser sandstone). At each locality, marked by a spot, the grain-size calculated is the mean of the grain-sizes of each 0.25 m interval in the 'vertical' sedimentological section. Contours on iterative-fit quadratic trend surfaces have been plotted.*

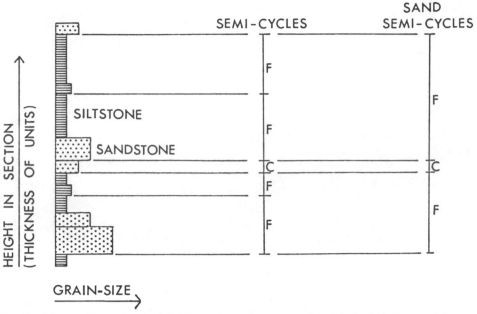

Fig. 12. *Diagram illustrating the definition of semi-cycles, and sand semi-cycles. F indicates a fining-up semi-cycle, C indicates a coarsening-up semi-cycle (in these cases, single sets).*

The explanation of this becomes clear when a plot of the percentages of sandstone in each semi-cycle is prepared (Fig. 14). The eastern areas (Austfjorden Sandstone Member) have a higher proportion of sandstone, and less siltstone than the western areas.

SAND SEMI-CYCLES

Definition

Fining-upwards cycles have generally been interpreted in terms of coarse member accumulation in a channel environment, and fine member accumulation in an overbank environment (e.g. ALLEN 1965b). The distinction between deposits of these two environments may, in some cases, be difficult to apply with certainty, especially where one of the two environments is not well developed, or where transitions occur (e.g. levée deposits, crevasse splay deposits). Because of this, we have found it helpful to make a distinction between bed-load deposits and suspended-load deposits. In the field, bed-load deposits are positively identified by their bed-form structures (asymmetrical ripples, cross-stratification, parting lineation).

FRIEND (1965 p. 46) published an analysis of the sedimentary structures in sets of different grain-size in some Wood Bay sections. This analysis suggests that a similar genetic distinction can be made, as a working approximation for regional analysis, between sandstone and siltstone sets. 80% of sets of medium siltstone grade were unlaminated or flat-laminated, and may, therefore, have been deposited from suspension. In contrast, 90% of sets of very-fine sandstone contain

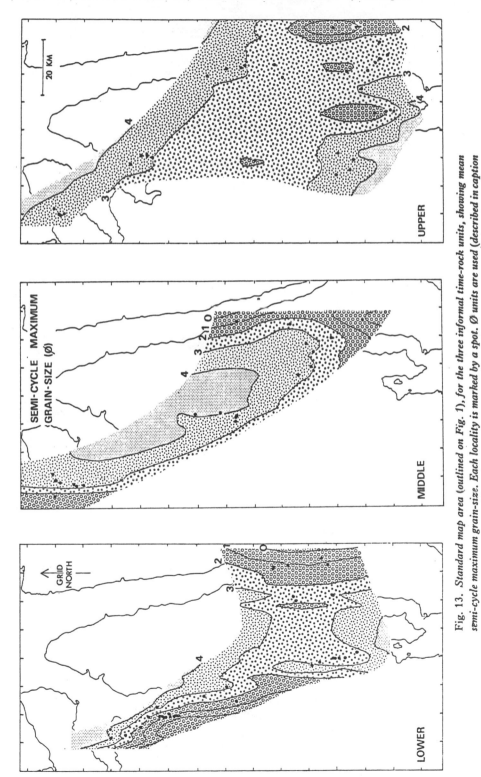

Fig. 13. Standard map area (outlined on Fig. 1), for the three informal time-rock units, showing mean semi-cycle maximum grain-size. Each locality is marked by a spot. Ø units are used (described in caption to Fig. 11). Contours on iterative-fit quadratic trend surfaces have been plotted.

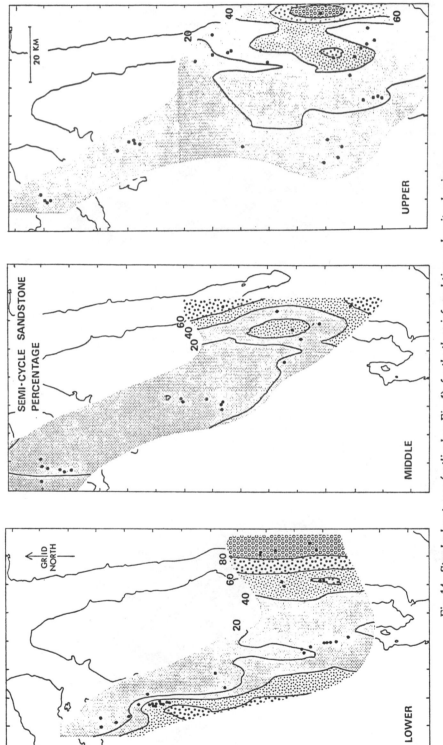

Fig. 14. Standard map area (outlined on Fig. 1), for the three informal time-rock units, showing mean semi-cycle sandstone percentage. Each locality is marked by a spot. Contours on iterative-fit quadratic trend surfaces have been plotted.

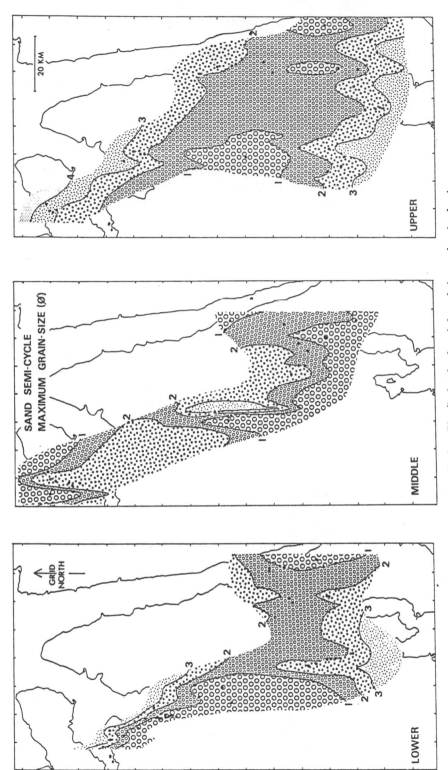

Fig. 15. *Standard map area (outlined on Fig. 1), for the three informal time-rock units showing mean sand-semi-cycle maximum grain-size. Each locality is marked by a spot. Ø units are used (described in caption to Fig. 11). Contours on iterative-fit quadratic trend surfaces have been plotted.*

small and large cross-stratification or parting lineation, and were, therefore, deposited from bed-load.

To use this distinction, we have defined another kind of semi-cycle, that which contains one or more sets of sand-grade (Fig. 12). These "sand semi-cycles" may either coincide with the semi-cycles considered above, or they may include two, or more of them. They have the special feature that it is reasonably certain that they included an episode of bed-load deposition.

Maximum grain-size

The pattern of variation of maximum grain-size of the sand semi-cycles is plotted on Fig. 15. The patterns for "Lower" and "Upper" times are very similar to those for the general semi-cycles (Fig. 13). However the pattern for "Middle" times is more complex, and particularly different in northern areas. This is because of the rareness of sand members in this region, and the variability of grain-size of the few that do occur.

Thickness

The pattern of thickness of the sand semi-cycles is plotted on Fig. 16. It shows a range of averages from 4 m (or less) to 16 m (or more). The sand semi-cycles become thicker in a generally downstream (Fig. 7) direction.

Thickness of sand member

The variation in thickness of the sand member is plotted on Fig. 17. In "Upper" rocks this variation is minor, and no overall pattern can be discerned. In "Lower" and "Middle" rocks, there is consistent thinning of sand members downstream (Fig. 7), from a thickness of 4 or 6 m (similar to that of the whole semi-cycle) to thicknesses considerably less than 1 m.

Check on downstream trends of variation

To check on the variation which we have just deduced from the patterns of iterative-fit trend-surfaces, we have (Fig. 18) selected groups of localities along palaeocurrent streamlines. We have then averaged for each group of localities, the characteristics of all the sand semi-cycles. On Fig. 18 we have plotted the actual average values against the values of the trend surfaces. It will be seen that these correspond closely.

Variation of individual sand semi-cycles

Any study of a feature which depends solely on the averages of individual characteristics, runs the risk that the average may conceal important variation. We have therefore plotted for four of the groups of localities, the characteristics of individual semi-cycles.(Fig. 19). Against semi-cycle maximum grain-size, we have plotted thickness of sandstone member, and thickness of siltstone member. The sum of the sandstone member and the siltstone member (if present), is the sand semi-cycle thickness which we analysed above.

This allows us to make a model of different sorts of sand semi-cycles, and their probable lateral variation (Fig. 20). But we would stress that in no case have we been able to follow an individual sand semi-cycle far enough to be able to see these variations.

In the "Lower" western streamline (LW on Figs. 18—20) the commonest cycles consist of less than 1 m of sand, either of very-fine or very-coarse (and pebbly) sand. These sand members sometimes rest directly on each other, but there is more often a siltstone member and these may be up to 10 m thick. A few kilometres downstream (LW2) the sand members are missing altogether, and the entire section is made of siltstone.

In the "Upper" eastern streamline (UE on Figs. 18—20), in the upstream area (UE1), cycles consist of up to 5 m of sandstone, varying from fine to very-coarse and pebbly in grade. No siltstone occurs. Further downstream (UE 2 and 3), the sandstone members tend to be thinner and siltstone members are increasingly common. More sand semi-cycles have finer sand grain-sizes, so we conclude that some of the sand members do become finer-grained downstream.

Alluvial environments

THE ALLUVIAL VARIABLES

KENNEDY (1971) analyses the very complex array of variables which make up an alluvial system. He distinguishes between *independent variables,* whose values are imposed externally on the system, and *dependent variables,* whose values are controlled, in turn, by the values of the independent variables. In our Spitsbergen study, we want ultimately to assess the independent variables.

Whether a variable is independent or dependent will depend, to some extent, on the nature of the system and on the period of time over which it is active. For instance, in some flume experiments, the slope is determined by the experimenter, i. e. externally and independently. In some natural streams, over a short period of time (days to months), the slope is also determined independently, because it has been inherited from the conditions of the previous major flood. However, over a longer period of time (several years), the slope of all natural alluvial streams is a dependent variable controlled by the independent variables, water and sediment discharge (KENNEDY 1971 p. 113). This long term adjustment of slope is the core of the idea of a graded river (MACKIN 1948).

In our regional work, our interest is mainly with the long-term balance of the variables of river systems. We shall therefore be concerned with the following pattern of independence and dependence (KENNEDY 1971 p. 113):

Independent variables: water discharge
                            sediment discharge
Dependent variables: alluvial grain-size
                            channel cross-section and plan
                            friction factor
                            strength of river flow
                            slope
We shall first consider the dependent variables, in turn, below.

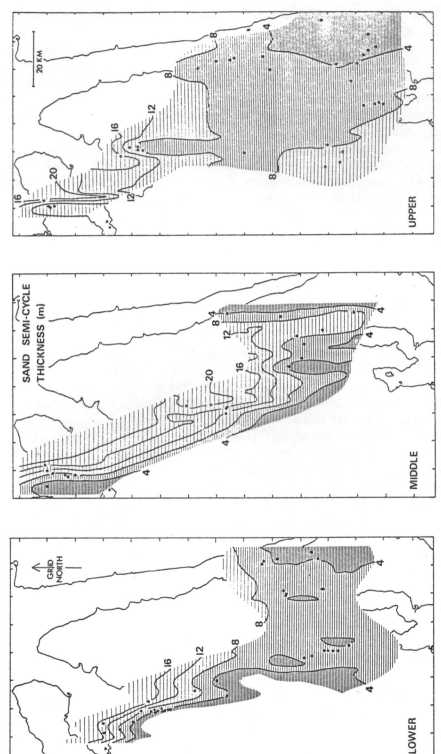

Fig. 16. *Standard map area (outlined on Fig. 1) for the three informal time-rock units, showing mean sand semi-cycle thickness in metres. Each locality is marked by a spot. Contours on iterative-fit quadratic trend surfaces have been plotted.*

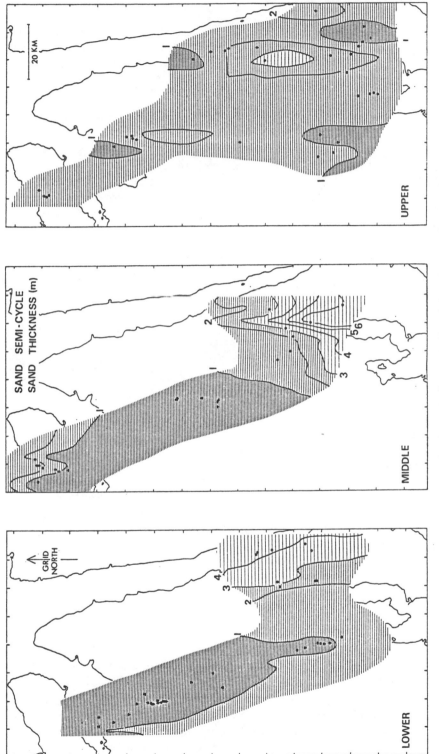

Fig. 17. *Standard map area (outlined on Fig. 1), for the three informal time-rock units, showing, in metres, the mean of the thickness of the sand units in each sand semi-cycle. Each locality is marked by a spot. Contours on iterative-fit quadratic trend surfaces have been plotted.*

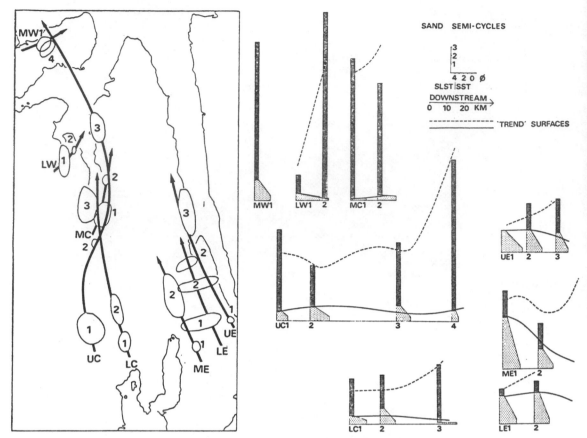

Fig. 18. *Study of downstream variation of sand semi-cycles. Data from localities is grouped from Lower (L), Middle (M) and Upper (U) informal time-rock units. Western (W), central (C) and eastern (E) streamlines are defined using the palaeocurrent trends of Fig. 7. From all the sand semi-cycles in each group, actual averages are plotted for total thickness, sand thickness and maximum grain-size. The corresponding trend surface values for the centre of each group are also plotted.*

ESTIMATING THE DEPENDENT VARIABLES
Alluvial grain-size

The grain-size of alluvium is partly controlled by processes acting in the source area. These processes determine the amount of sediment supplied and its maximum grain-size. In the alluvial area the grain-size is then further influenced by processes of local sorting, often involving many cycles of transport and deposition.

We have already distinguished between siltstones, mainly deposited from suspension, and sandstones, mainly deposited from bed-load. It is a remarkable feature of the bed-load deposits of the Wood Bay Formation that they have so limited a range of grain-size (very fine to medium sand grades, Fig. 15). Pebbles are rare and usually of locally derived, intraformational material. Conglomerates hardly exist. Our knowledge of modern rivers allows us to make a (general) estimate of stream strength based on this fact. This will be done below.

More subtle distinctions of grain-size, within the sand grade, may be due to a

[*Editor's Note:* Figure 19 is not reproduced here.]

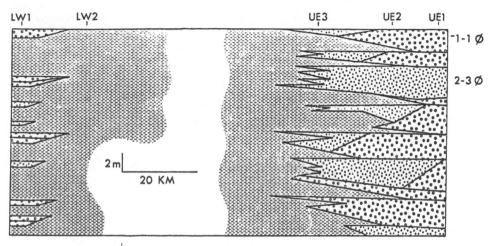

Fig. 20. *Hypothetical vertical-plane section showing lateral variation of sandstone units, and their relationships to siltstone units, in the western area groups (LW1 and LW2) and the eastern groups (UE1, UE2 and UE3) which are located in Fig. 18. These relations are characteristic of the eastern and western river systems.*

number of local processes of selective erosion, transport and deposition of the hydrodynamically different particle fractions.

Possible processes are:

a) sorting on river bar faces. Lateral components of river flow, weaker in shallow water, balance downslope components of particle weight (ALLEN 1970).

b) downstream sorting. Different size-fractions of bed-load travel at different rates, and are buried differentially in ripples (MELAND and NORRMAN 1969). Some fractions may temporarily travel as suspended load, during peak floods. They will therefore travel faster and further than coarser fractions which travel entirely as bed-load (BRUSH 1965).

Channel measurements

Channel type. — In his work on present-day rivers and their alluvium, SCHUMM (1968a, p. 1579) distinguished a *bed-load* type of channel. He defined channels of this type as those having less than 5% of silt or clay (suspended-load deposits) in a systematic sampling of their walls and floor (their perimeters). In terms of the rivers, rather than their perimeters, SCHUMM thought that they carry at least 11% of their total sediment load as bed-load. It is because of the difficulty of measuring bed-load, that SCHUMM's definition must, in practice, be based on perimeter sampling.

SCHUMM was able to generalise about channel cross-section and plan (1968a, p. 1580). Bed-load channels are characterised by great width compared with their depth (ratio about 60). They are almost straight (sinuosity 1.0 to 1.1) and are often braided, having extensive alluvial islands at low stage.

The deposits of a series of these bed-load channels accumulate as thick successions of sand or gravel (SCHUMM 1968, p. 1582) with only rare and thin beds of suspended load material. Although suspended-load sediments may accumulate as superficial veneers in years of low peak discharge, they tend to be destroyed in the wholesale reworking of the underlying cohesionless bed-load material which occurs during major floods.

On the basis of the grain-size of its channel perimeter, we would regard the Brahmaputra River (COLEMAN 1969) as of bed-load channel type. This is supported by its low sinuosity and braiding. Although there are extensive "floodbasins" adjacent to its braided channels, COLEMAN (1969, p. 233) stated that "Because of the large number of major and minor channels within the Bengal Basin and their active migrating nature, floodplain deposits form only thin veneers capping channel sands and silts".

McGOWEN and GARNER (1970) classify the Amite River (Louisiana) and the Colorado River (Texas) in SCHUMM's bed-load type. The banks of these rivers are unusually strongly stabilized by vegetation, so they are not braided, and have sinuosities of 1.4 to 1.75. The floodplain deposits are mainly of bed-load material (p. 89), like those of Bijou Creek (McKEE, CROSBY and BERRYHILL 1967).

It seems clear to us that the dominantly sandy deposits of the south-east (Aust-fjorden Sandstone Member) of the Wood Bay Formation were deposited by rivers of SCHUMM's bed-load type. This carries with it implications about channel cross-section and plan.

SCHUMM (1968a, p. 1579), also defined *mixed* and *suspended-load* types of channels. These have respectively between 5 and 20%, and more than 20%, of suspended-load sediments in their channel perimeters. He suggested that these channel perimeter features correspond to 3 to 11%, and less than 3% bed-load in the total sediment load.

As examples of the "valley fill" or floodplain stratigraphy of a mixed-load channel, SCHUMM (1968a, p. 1582) gave the "prior stream" channels of the Murrumbidgee River. Channel forms were cut at different times in a Pleistocene fluvial and aeolian formation, and were then plugged by the aggrading streams. Each plug consists of a lens of sand or gravel, decreasing in width upwards from its maximum, because the depositing channel decreased in size and deposited increasing amounts of cohesive suspended-load material. During its earlier and larger stages, each channel was of bed-load type (SCHUMM, 1968b, p. 36), but as time passed, it became increasingly a suspended-load channel, presumably passing through a mixed-load stage.

The present Murrumbidgee provides SCHUMM's example of the sort of alluvial succession built by a suspended-load type of channel. This is depositing an alternation of sand and clay members. "Lateral migration of this suspended-load channel leaves a sheet of channel sand which is overlain by point-bar deposits of lateral accretion and overbank deposits. Only about one-third or less of this deposit will be composed of sand or coarse sediments" (SCHUMM 1968a, p. 1582).

We do not feel confident that we can distinguish between the deposits of mixed and suspended-load types in the Wood Bay Formation. SCHUMM's examples allow us to feel sure that any section composed of sand and silt semi-cycles, was formed by mixed or suspended-load channel activity, compared with a sand only section which was formed by bed-load channel activity. Within the sand and silt sections around Woodfjorden, MOODY-STUART (1966) distinguished between high and low sinuosity channel deposits, and Schumm (1968a, p. 1582) suggested that they might correspond to suspended and mixed-load channel types. In our regional study we are not able to apply this distinction widely to our body of general data. We therefore group together *mixed*-or-*suspended-load* channel deposits, and characterise them by single values of the channel measurements. In contrast to the bed-load channels, they have a smaller width/depth ratio (10), and higher sinuosity (2) (Schumm 1968a, p. 1580).

Cycles. — We have explained above our decision to define a "semi-cycle" for analytical purposes, purely in terms of vertical variations of grain-size. We now come to consider possible origins for these semi-cycles.

Beerbower (1964) described *allocyclic* sequences in which the cyclicity results from disturbances outside the sedimentary system (alluvial part of system in our terminology) "by changes in discharge, load and slope". All these changes would produce characteristically widespread changes in the sedimentary basin.

(1) *Changes of water discharge*. The "prior streams" of the Murrumbidgee River produced fining-upwards semi-cycles (Schumm 1968b, p. 30). The fining-upwards was partly the result of decreasing peak discharge and reduction of sand load due to change of climate.

(2) *Change of sediment-load*. Schumm (1968a) explained how sediment may accumulate in source area valleys, to be flushed out intermittently by exceptional (say 1000 year) floods and deposited in an alluvial area.

(3) *Change of slope* (or base level). A rise in base level first decreases the slope nearest to that level. This results in a decrease in flow strength, and deposition of sediment. The Recent deposits of the Mississippi River are examples of a large single fining-upward semi-cycle due to this cause. In the case of the Mississippi, the change from erosion to gravel, and then sand, deposition has taken place over about 30,000 years (Fisk 1952, p. 66). The river Cam, near Cambridge, has a similar sequence: erosion surface — sand and gravel — silty clay, which again reflects post-glacial sea level rise (Sparks and West 1965, p. 35).

If the cycles can be shown to be local, not reflecting basin-wide changes, they are of Beerbower's *autocyclic* type. These are "generated purely within the sedimentary prism". Subsurface plots of the cyclic deposits of the alluvial Rechna Doab, West Pakistan, provide a clear example of local cycles which obviously result from movements of the rivers of the area (Kazmi 1964). We have no doubt that the cycles of the Wood Bay Formation are of this local type. Fig. 21 presents a rare example of sections which are known to be lateral equivalents (i.e. not separated by faulting). The cycles are highly impersistent.

We shall therefore consider interpretation of Wood Bay cycles in terms of local factors. In Fig. 22, we present a range of typical successions in the Wood Bay Formation. We have classified the semi-cycles in three categories (A, B, C below and in Fig. 22).

(A) *Single-set sandstone and siltstone*. The presence of a silt member, assuming it to be thick, or laterally persistent, is evidence for a significant episode of deposition from suspension in fluvial overbank, or lacustrine conditions. The fact that, in this category, the episode of bed-load deposition resulted in a single set of one

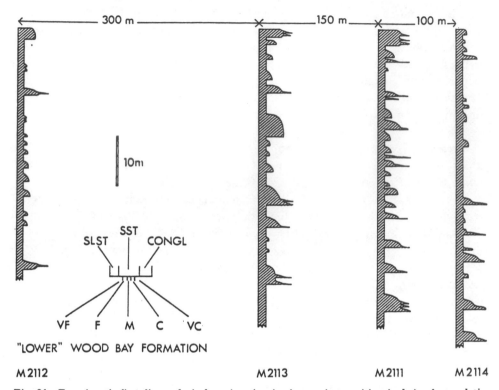

Fig. 21. *Four 'vertical' sedimentological sections (grain-size against position in log) whose relative positions are known. No significant faults separate these sections, so they provide some idea of the lateral variation in number and grain-size of sandstone members.*

grain-size and one type of internal structure, means that the episode was short-lived, probably the duration of a single flood. Single sets of this sort have been deposited by particular floods of the Connecticut River (JAHNS 1947). They are a feature of the lower parts of the fan deposits of the northern Sahara (WILLIAMS 1970). They are also a feature of many levée deposits, in which sand members accumulate during one or more floods as crevasse splays. In the deposits of the Brahmaputra (COLEMAN 1969 p. 230), the levée sand members are thickest close to the main channel, and thin progressively away from it.

(B) *Multiple-set sandstone and siltstone.* In this category the presence of a silt-member again points to a significant episode of suspended-load accumulation. However it differs in that the sand member consists of more than one sedimentation event. By definition the sets become finer in grain upwards, and are similar to the point bar deposits of some meandering streams (e.g. BERNARD and others 1970). This fining upwards has been explained by ALLEN (1970). It results from the balancing of the component of weight of the various particles down the bar surface, against an upsurface component of river flow which decreases as the water becomes shallower. The sand-member thickness corresponds directly to the bank-full depth of the channel.

In a cyclic sequence of this sort, the change from an episode of major suspended-

P. F. Friend and M. Moody-Stuart

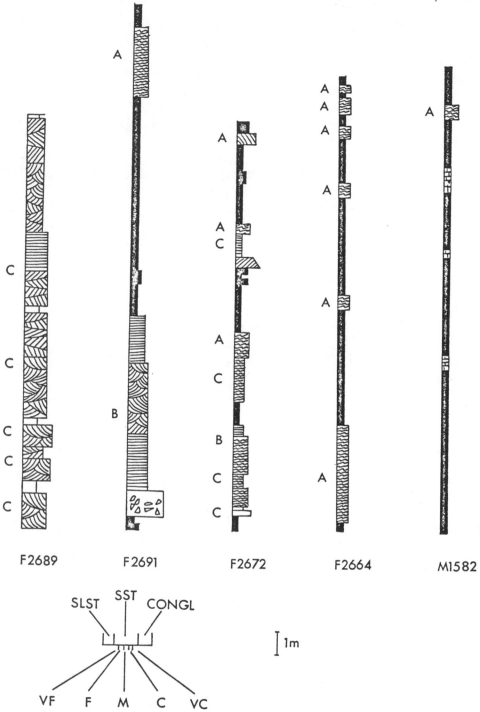

F2689   F2691   F2672   F2664   M1582

Fig. 22. *Parts of four 'vertical' sedimentological sections, deliberately chosen to illustrate four different patterns of grain-size variation which occur in the Wood Bay Formation. The letters indicate semi-cycles consisting of A) single-set sandstone and siltstone, B) multiple-set sandstone and siltstone, and C) no siltstone.*

load accumulation, to an episode of composite bed-load sedimentation, and back again, must result from important lateral movements of the river channel complex. Steady meander migration within the channel complex will not be enough. In two sections in the Recent deposits of the Murrumbidgee River (SCHUMM 1968b, p. 27), three sand members alternate with silt members. These sand members are quite discrete bodies, and therefore represent distinct episodes of channel location, rather than stages in a continuous and gradual sequence of lateral movements. The Mississippi River meander belt has moved as a whole by distinct steps four times over the last 2,000 years (FISK 1952).

(C) *No siltstone.* In this category the siltstone is either absent altogether, or impersistent. These semi-cycles occur in two main situations.

Firstly they may form the completely sandy sections (e.g. F2689 of Fig. 22) which are typical of areas of bed-load channel deposition. These semi-cycles (C1) are the deposits of the bars of braided rivers. Any tendency to become finer in grain-size upwards, may be due to the mechanism proposed by ALLEN (1970) and described above, or it may result from fluctuations of local flow strength or sediment supply.

Secondly they may occur below other sand semi-cycles in silt-bearing successions (e.g. F2672 of Fig. 22), where they form the lower parts of complex sand units. These complex units could arise in two ways:

(C2) by superposition of two semi-cycles of types A or B, with intervening erosion of silt that had been deposited.

(C3) by complex deposition during an interval of bed-load channel activity. This implies a distinct period of complex channel activity between periods of prolonged suspended-load (overbank or lacustrine) deposition. This is what one of us (MOODY-STUART 1966) had in mind in describing his "low sinuosity" model.

The general sediment pattern of this low sinuosity model is regarded by SCHUMM (1968a, p. 1582) as typical of his mixed-load type. ALLEN (1970) has doubted its existence in nature. We have certainly not found clear published descriptions of Recent deposits of this type formed by bed-load or low-sinuosity rivers. The few descriptions available of bed-load channels with overbank suspended-load deposits, stress that suspended load deposits are usually destroyed by channel movements (SCHUMM 1968a; COLEMAN 1969). However, MOODY-STUART did establish the distinctiveness of the sedimentary association in the Wood Bay Formation. The absence of modern examples may result from a natural tendency to select, for examination, systems which appear to conform with existing knowledge.

Channel depth. — The estimation of depth is probably the most critical and difficult step in our consideration of the alluvial variables. We would stress that any use we make of these alluvial variables will be based on the average properties of a large number of semi-cycles. We are anxious, therefore, to establish that the averages of certain properties are likely to be related to each other, rather than that individual measurements necessarily correspond.

We intend to use the average thickness of the sand members of our semi-cycles, as a measure of average channel flood depth.

70

We shall now attempt to justify this for the various semi-cycle types described in the previous section:

(A) *single-set sandstone and siltstone*. The amount of bed-load sediment being moved at any one time depends on the stream power (BAGNOLD 1966), which depends on water depth, amongst other variables. All other factors being equal, a thicker sandstone set is likely, therefore, to have been deposited from a deeper water flow.

(B) *multiple-set sandstone and siltstone*. Bank-full (flood) depth is generally equivalent to sand-member thickness, because the sand-member is interpreted as a complete bar deposit (as discussed above).

(C) *no siltstone*. In all types (C1, C2, C3) the semi-cycles are relict fragments of sand-members equivalent to those of types A and B. The thickness of these semi-cycles is therefore a minimum measure of bar, and channel depth.

In general we shall use the thickness of the sand members of our semi-cycles, as statistical measures of mean minimum flood depth.

An independent indicator of water depth is the height of sets of cross-stratification. The heights of large-scale ripples (dunes) with relatively straight crests have been plotted by ALLEN (1968, p. 139) against the depth of the water in which they formed. In his examples the depths varied from 0.5 to 38 m. The ratio of depth of water to height of ripple ranges from 3 to 22, with a mean at 8. ALLEN points out, however, that many of the small and large scale ripples which have strongly sinuous crests form in water only slightly deeper than ripple height. Our cross-bedding sets are the relict fragments of ripples (dunes), some with obviously sinuous crests, and some nearly straight. We feel that it is reasonable to base an assertion about water depth on our statistical observations of cross-bedding set thickness.

The north-western group of sections (LW1 in Fig. 20) have an average sand member thickness of 0.8 m, which we shall use as a measure of mean minimum flood depth. The mean cross-stratification set-thickness for the same sections is 0.2 m (Fig. 23). In contrast the south-eastern groups of sections (UE2 and 3) would have mean minimum flood depths of 2.2 and 2.6 m, and mean cross-stratification set thicknesses of 0.5 m (Fig. 23). This internal consistency encourages us to feel that our assumptions are reasonable.

Friction factor

Hydraulic engineers use a number of different equations to relate mean flow velocity to depth and slope. These equations can be regarded as relating the kinetic energy of the stream flow to the energy loss resulting from frictional drag at the stream boundary. The successful use of these equations depends in practice, on the estimation of the friction or resistance factor, or coefficient. The DARCY-WEISBACH equation is recommended by CARTER and others (1963, p. 99) in their progress report to the American Society of Civil Engineers on friction factors in open channels. They recommend it, in slight preference to the similar CHEZY-MANNING equations, which are rather less widely applicable and less generally used.

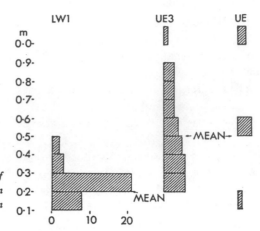

Fig. 23. *Histograms showing the occurrence of cross-bedding sets of various thicknesses in certain groups of localities (positions of groups shown on Fig. 18).*

$$U = \left(\frac{8g}{f}\right)^{\frac{1}{2}} (DS)^{\frac{1}{2}} \quad \text{Darcy-Weisbach,} \tag{1}$$

where f = friction factor or resistance coefficient, U = mean velocity, D = depth, S = slope.

In alluvial channels, frictional losses depend on (a) grain-size of the sediment, (b) nature and size of bed forms (e.g. ripples, plane beds etc.), (c) form of channels (sinuosity, size etc.).

It remains to consider the actual values of the friction factor or resistance coefficient (f). Bagnold (1966, Table 1) has published a list of observations on certain rivers, mainly in south-western North America. These observations include mean velocity, depth and slope, and it is therefore possible, using the above equation, to calculate 'f' for each set of observations. Fig. 24 shows the variation of values of "f". It is found that 95% of the 146 values lie between .01 and .15, and that the mode is .04. These values are therefore used in this investigation.

Many of the rivers which make up Bagnold's list are large. They are probably larger than the rivers responsible for much ancient river sedimentation. At the other extreme of size are the laboratory flume runs reviewed by Guy, Simons and Richardson (1966). Of the 8' runs involving 0.19 mm, 0.45 mm, and 0.93 mm sand, 100% of the "f" determinations lie within the range selected here (0.01 to 0.15). Allen (1970, Fig. 14, p. 316) uses the same flume data to show a distinction between the friction factor for plane beds and antidunes (f = 0.02), and for ripples and dunes (large-scale ripples) (f = 0.08). In our Spitsbergen regional work, cross-strata formed in ripples or dunes are much the commonest structures. As we shall show below, the assumption of a single modal friction factor does not seem to be a serious source of error in our sort of work.

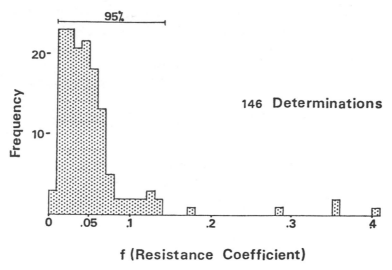

f (Resistance Coefficient)

Fig. 24. *Histograms showing the occurrence of different values of the Resistance coefficient (f) for the 146 present-day measurements of rivers listed by* BAGNOLD (*1966, Table 1*).

Strength of River Flow

Arbitrary choice of index. — HJULSTRÖM's well-known diagram (1935, p. 298) used water flow velocity as an index of the erosion, transportation and deposition of sediment of different grain-sizes. Some such index of stream strength seems a valuable measure of the processes which do much to control sediment movement. Many other measures of stream strength have been used. For instance, in addition to velocity, WILLIAMS (1967, p. 7) gives four measures (stream power, shear stress using depth, shear stress using hydraulic radius, regime theory bed factor). He comments, "No single measure of flow strength has yet gained wide-spread acceptance".

Since that was written, the work of BAGNOLD (1966) has done much to establish the use of stream power (defined below) as an index of stream strength, for portraying the ability of a stream to move sediment. BAGNOLD regards the transporting of sediment as a form of work carried out by the stream, and its power as a measure of the rate at which this work can be performed. Having developed this idea theoretically, BAGNOLD supported it by presenting detailed measurements made in laboratory flumes and natural rivers. We have decided to use stream power as one index of river strength.

The definition of stream-power involves shear stress, which itself involves slope, as will be described below. KENNEDY (1971, p. 132) points out that "sediment transport rate and flow velocity are not particularly sensitive to slope and shear stress". In addition, single values of slope sometimes correspond to three different values of mean velocity, even with constant discharge (because of variation in bed-form development). "Depth-discharge predictors which use slope or any quantity including slope as an independent variable, are inherently inferior in at least two respects (uniqueness and sensitivity) to those which treat S as a dependent variable". In our work, we use slope, depth and a highly generalised friction factor to estimate a stream-power for modern rivers. This use is clearly open to criticism on the grounds put forward by KENNEDY. However we feel that our use of stream power is justifiable over the range of river strengths which concerns us, which is much larger than that being considered by KENNEDY. Our interest is the distinction between silt, sand and pebble movement. We also have an interest in utilising stream power as an index, because of its value in measuring the rate of river work. We shall, however, also quote mean velocity as a strength index.

Relationship of velocity and power to depth and slope. — The relationship between velocity and depth and slope has been considered above.

The power of stream flow (ω), per unit area of stream bed, is defined as the product of the mean velocity (U), and the bed shear stress per unit area (τ).

$$\omega = U\tau \tag{2}$$

The shear stress (τ) is then defined as follows, where p = density, g = acceleration due to gravity, D = depth of water, S = slope of water:

$$\tau = pg\, DS \tag{3}$$

This stress is the downslope component of the weight of the water. The relationship is approximately true only if 1) the slope, S, is a low angle, 2) the flow is uniform, 3) the depth, D, is small compared with the width (WILLIAMS 1970). Combining equations 1, 2, 3

$$\omega = pg\,(DS)^{3/2}\left(\frac{(8g)}{f}\right)^{\frac{1}{2}} \tag{4}$$

We assume that p and g are constant and fix f at 0.04 (see above) in our particular situation. On plots of log D against log S therefore a series of straight and parallel lines will represent different values of mean velocity and power. (Figs. 25, 26).

$$\log D = \text{constant} - \log S \tag{5}$$

Velocity and power increase with increases of depth or of slope. To achieve fixed velocity or fixed power, depth and slope must behave inversely.

Sand bed-forms as indicators of local stream strength. — BAGNOLD (1966) and GUY, SIMONS and RICHARDSON (1966) showed that a systematic change of bed forms occurs in the sand of experimental flumes when the stream power and the transport rate increase. This relationship has been examined in more detail by ALLEN (1968, Fig. 6.9; 1969 Figs. 1 and 2; 1970, Fig. 2.6) who particularly concentrated on the variation of bed forms which is found with differing mean grain-size. He used the experimental data of GUY, SIMONS and RICHARDSON (1966), and WILLIAMS (1967).

We have used these same data, and some provided by LAURSEN (1958), to plot stability ranges of different bed forms against power and mean grain-size, in S. I.

74

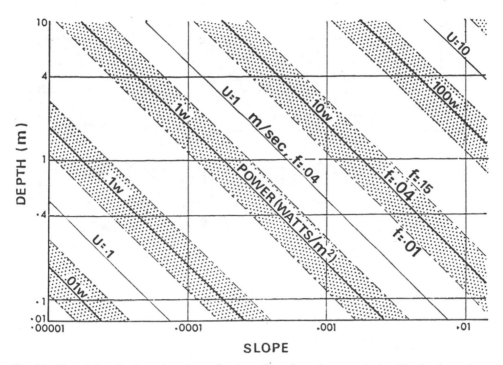

Fig. 25. *Plot of flow depth against slope, showing a line of equal mean velocity (U=1 m/sec, where Resistance coefficient (f) = 0.4), and lines of equal power with Resistance coefficient (f) = 0.01, 0.04 and 0.15. This plot is based on equations 1 and 4.*

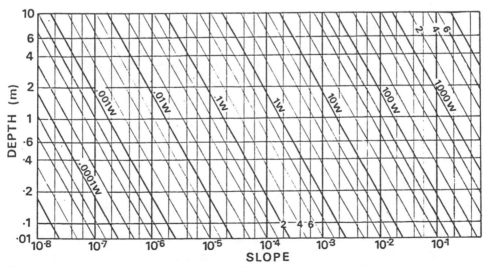

Fig. 26. *Plot of flow depth against slope, showing lines of equal stream power (watts). This plot is based on equation 4, assuming Resistance coefficient (f) = 0.04.*

75

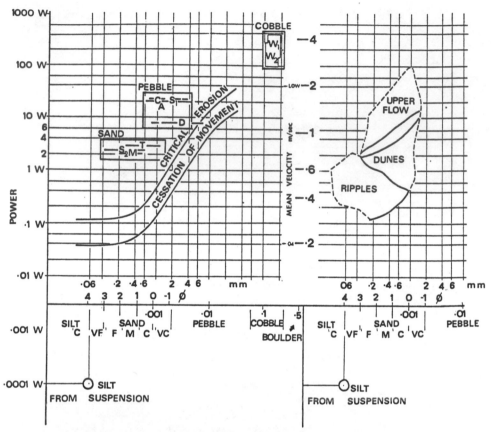

Fig. 27. *Plot of grain-size against stream power and mean velocity, showing 1) cessation of movement and critical movement curves (SUNDBORG), 2) alluvial grain-size (flood power data from some present-day rivers (Table 1), 3) grain-size power fields for sand, pebble and cobble rivers and 4) the fields of occurence of the most important sand bed-forms (GUY, SIMONS and RICHARDSON, 1966).*

units (Fig. 27). This plot shows a clear sequence from ripples (or lower-phase flat beds) to dunes (not always represented) to upper phase flat beds, with increasing power. ALLEN (1970) used variations in power and in grain-size as variables controlling the formation of one or other of the bed forms.

There are limitations to this method. Firstly, when the water is shallow, the general relationship between sand bed-load transport and power will break down completely. BAGNOLD suggested (1966, p. 19) that this will occur with depths of 2 to 3 centimetres. Secondly, the actual measurements reported by GUY, SIMONS and RICHARDSON (1966) may refer to conditions rather different from those in an aggrading stream. Equilibrium, rather than aggradation, was the experimental situation, and the power was generally raised to the equilibrium level, rather than dropped to it after a flood peak.

Finally we know so little about the significance, in terms of flood stage (or depth), of individual sets, that we prefer a method which uses a feature of a complete sand member to indicate the flood power of the river. This we shall consider below.

Grain-size generally as an indicator of general stream strength. — It is common knowledge that boulder beds are deposited by stronger streams than silt beds. It is our object in this section to consider this quantitatively.

For over 150 years, hydraulic engineers have been using empirical, and partly empirical, relationships between grain-size and velocity of flow (LELIAVSKY 1966, p. 34). Their work has often been concerned with problems of scour, so that "pick up" velocities ("critical erosion velocities") are usually quoted. However velocities for the cessation of sediment movement are also important in the sedimentation process. For any particular grain-size, cessation of sediment movement occurs at about 2/3 of the critical erosion velocity (HJULSTRÖM 1935, p. 320; MENARD 1950, p. 151; SUNDBORG 1956, p. 181).

For our purposes we have used the plots of velocity against grain-size published by SUNDBORG (1967, p. 337). He quoted velocities 1 m above the sediment surface, and we have converted these to mean velocities by multiplying them by 0.8 (HJULSTRÖM 1935, p. 294; SUNDBORG 1956, p. 175). We have then converted these mean velocities to powers using a combination of equations 1, 2 and 3:

$$\omega = \frac{U^3}{8} pf \qquad (6)$$

where ω = power, U = mean velocity, p = density, and f = friction factor.

This becomes $\omega = 5U^3$, in S. I. units, assuming standard friction factor, $f = 0.04$.

Curves for critical erosion, and for cessation of movement, are then plotted on Fig. 27, against axes of mean velocity, power and grain-size. Assuming that clasts of all sizes of material were available in the source areas, we would expect the flood velocities and powers of rivers to be reflected in the grain-size of their alluvium.

We have already pointed out that many local processes may influence the grain-size of particular beds. These include 1) processes influencing the grain-size of material supplied to the alluvial area, 2) processes of local hydrodynamic sorting, 3) processes causing local fluctuations of velocity or power, due to bed forms, channel shape etc.

We hope to avoid some of these difficult problems by basing our method on a general range of grain size for our alluvial systems, rather than the actual grain-size of local features. On Fig. 27 we have plotted ranges of grain size against power or mean velocity estimates for a number of modern rivers (Table 1), from the fine and medium sand of the lower part of the Mississippi to the cobbles of the White River. The estimate of river strength are flood estimates using the relationships to depth and slope (equations 1 and 4), and a constant friction factor of $f = 0.04$. Estimates of flood depth, particularly, are often difficult to make from the published descriptions, but the possible errors are not serious for our purposes at this stage.

On the basis of the grouping of these data, we distinguish:

1) *sand rivers,* with a median flood power of about 2W
2) *pebble rivers,* with a median flood power of about 15W
3) *cobble rivers,* with a median flood power of about 200W

Table 1
Grain-size and flood power of some present day rivers

Mississippi, U.S.A. (M) (FRAZIER and OSANIK 1961; KOLB 1963; LEOPOLD, WOLMAN and MILLER 1964; MOORE 1970), *alluvial tract.* fine-medium sand; flood depth, 30–34 m; slope 8 cm/km. Power, therefore 2W.

Slims River, Canada (S) (FAHNESTOCK 1969)

Intermediate (S$_1$), medium gravel, sand; flood depth, 1.3 m; slope 100–330 cm/km. Power, therefore, 20W.

Downstream (S$_2$), sand, silt; flood depth, 1 m; slope 30 cm/km. Power, therefore, 2W.

Tana River, Norway (T) (COLLINSON 1970), medium to coarse sand; flood depth (based on our calculations from discharge figures), 2.5 m.; slope 20 cm/km. Power, therefore, 3W.

Amite (A), Louisiana, U.S.A. (McGOWEN and GARNER 1970), coarse sand and pebble gravel; flood depth, 7 m; slope 60 cm/km. Power, therefore, 20W.

Colorado (C), Texas, U.S.A. (McGOWEN and GARNER 1970), coarse sand and pebble gravel; flood depth, 11.5 m; slope 30 cm/km. Power, therefore, 20 W.

White River (W), Washington, U.S.A. (FAHNESTOCK 1963, p. 26 etc.). Above moraine (W$_1$) cobbles (.12 – .18 m), flood depth about 0.8 m; slope 3000 – 8000 cm/km. Power, therefore, 150 – 400W. Below moraine (W$_2$), cobbles (.1 – .16 m), flood depth, about 0.8 m; slope 2000 – 6000 cm/km. Power, therefore, 100 – 300W.

Donjek River (D), Canada. (WILLIAMS and RUST 1969; RUST personal communication). Gravel most abundant, sand; flood depth about 2.8 m (our calculation using flood discharge estimate); slope 60 cm/km. Power, therefore, 8W.

All the Wood Bay rivers deposited alluvium of very-fine to coarse sand. They were all "sand rivers", and as the curves of Fig. 27 show, changes within the sand grade appear to make little difference to the power ratings. We therefore suggest that all the Wood Bay rivers were characterised by flood velocities of about 0.7 m/sec., and flood powers of about 2W. The presence of conglomerates would greatly extend the range of flow-strengths implied.

Slope

Slope, the last of our dependent variables, is estimated by using the depth-slope-power relationships of equation 4, which we have plotted graphically in Figs. 25 and 26.

Assuming a general friction factor of 0.04, and a general flood power of 2W, then slope will depend on fluctuations of flood-depth. The local slope of the channels is converted to a more regional slope by multiplying it by the sinuosity.

[*Editor's Note:* Material has been omitted at this point.]

References

ALEXANDER-MARRACK, P. D., P. F. FRIEND, and A. K. YEATS, 1971: 1. Mark Sensing for Recording and Analysis of Sedimentological Data. In: *Data Processing in Biology and Geology*, (J. L. CUTBILL ed.), Systematics Association, Special. 3. London.

ALLEN, J. R. L., 1963: The classification of cross-stratified units, with notes on their origin. *Sedimentology*. 2, 93–114.

— 1964: Studies in fluviatile sedimentation: six cyclothems from the Lower Old Red Sandstone, Anglo-Welsh Basin. *Sedimentology*. 3, 163–198.

— 1965a: Sedimentology and palaeogeography of the Old Red Sandstone of Anglesey, North Wales. *Proc. Yorks. Geol. Soc.* 35, 139–185.

— 1965b: Fining-upwards cycles in alluvial successions. *Geol. Jour.* 4, 229–246.

— 1966: On bed forms and palaeocurrents. *Sedimentology*. 6, 153–190.

— 1968: *Current Ripples*. North Holland, Amsterdam. 433p.

— 1969: Some Recent Advances in the Physics of Sedimentation. *Proc. geol. Ass.* 80(1), 1–42.

— 1970: Studies in fluviatile sedimentation: A comparison of fining-upwards cyclothems, with special reference to coarse member composition and interpretation. *Jour. sed. Pet.* 40, 298–323.

ALLEN. J. R. L. and P. F. FRIEND, 1968: Deposition of the Catskill Facies, Appalachian region: with notes on some other Old Red Sandstone basins. In: Late Paleozoic and Mesozoic Continental Sedimentation, northeastern North America, (G. de V. KLEIN ed.), *Spec. Paper Geol. Soc. Amer.* 106, 21–74.

BAGNOLD, R. A., 1966: An approach to the sediment transport problem from general physics *U.S. geol. Surv. Prof. Pap.* 422 (I), 1–37.

BEERBOWER, J. R., 1964: Cyclothems and Cyclic Depositional Mechanisms in Alluvial Plain Sedimentation. *Bull. Kansas geol. Surv.* 169, 31–42.

BERNARD, H. A., C. F. MAJOR JR., B. S. PARROTT, and R. J. LE BLANC SR., 1970: Recent sediments of southeast Texas: A Field Guide to the Brazos Alluvial and Deltaic Plains and the Galveston Barrier Island Complex. *Texas, Bureau of Economic Geology, Guidebook.* 11, 16 p.

BIRKENMAJER, K., 1964: Devonian, Carboniferous and Permian Formations of Hornsund, Vestspitsbergen. *Studia Geol. Polon.* VII, 47–123. Warsaw.

— 1965: Some sedimentological observations in the Old Red Sandstone at Lykta, Vestspitsbergen. *Norsk Polarinst. Årb.* 1963. 137–150. Oslo.

BLUCK, B. J., 1971: Sedimentation in the meandering river Endrick. *Scottish Jour. Geology.* 7, 93-187.

BRUSH, L. M., 1965: Sediment sorting in alluvial channels. In: Primary sedimentary structures and their hydrographic interpretation (G. V. MIDDLETON ed.). *Society of Economic Paleontologists and Mineralogists, Special Publication,* No. 12, 25–33.

CARTER, R. W., H. A. EINSTEIN, J. HINDS, R. W. POWELL, and E. SILBERMAN, 1963: Friction Factors in open channels. *Jour. Hydr. Div. Proc. Amer. Soc. Civil Eng.* 89, (HY2), 97–143.

COLE, A. J., 1968: Algorithm for the Production of Contour Maps from Scattered Data. *Nature*. 220, 92–94.

— 1969: An iterative approach to the fitting of trend surfaces. *Computer contributions Kans. Univ. geol. Surv.* No. 37.

COLEMAN, J. M., 1969: Brahmaputra River: Channel processes and sedimentation. *Sedimentary Geology*. 3, (2/3), 129–239.

COLLINSON, J. D., 1970: Bedforms of the Tana River. *Geogr. Annaler.* 52a, 31–56.

CURRAY, J. R., 1956: The analysis of two-dimensional orientation data. *Jour. Geol.* 64, 117–131.

CUTBILL, J. L. and A. CHALLINOR, 1965: Revision of the stratigraphical scheme for the Carboniferous and Permian rocks of Spitsbergen and Bjørnøya. *Geol. Mag.* 102, 418–439.

DENISON, R. H., 1970: Revised classification of Pteraspidae with Description of New Forms from Wyoming. *Fieldiana, Geology*. 20, 1–41.

DINELEY, D. L., 1960: The Old Red Sandstone of eastern Ekmanfjorden, Vestspitsbergen. *Geol. Mag.* 97, 18–32.

FAHNESTOCK, R. K., 1963: Morphology and hydrology of a Glacial Stream – White River, Mount Rainier, Washington. *U. S. geol. Surv. Prof. Paper.* **422A**, 70 p.

— 1969: Morphology of the Slims River. *Icefield Ranges Research Project, Scientific Results.* **1**, 161–172.

FISK, H. N., 1952: Geological investigation of the Atchafalaya basin and the problem of Mississippi River diversion. *Vicksburg, Mississippi River Commission*, 145 p.

FLOOD, B., D. G. GEE, A. HJELLE, T. SIGGERUD, and T. S. WINSNES, 1969: The geology of Nordaustlandet, northern and central parts. *Norsk Polarinst. Skr.* Nr. 146. 139 p. Oslo.

FØYN, S. and A. HEINTZ, 1943: the Downtonian and Devonian vertebrates of Spitsbergen VIII. The English-Norwegian-Swedish Expedition, 1939. Geological results. *Skr. om Svalbard og Ish.* Nr. 85, 1–51. Oslo.

FRAZIER, D. E. and A. OSANIK, 1961: Point-bar deposits, Old River Locksite, Louisiana. *Trans. Gulf Coast Assoc. Geol. Soc.* **11**, 121–137.

FRIEND, P. F., 1958: Cambridge University Expedition to Spitsbergen, 1957. *Polar Record.* **9**, 141.

— 1959: Cambridge Spitsbergen Expedition, 1958. *Polar Record.* **9**, 463–464.

— 1961: The Devonian stratigraphy of north and central Vestspitsbergen. *Proc. Yorks. geol. Soc.* **33**, 77–118.

— 1962: Cambridge Spitsbergen Expedition, 1961. *Polar Record.* **11**, 44–46.

— 1965: Fluviatile sedimentary structures in the Wood Bay Series (Devonian) of Spitsbergen. *Sedimentology.* **5**, 39–68.

— 1966: Clay fractions and colours of some Devonian red beds in the Catskill Mountains, U.S.A. *Quart. Jour. Geol Soc.* **122**, 273–292. London.

— 1967: Tectonic implications of Sedimentation in Spitsbergen and midland Scotland. In: *International Symposium on the Devonian System, Calgary, 1967*, (D. H. OSWALD ed.). **2**, 1141–1147.

— 1969: Old Red Land of the Atlantic. *Geographical Mag.* **41**, 689–694. London.

— and M. R. HOUSE, 1964: The Devonian period. In: The Phanerozonic Time-scale, *Quart. Jour. Geol. Soc.* **120**, special, 233–236. London.

— N. HEINTZ, and M. MOODY-STUART, 1966: New unit terms for the Devonian of Spitsbergen, and a new stratigraphical scheme for the Wood Bay Formation. *Norsk Polarinst. Årb.* 1965. 59–64. Oslo.

— and M. MOODY-STUART, 1970: Carbonate deposition on the river floodplains of the Wood Bay Formation (Devonian) of Spitsbergen. *Geol. Mag.* **107**, (3), 181–195.

GEE, D. G. and A. HJELLE, 1966: On the crystalline rocks of north-west Spitsbergen. *Norsk Polarinst. Årb.* 1964. 31–46. Oslo.

GEDDES, A., 1960: The alluvial morphology of the Indo-Gangetic plain: its mapping and geographical significance. *Inst. British Geographers, Trans.* 253–276.

GILBERT, G. K., 1917: Hydraulic mining debris in the Sierra Nevada. *U.S. geol. Surv. Prof. Pap.* **105**, 1–154.

GROVE, A. T. and A. S. GOUDIE, 1971: Secrets of Lake Stefanie's Past. *Geographical Mag.* **43** (8), 542–567. London.

GUY, H. P., D. B. SIMONS, and E. V. RICHARDSON, 1966: Summary of alluvial channel data from flume experiments, 1956–61. *U. S. geol. Surv. Prof. Pap.* **462**, (1), 1–96.

HARLAND, W. B., 1960: The Cambridge Svalbard Expedition, 1959. *Polar Record.* **10**, 40–44.

— 1961: An outline structural history of Spitsbergen. In: *Geology of the Arctic* (G. O. RAASCH ed.). Univ. of Toronto Press. **1**, 68–132.

— 1963: Cambridge Spitsbergen Expedition, 1962. *Polar Record.* **11**, 435–438.

— 1964: Cambridge Spitsbergen Expedition, 1963. *Polar Record.* **12**, 303–304.

— 1965a: Cambridge Spitsbergen Expedition, 1964. *Polar Record.* **12**, 589–591.

— 1965b: The tectonic evolution of the Arctic – North Atlantic region: *Roy. Soc. London. Philos. Trans.* **258**, (set. A.), 59–75.

— 1969: Contribution of Spitsbergen to understanding of Tectonic evolution of North Atlantic region. In: North Atlantic Geology and Continental Drift (a symposium, M. KAY ed.). *Am. Ass. Petrol. Geol., Mem.* **12**. 817–851. Tulsa.

— 1971: Tectonic Transpression in Caledonian Spitsbergen. *Geol. Mag.* **108**, 27–41.

HARLAND, W. B. and D. MASSON-SMITH, 1962: Cambridge survey of central Vestspitsbergen. *Geogr. Jour.* **128**, 58–70.

—   and R. H. WALLIS, 1966: Cambridge Spitsbergen Expedition, 1965. *Polar Record.* **13**, 192–194.

HAYES, J. R. and M. A. KLUGMAN, 1959: Feldspar Staining Methods. *Jour. sed. Pet.* **29**, 227–232.

HEINTZ, N., 1962: *Gigantaspis* – a new genus of fam. Pteraspidae from Spitsbergen. A preliminary note. *Norsk Polarinst. Årb.* 1960. 22-27.

—   1968: The pteraspid *Lyktaspis* n.g. from the Devonian of Vestspitsbergen. In: Current problems of lower vertebrate phylogeny (T. ØRVIG ed.). *Proc. 4th Nobel Symp.* 73–80. Stockholm 1967.

HJULSTRÖM, F., 1935: Studies of the morphological activity of rivers as illustrated by the River Fyris. *Bull. geol. Inst. Univ. Upsala.* **25**, 221–527.

HOEL, P. G., 1960: *Elementary Statistics,* Wiley, New York. 261 p.

HOLTEDAHL, O., 1914: On the Old Red Sandstone of northwestern Spitsbergen. *C. R. XIIth Session Int. Geol. Cong.* Canada, 1913, 707–712.

JAHNS, R. H., 1947: Geologic Features of the Connecticut Valley, Massachusetts, as related to recent floods. *U.S. geol. Surv., Water Supply Papers.* **996**, 158 p.

KAZMI, A. H., 1964: Report on the Geology and Ground-water Investigations in Rechna Doab, West Pakistan. *Records Geol. Surv. Pakistan.* **X**, (3), 26 p.

KENNEDY, J. F., 1971: Sediment Transportation Mechanics: F. Hydraulic Relations for Alluvial Streams, Task Committee for Preparation of the Sedimentation Manual, (V. A. VANONI, chairman). *Jour. Hydr. Div., Proc. Amer. Soc. Civil Eng.* **97** (HY1), 101–141.

KOLB, C. R., 1963: Sediments forming the bed and banks of the lower Mississippi River, and their effect on River Migration. *Sedimentology.* **2**, 227–234.

KRINSLEY, D. B., 1970: *A geomorphological and paleoclimatological study of the playas of Iran.* 486 p. Superintendent of Documents, U.S. Govt. Printing Office, Washington D.C. 20402, U.S.A.

LANGBEIN, W. B. and S. A. SCHUMM, 1958: Yield of sediment in relation to mean annual precipitation. *Am. Geophys. Union Trans.* **39**, 1076–1084.

LANKESTER, E. R., 1884: Report on fragments of fossil fishes from the Palaeozoic strata of Spitzbergen. *K. Svenka Vet. Akad. Handl.* **20**, (9), 9 p.

LAURSEN, E. M., 1958: The total sediment load of streams. *Jour. Hydr. Div., Proc. Amer. Soc. Civil Eng.* **84** (HY1), 36p.

LELIAVSKY, S., 1966: *An introduction to Fluvial Hydraulics.* Dover, New York, 257 p. (republication of 1955 edition).

LEOPOLD, L. B., M. G. WOLMAN, and J. P. MILLER, 1964: *Fluvial Processes in Geomorphology.* Freeman, San Francisco. 522p.

LOBBAN, M. C., 1956: Cambridge Spitsbergen Physiological Expedition, 1955. *Polar Record.* **8**, 253–254.

MACKIN, J. H., 1948: Concept of the graded stream. *Bull. geol Soc. Amer.* **59**, 463–512.

MELAND, D. and J. O. NORRMAN, 1969: Transport velocities of individual size fractions in heterogeneous bed load. *Geog. Ann.* **51A**, 127–144.

McGOWEN, J. H. and L. E. GARNER, 1970: Physiographic Features and Stratification Types of coarse-grained point bars; modern and ancient examples. *Sedimentology.* **14** (1/2), 77–112.

McKEE, E. D., E. J. CROSBY, and H. L. BERRYHILL JR., 1967: Flood deposits, Bijou Creek, Colorado, June 1965. *Jour. sed. Pet.* **37**, 829–851.

MENARD, H. W. 1950: Sediment movement in relation to current velocity. *Jour. sed. Pet.* **20**, 148–160.

MILLER, C. R. and R. F. PIEST, 1970: Chapter IV. Sediment Sources and Sediment Yields. *Jour. Hydr. Div., Proc. Amer. Soc. Civil. Eng.* **96** (HY6), 1283–1329.

MOODY-STUART, M., 1966: High- and Low-Sinuosity Stream Deposits, with Examples from the Devonian of Spitsbergen. *Jour. sed. Pet.* **36**, (4), 1102–1117.

MOORE, G. T., 1970: Role of Salt Wedge in Bar Finger Sands and Delta Development. *Bull. Amer. Ass. Petrol. Geol.* **54**, (2), 326–333.

PETTIJOHN, F. J., 1957: *Sedimentary Rocks.* Second Edition, Harper, New York. 718p.

READ, W. A., J. M. DEAN, and A. J. COLE, 1971: Some Namurian (E_2) paralic sediments in central Scotland: an investigation of depositional and facies change using iterative-fit trend-surface analysis. *Jour. geol. Soc. London.* 127, 137–176.

SCHUMM, S. A., 1963: A tentative classification of Alluvial River Channels. *U. S. geol. Surv. Circular.* 477, 10p.

— 1968a: Speculations Concerning Paleohydrologic Controls of Terrestrial Sedimentation. *Bull geol. Soc. Amer.* 79, 1573–1588.

— 1968b: River Adjustment to Altered Hydrologic Regimen – Murrumbidgee River and Paleochannels, Australia. *U. S. geol. Survey Prof. Paper.* 598, 65p.

— 1969: River metamorphosis. *Jour. Hydr. Div., Proc. Amer. Soc. Civil Eng.* 95 (HY1), 255–273.

SPARKS, B. W. and R. G. WEST, 1965: The relief and drift deposits. In: *The Cambridge Region 1965* (J. A. STEERS ed.). The British Association for the Advancement of Science. 18–41.

SUNDBORG, Å., 1956: The River Klarälven, A study of fluvial processes. *Geog. Ann.* 38, 125–316.

— 1967: Some aspects on fluvial sediments and fluvial morphology, I General views and graphic methods. *Geog. Ann.* 49A, 333–343.

VOGT, T., 1929: Fra en Spitsbergen-ekspedition i 1928. *Årb. norske Vidensk. Akad. (Nat. Vid. Kl.)* 11, 10–12.

WALCOTT, R. I., 1970: An isostatic origin for basement uplifts. *Canadian Journal of Earth Sciences.* 7, 931–937.

WHITE, E. I., 1935: The ostracoderm *Pteraspis* Kner and the relationships of agnathous vertebrates. *Phil. Trans. Roy. Soc. London.* (B) 225, 381–457.

WILLIAMS, G. E., 1970: Piedmont sedimentation and late Quaternary chronology in the Biskra region of northern Sahara. *Zeitschrift für Geomorphologie.* Supplementband 10, 40–63.

WILLIAMS, G. P., 1967: Flume experiments on the Transport of a coarse sand. *U.S. geol. Surv. Prof. Paper.* 562-B, 1–31.

— 1970: Flume width and water depth effects in Sediment Transport experiments. *U. S. geol. Surv. Prof. Paper.* 562-H, 1–37.

WILLIAMS, P. F. and B. R. RUST, 1969: The sedimentology of a Braided River. *Jour. sed. Pet.* 39, 649–679.

ØRVIG, T., 1957: Notes on some Paleozoic lower vertebrates from Spitsbergen and North America. *Norsk Geol. Tidsskr.* 37, 285–353. Oslo.

— 1969a: Vertebrates from the Wood Bay Group and the position of the Emsian-Eifelian boundary in the Devonian of Vestspitsbergen. *Lethaia.* 2, 273–328. Oslo.

— 1969b: Thelodont scales from the Grey Hoek Formation of Andrée Land, Spitsbergen. *Norsk Geol. Tidsskr.* 49, 387–401. Oslo.

5

Reprinted from *Geol. Jour.* **9**, Pt. 1:1–16 (1974)

Studies in fluviatile sedimentation: lateral variation in some fining-upwards cyclothems from the Red Marls, Pembrokeshire

J. R. L. Allen

The lateral lithological variability of 18 stratigraphically close fluviatile cyclothems is analysed in three ways: (1) qualitatively, (2) by means of the upwards facies transition probability matrix, and (3) using the non-parametric Mann-Whitney U statistic. Some cyclothems are relatively very uniform over lateral distances comparable with 100 cyclothem thicknesses, whereas others have a lateral variability similar to the variability characteristic of the set of cyclothems. Hence some amongst stratigraphically close cyclothems are considered to show real differences, which may be attributed to allocyclic environmental factors operating on a relatively large time scale. The variation within cyclothems can be explained by shorter term autocyclic factors.

[*Editor's Note:* The introduction and Table 1 have been omitted at this point.]

2. Sources and nature of data

The data represent profiles measured normal to bedding in each of 18 fining-upwards cyclothems in the Red Marls (L. Devonian) of the Pembrokeshire coast (Table 1). Cyclothems were chosen on the grounds that (1) each was confidently traceable over a sufficient distance along strike, (2) each was for practical purposes fully exposed, and

(3) weathering did not obscure the lithologies and sedimentary structures. Hence most lay between mid-tide level and a few metres above high water mark.

The stratigraphical background is as follows. Cyclothem *1* (SM 801110) lies in the Winsle anticline 200–300 metres above the base of the Red Marls (Cantrill *et al.* 1916 p. 91). Cyclothems *2–4* crop out 14 km to the southeast, in Gravel Bay, Freshwater West, on the Castlemartin Corse anticline (SM 878007). They occur in Dixon's (1921 p. 33) Lower Marl Group, and therefore compare in horizon to cyclothem *1*. The younger cyclothems *5–11* come from the lower and middle Sandstone and Marl Group (Dixon 1921 p. 34) east of the main north-south fault in Freshwater West (SR 8898). Higher cyclothems, listed as *12–15*, were measured in the centre of Manorbier Bay (SS 057976) 17 km to the east in the Freshwater East anticline. Cyclothem *16* is from the upper Sandstone and Marl Group on the eastern cliffs in Manorbier Bay (SS 059973). In neighbouring Swanlake Bay, cyclothems *17* and *18* (SS 043981) were measured immediately west of the beach. They also lie in the upper Sandstone and Marl Group.

The cyclothems are represented by 2–11 measured profiles each, the lateral spacing of which is given non-dimensionally, as the ratio of the measured spacing along strike to the mean of the thicknesses of the adjacent two profiles (Table 1; Fig. 1). The lithologies found were classified into sedimentation units which, singly or contiguously in groups, were divided between the following six facies states widely recognized in fining-upwards cyclothems (Allen 1970 p. 300).

> State A—Conglomeratic Facies. Intraformational conglomerate, usually a sandstone matrix, cross-bedded or without internal bedding. Exotic gravel subordinate or absent.
>
> State B_1—Cross-bedded Sandstone Facies. Relatively coarse-grained sandstones, usually in more than one successive set.
>
> State B_2—Flat-bedded Sandstone Facies. Parallel laminated sandstones of variable grain size, often in successive, erosively related sets.
>
> State B_3—Cross-laminated Sandstone Facies. Relatively fine grained sandstones commonly with climbing-ripple structures.
>
> State C—Alternating Beds Facies. Siltstones, often mudcracked, interbedded on a centimetre or decimetre scale with relatively fine grained sandstones, commonly graded and with erosive bases. Limestone concretions (calcrete) may be present.
>
> State D—Siltstones Facies. Relatively thick, poorly-bedded siltstones commonly with limestone concretions (calcrete).

The data relevant to each cyclothem therefore are (1) the number of profiles measured, (2) the relative profile spacing, (3) the total thickness of each profile, (4) the number of facies states in each profile (5) the number of facies representatives in each profile, (6) the thicknesses of the individual facies representatives in each profile, and (7) the upward sequence of facies representatives in each profile. Directional sedimentary structures were measured wherever possible, so that the lines of profiles could be related to the palaeocurrents.

3. Approaches to the problem of lateral variation in the cyclothems

The problem of lateral variation may be approached by analysing some or all of the listed characters either (1) qualitatively (subjectively at all stages of the analysis), (2) semi-quantitatively (objective first stages but empirical final stage), or (3) quantitatively (objectively at all stages).

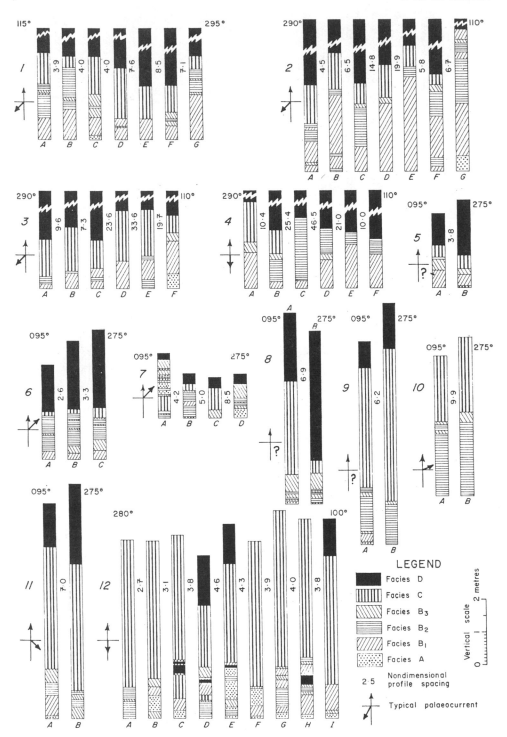

Fig. 1. Vertical sequences of facies states observed in profiles through the 18 fining-upwards cyclothems.

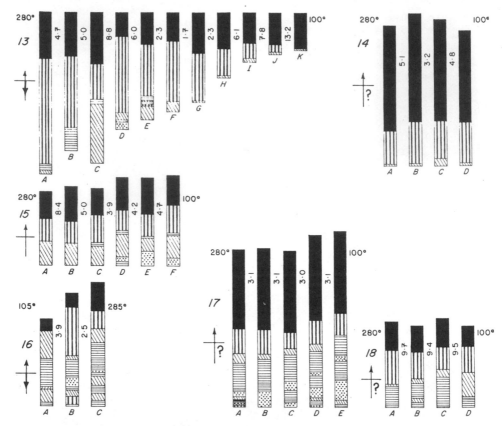

Fig. 1.—Continued from previous page.

The qualitative approach has the advantage that all of the characters can be assessed together. Its main weakness is that different workers, because of their differing experience and views, will choose a different balance of importance between the various characters of the cyclothems. Hence statements based on such an approach about whether or not there is little or much lateral variation become, in the final analysis, expressions of opinion.

The strength of the second and third modes is that some or all statements and inferences regarding lateral variation can be justified objectively. The weaknesses of a wholly or partly quantified approach are that (1) the data group into small samples, and (2) there are very few powerful ways of comparing paired profiles even in respect of one or two characters. The choice of quantitative methods should, however, be partly related to previous ways of analysing fining-upwards cyclothems (Allen 1970). The earlier study considered *one* profile from each of numerous cyclothems; we now examine the variation between *numerous* profiles existing in cyclothems taken one at a time.

The final choice determined by these considerations is the qualitative approach, a semi-quantitative attack in which facies transitions are studied in the context of a first-order Markov process, and a fully quantitative analysis employing the Mann-Whitney *U* statistic. This is not the only choice available, and other workers may have reasons for a different preference.

4. Qualitative assessment

It is possible with confidence to indicate the least and most variable cyclothems (Fig. 1), and to assign the remainder without ranking to an intermediate group. This group comprises cyclothems 2–6, 8–10, and 15, together with 7, 11 and 16 which are singled out because on field evidence they include a lateral deposit.

Cyclothem 14 is least variable. Although the profiles differ markedly in total thickness, every profile comprises the same three facies states which, except for the lowest state B_3, are similarly developed. Another comparatively uniform cyclothem is 18, represented by four relatively widely spaced profiles of similar thickness. Profiles A, C and D are similar in having four states expressed each by one representative, but profile B differs from them in that two states are represented twice. Cyclothem 17 is more variable than either 14 or 18, but is placed amongst the relatively uniform. Although the profiles differ more in the number of facies representatives than in total thickness, the variation is largely due to the number of times state A is represented.

Three cyclothems vie as the most variable. Cyclothem 13 has a flat top and concave-up base, and its coarse member fingers out eastwards amongst massive and concretionary siltstones. There is an almost five-fold eastward decrease of profile thickness but a much weaker if consistent trend in the number of facies representatives. Profiles A-C are similar in kinds of facies states, as also are profiles F-K. Profile D differs from A-C in possessing state A, and profile E resembles no other in the cyclothem. Cyclothem 1 is also amongst the most variable, showing substantial lateral variations of total thickness, number of facies states, and number of facies representatives. A persisting motif, however, is a substantial thickness of state B_1 in the lower part of the coarse member. Cyclothem 12 is also comparatively variable, especially in terms of the number and sequence of facies states.

The coarse members of cyclothems 7, 11 and 16 in the field visibly have the structure of a lateral deposit (Allen 1965 fig. 2). Of these only 7 can be compared in variability with cyclothems 13, 1 and 12. The others are definitely of intermediate variability.

5. Facies transitions

A vertical lithological sequence of several different facies states expressed as a number of facies representatives may be analysed for pattern (cyclicity) by calculating the first-order upwards facies transition probability matrix (Potter and Blakely 1968). The matrix, which may alternatively be shown as a path diagram, lists the empirical probability that a given facies state will be succeeded by any other. In the study of variation between cyclothems (Allen 1970), for example, the probability in each vertical sequence that facies A will be overlain by facies B_1 is the ratio of the total observed 'transitions' (erosional as well as continuous relationships are possible) from A to B_1 to the total observed occurrences of A. Hence many of the transitions will carry a relatively high probability value if there is a strong vertical facies pattern. If the pattern is weak, and the sequence relatively disordered, most transitions will carry uniformly low probabilities.

Thus the transition probability matrix denotes not only the type and strength of pattern but also the variability of the sequence, since a strong pattern demands little variation in the relative arrangement of states. The variability may be assessed empirically by (1) the number of unoccupied transitions in the matrix, (2) the number of alternative paths from a given facies state back to the same state, assuming an arbitary threshold probability, and (3) the total number of possible steps (transitions) in the

Table 2. Upwards facies transition probability matrices for the fining-upward cyclothems (See also figure 2)

1

	A	B₁	B₂	B₃	C	D
A		1.00	–	–	–	–
B₁	–		0.13	0.62	0.25	–
B₂	–	–		0.62	0.38	–
B₃	–	0.10	0.70		0.20	–
C	–	–	–	–		1.00
D	0.14	0.86	–	–	–	

2

	A	B₁	B₂	B₃	C	D
A		1.00	–	–	–	–
B₁	–		0.77	0.15	–	0.08
B₂	–	0.46		0.27	0.27	–
B₃	–	0.40	–		0.60	–
C	–	–	–	–		1.00
D	0.14	0.72	0.14	–	–	

3

	A	B₁	B₂	B₃	C	D
A		1.00	–	–	–	–
B₁	–		0.57	0.14	0.29	–
B₂	–	0.40		–	0.60	–
B₃	–	–	–		1.00	–
C	–	–	–	–		1.00
D	0.17	0.66	0.17	–	–	

4

	A	B₁	B₂	B₃	C	D
A		1.00	–	–	–	–
B₁	–		0.67	0.33	–	–
B₂	–	–		0.20	–	0.80
B₃	–	–	0.33		0.67	–
C	–	–	–	–		1.00
D	–	1.00	–	–	–	

6

	A	B₁	B₂	B₃	C	D
A		–	–	–	–	–
B₁	–		0.33	0.64	–	–
B₂	–	–		0.71	0.29	–
B₃	–	–	0.86		0.14	–
C	–	–	–	–		1.00
D	1.00	–	–	–	–	

7

	A	B₁	B₂	B₃	C	D
A		–	0.20	0.33	–	–
B₁	–		0.20	0.67	–	–
B₂	–	–		0.40	0.20	–
B₃	–	–	0.12		0.25	0.13
C	–	–	–	–		–
D	0.33	–	0.50	–	–	

12

	A	B₁	B₂	B₃	C	D
A		0.44	0.14	0.14	0.07	0.21
B₁	0.07		0.08	0.46	0.30	0.08
B₂	0.25	0.13		0.37	0.25	–
B₃	0.17	0.17	0.25		0.33	0.08
C	0.42	0.08	0.17	0.08		0.25
D	0.49	0.38	–	–	0.13	

13

	A	B₁	B₂	B₃	C	D
A		1.00	–	–	–	–
B₁	–		1.00	–	–	–
B₂	–	–		0.33	0.67	–
B₃	0.09	–	0.27		0.55	0.09
C	–	–	–	–		1.00
D	0.09	–	0.09	0.82	–	

14

	A	B₁	B₂	B₃	C	D
A		–	–	–	–	–
B₁	–		1.00	–	–	–
B₂	–	–		1.00	–	–
B₃	–	–	–		1.00	–
C	–	–	–	–		1.00
D	1.00	–	–	–	–	

15

	A	B₁	B₂	B₃	C	D
A		–	0.20	–	–	–
B₁	–		–	1.00	–	–
B₂	–	–		0.67	0.80	–
B₃	–	–	–		0.33	–
C	–	–	0.17	0.50		1.00
D	0.33	–	–	–	–	

16

	A	B₁	B₂	B₃	C	D
A		1.00	–	–	–	–
B₁	–		1.00	–	–	–
B₂	0.22	–		0.67	0.11	–
B₃	–	–	0.43		0.29	0.14
C	–	0.14	–	0.33		0.67
D	–	–	1.00	–	–	

17

	A	B₁	B₂	B₃	C	D
A		0.91	–	0.09	–	–
B₁	0.27		–	0.64	–	0.09
B₂	0.44	–		–	0.56	–
B₃	–	–	0.17	0.17	–	–
C	0.66	–	–	–	–	1.00
D						

18

	A	B₁	B₂	B₃	C	D
A		–	–	–	–	–
B₁	–		–	1.00	–	–
B₂	–	–		–	0.80	1.00
B₃	–	–	0.20		–	–
C	–	–	–	–		1.00
D	–	–	1.00	–	–	

88

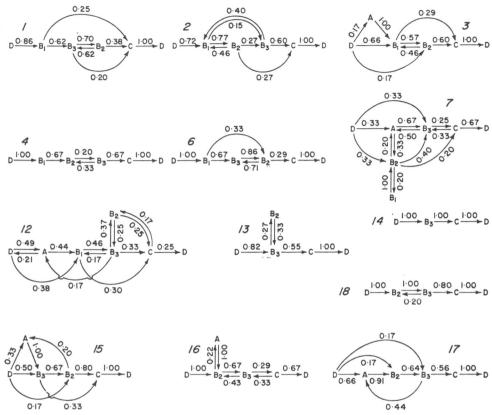

Fig. 2. Path diagrams (p ⩾ 0·15) for the fining-upwards cyclothems based on the upwards facies transition probability matrices (see table 2).

path diagram, again for some threshold probability. The number of unoccupied transitions should decrease as sequences become more variable. Clearly the other two indices will increase with increasing diversification. Read and Merriam (1972) have independently also used the transition probability matrix as the basis for comparing cyclical sequences, in this case of the coarsening-upwards type. The matrices provide for them a set of vectors which are then separated and compared by cluster analysis.

In the present study of lateral variation, a transition matrix for each cyclothem may be calculated by the device of treating the lateral sequence of profiles as a vertical one. For example, referring to Figure 1, the lowest state in profile B is arbitrarily taken as transitional from the highest state in profile A, and so on through the lateral sequence. The equally arbitrary alternative, yielding nearly or exactly identical results, is to work from the last lettered profile.

Table 2 and Figure 2 give respectively the transition matrixes and path diagrams (beginning in state D) for 13 of the cyclothems, starting in each case with profile A. Cyclothems 5 and 8–11 were excluded because they gave too few transitions. Listed in Table 3 are values for the above three empirical indices of variation. The number of paths and steps has in each case been normalized by the number of facies states in the path diagram, the selected threshold probability being $p = 0.15$.

The indices suggest that the cyclothems differ greatly in variability. Unoccupied transitions range between 27 and 3, out of a possible total of 30 for the group of cyclothems as a whole. Correspondingly, the number of paths per state varies between 0·20 and 5·67, and the number of steps per state between 1·00 (the smallest possible value) and 2·50. The cyclothems ranked in order of increasing variability and with ties bracketed are:

14, 18, 4, (6, 15), (3, 13), (1, 16, 17), 2, 7 and 12

in terms of the number of unoccupied transitions;

4, 14, (13, 18), (1, 6), 17, 3, 2, 16, 15, 7 and 12

in terms of the number of paths per state; and

14, 4, (13, 18), 6, (1, 16, 17), (2, 3, 15), 7 and 12

in terms of the number of steps per state. These rankings are broadly consistent, and there is especially good agreement in terms of the numbers of paths and steps per state. Thus all three indices denote cyclothem 12 as the most variable, and two of the three indicate 14 to be the most uniform. Each index shows cyclothems 3, 6, and 17, for example, to have an intermediate variability. Cyclothem 7, with a lateral deposit, is the second most variable by all three indices.

6. Mann-Whitney statistic

The Mann-Whitney U test (Siegel 1956 p. 116; Owen 1962 p. 340) is a powerful non-parametric test of whether two independent groups of numerical data come from the same population. Let there be a sample drawn from each of populations X and Y. The null hypothesis is that X and Y have the same distributions, and the alternative hypothesis, making the test two-tailed, is that X is stochastically either larger or smaller than Y. The test involves (1) ranking the numerical observations in the combined samples in order of increasing size, and (2) then counting the number of observations representing, say, population X that precede each observation drawn from population Y. The total of these counts is the Mann-Whitney statistic U, the sampling distribution of which, under the null hypothesis, is tabled in detail (Siegel 1956; Owen 1962) for sample sizes up to 10. Critical values of U only are available for larger samples.

This test was applied to the cyclothems in a manner allowing their variability to be assessed in terms of the combined kinds and thicknesses of the facies. The states were ranked from 1 through to 6, beginning with state A and ending with state D. Each profile was divided into facies representatives, and the thickness in metres of each representative was multiplied by the rank number of the appropriate state, to form the rank-thickness product. Thus every profile was described by a group of rank-thickness products. These groups were the raw data for the comparison by the Mann-Whitney test of the profiles two at a time within each cyclothem. The test therefore establishes the likelihood that each two groups of rank-thickness products had come from a single population. Hence as the cyclothems grow more uniform laterally in lithological composition, the inferred probabilities under the null hypothesis become larger, and the mean of the probabilities for each cyclothem may serve as an index to its lateral variability.

Table 4 shows for each cyclothem the matrix of probability values under the null hypothesis obtained when the constituent profiles are compared two at a time. Mean values appear in Table 3. The matrices show generally high or intermediate probabilities. Only cyclothem 7 has profiles differing at the 5 percent level, and differences at the 10 per cent level are restricted to 12, 13 and 17. Using the mean probability, the cyclothems may be ranked in order of increasing variability as follows:

14, 4, 16, 18, (3, 15,) 13, 1, 6, 2

Table 3. Summary of quantitative indices of variation (transition matrix and Mann-Whitney test) for the 18 fining-upwards cyclothems.

Cyclothem	Number of unoccupied transitions (max. = 30)	Paths per state (p ⩾ 0·15)	Steps per state (p ⩾ 0·15)	Mean probability in Mann-Whitney test
1	18	0·60	1·60	0·62
2	17	1·40	1·80	0·44
3	19	1·20	1·80	0·66
4	22	0·20	1·20	0·82
5	—	—	—	0·84*
6	21	0·60	1·40	0·61
7	14	3·67	2·33	—
8	—	—	—	0·80*
9	—	—		0·28*
10	—	—	—	0·25*
11	—	—	—	0·53*
12	3	5·67	2·50	—
13	19	0·50	1·25	0·63
14	27	0·33	1·00	0·95
15	21	1·80	1·80	0·66
16	18	1·60	1·60	0·75
17	18	1·00	1·60	—
18	25	0·50	1·25	0·73

*based on only one pair of profiles

with the ties bracketed. Cyclothems 7, 12 and 17 are excluded because they have some sample groups too large to appear in the detailed tabulation of the U statistic. Cyclothems 5 and 8–11 have only two profiles each, and are also excluded. The mean probabilities associated with the ranked cyclothems suggest noticeable but not great differences of lateral variability. Cyclothem 14, for example, yielded a mean of 0·95, whereas cyclothem 2, the most variable, gave a mean of 0·44.

Since the profiles of Figure 1 are spatially ordered sequences, it is worth examining Table 4 to see if they become less alike as their lateral spacing grows. This would be indicated by a steady decrease with increased spacing, or some other systematic variation, of the probability under the null hypothesis. The matrices for several cyclothems, for example 2, 3, 14 and 16, show runs of decreasing probability values involving 3 or more consecutive items. Significantly, the opposite occurs but once

92

Table 4. Probability matrices comparing the profiles two at a time within each of the 18 cyclothems. The probability shown is that under the null hypothesis in the Mann-Whitney test. See Figure 1 for a summary of the lithological compositions of the profiles compared.

1

	A	B	C	D	E	F	G
A		0.79	0.54	0.56	0.38	0.69	0.25
B	0.79		0.70	0.93	0.43	0.93	0.94
C	0.54	0.70		1.00	0.43	0.66	1.00
D	0.56	0.93	1.00		0.53	0.46	0.76
E	0.38	0.43	0.43	0.53		0.18	0.29
F	0.69	0.93	0.66	0.46	0.18		0.66
G	0.25	0.94	1.00	0.76	0.29	0.66	

2

	A	B	C	D	E	F	G
A		0.63	0.16	0.27	0.27	0.37	0.46
B	0.63		0.76	0.38	0.26	0.59	1.00
C	0.16	0.76		0.40	0.63	0.48	0.11
D	0.27	0.38	0.40		1.00	0.16	0.12
E	0.27	0.26	0.63	1.00		0.16	0.18
F	0.37	0.59	0.48	0.16	0.16		0.73
G	0.46	1.00	0.11	0.12	0.18	0.73	

3

	A	B	C	D	E	F
A		1.00	0.63	0.40	1.00	1.00
B	1.00		0.86	0.40	0.34	1.00
C	0.63	0.86		0.27	0.34	0.34
D	0.40	0.40	0.27		0.27	0.53
E	1.00	0.34	0.34	0.27		0.87
F	1.00	1.00	0.34	0.53	0.87	

4

	A	B	C	D	E	F
A		0.40	1.00	0.70	1.00	1.00
B	0.40		0.80	1.00	0.80	0.80
C	1.00	0.80		1.00	0.67	0.67
D	0.70	1.00	1.00		0.80	1.00
E	1.00	0.80	0.67	0.80		0.67
F	1.00	0.80	0.67	1.00	0.67	

5

	A	B
A		0.84
B	0.84	

6

	A	B	C
A		0.52	0.57
B	0.52		0.74
C	0.57	0.74	

7

	A	B	C	D
A		>0.10	<0.05	>0.10
B	>0.10		0.12	1.00
C	<0.05	0.12		0.12
D	>0.10	1.00	0.12	

8

	A	B
A		0.50
B	0.50	

9

	A	B
A		0.28
B	0.28	

10

	A	B
A		0.25
B	0.25	

11

	A	B
A		0.53
B	0.53	

12

	A	B	C	D	E	F	G	H	I
A		0.73	0.33	0.74	>0.10	0.41	0.46	0.45	0.95
B	0.73		0.60	1.00	>0.10	0.84	0.22	0.21	1.00
C	0.33	0.60		1.00	>0.10	0.52	0.17	1.00	0.30
D	0.74	1.00	1.00		>0.10	1.00	1.00	0.35	1.00
E	>0.10	>0.10	>0.10	>0.10		>0.10	>0.10	>0.10	>0.10
F	0.41	0.84	0.52	1.00	>0.10		0.59	0.72	0.22
G	0.46	0.22	0.17	1.00	>0.10	0.59		0.99	0.68
H	0.45	0.21	1.00	0.35	>0.10	0.72	0.99		0.62
I	0.95	1.00	0.30	1.00	>0.10	0.22	0.68	0.62	

13

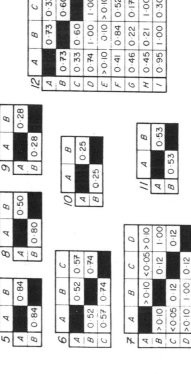

	A	B	C	D	E	F	G	H	I	J	K
A		0.86	0.87	0.56	0.16	0.86	0.63	0.63	0.86	0.63	0.53
B	0.86		0.63	0.39	0.07	0.40	0.40	0.40	0.40	0.20	0.40
C	0.87	0.63		0.41	0.16	0.86	0.40	0.40	0.63	0.40	0.53
D	0.56	0.39	0.41		0.53	0.39	0.86	0.79	0.79	0.79	0.86
E	0.16	0.07	0.16	0.53		0.27	0.39	0.27	0.67	0.37	0.89
F	0.86	0.40	0.86	0.39	0.27		1.00	0.70	0.70	0.40	0.80
G	0.63	0.40	0.40	0.86	0.39	1.00		0.70	0.70	0.40	0.80
H	0.63	0.40	0.40	0.79	0.27	0.70	0.70		1.00	0.70	0.80
I	0.86	0.40	0.63	0.79	0.67	0.70	0.70	1.00		1.00	1.00
J	0.63	0.20	0.40	0.79	0.37	0.40	0.40	0.70	1.00		1.00
K	0.53	0.40	0.53	0.86	0.89	0.80	0.80	0.80	1.00	1.00	

14

	A	B	C	D
A		0.70	1.00	1.00
B	0.70		1.00	1.00
C	1.00	1.00		1.00
D	1.00	1.00	1.00	

15

	A	B	C	D	E	F
A		1.00	0.96	0.26	0.29	0.57
B	1.00		0.63	0.17	0.29	0.57
C	0.96	0.63		0.76	0.73	0.95
D	0.26	0.17	0.76		0.66	1.00
E	0.29	0.29	0.73	0.66		1.00
F	0.57	0.57	0.95	1.00	1.00	

16

	A	B	C
A		0.53	0.71
B	0.53		1.00
C	0.71	1.00	

17

	A	B	C	D	E
A		>0.05	>0.10	>0.10	>0.10
B	>0.05		0.53	0.21	0.44
C	>0.10	0.53		0.60	0.92
D	>0.10	0.21	0.60		0.66
E	>0.10	0.44	0.92	0.66	

18

	A	B	C	D
A		0.48	1.00	0.87
B	0.48		0.35	0.76
C	1.00	0.35		0.89
D	0.87	0.76	0.89	

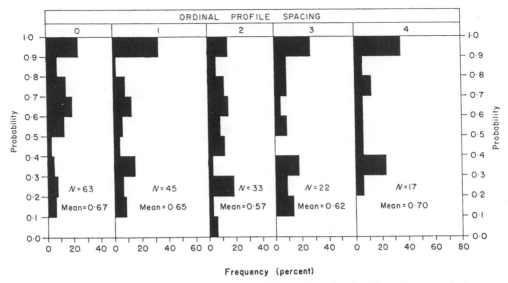

Fig. 3. Frequencies of probabilities under the null hypothesis in the Mann-Whitney test as a function of the ordinal spacing of the compared cyclothem profiles.

(cyclothem *15*). The impression that the profiles might vary in similarity with increased spacing is strengthened a little by compiling over the cyclothems as a whole the frequencies of probabilities under the null hypothesis for paired profiles 0, 1, 2, 3 and 4 profiles apart (Fig. 3). The largest proportions of high probability values are found for pairs separated by 0 and 4 profiles. Pairs 5 profiles apart, of which there are only 10, yield a mean probability of 0·50 and hence suggest a return at this spacing to the dominance of low probabilities. The lateral variation might therefore be periodic, with a repetition distance comparable with a few tens of profile thicknesses (Table 1).

7. Discussion

The three modes of analysis lead to different statements about the lateral variability of the cyclothems, because they depend on different character groupings and are unequally incisive. The qualitative approach uses all characters simultaneously, to summarize, the profile thickness, lithological content, and vertical facies pattern. Attention when using the facies transition matrix is directed primarily to lithological pattern, the thickness content being ignored. A comparison of lithological contents is, however, afforded by the Mann-Whitney test, but at the cost of ignoring pattern. Moreover, this test has an ambiguity, since closely similar rank-thickness product values may arise from markedly different combinations of rank number and representative thickness. To have used rank alone, however, would have reduced considerably the effectiveness of the test, on account of the consequently large numbers of ties.

Qualitatively, many cyclothems are so variable laterally that profiles taken from them are not recognizably related, except generically. Again qualitatively, a number of cyclothems seem laterally very uniform. The three empirical indices of lateral variability derived from the transition matrices also suggest marked differences in lateral variability, with some cyclothems being very uniform and others very variable. Neither mode of analysis affords an objective statement as to how variable is a particular cyclothem. By contrast, the Mann-Whitney test gives for each cyclothem a probability matrix objectively defining the lateral variability. The test reveals the cyclothems as relatively homogeneous, and certainly more homogeneous than the other tests, although some are less uniform than others. Very few profiles differ even at the 10 per cent level.

It is interesting to compare the rankings based on the different modes. All three place cyclothem *14* amongst the most uniform of the set; the transition matrix and Mann-Whitney test also place cyclothem *4* amongst the most uniform, though the qualitative approach assigns to it an intermediate variability. Cyclothem *12* is amongst the most variable in terms of the qualitative approach and the transition matrix. Its position by the Mann-Whitney test is uncertain, because of the incomplete probability data, but an intermediate ranking seems probable. Cyclothem *1*, regarded qualitatively as very variable, has a high variability according to the Mann-Whitney test but an intermediate variability by the transition matrix. These examples show that there is a broad agreement between the rankings achieved by the different approaches. Hence increasing variation in one or a small group of characters is associated with a growing variation in all. Finally, no mode justifies the view that the cyclothems with visible lateral deposits (*7, 11, 16*) are unduly variable compared with those apparently lacking this feature.

Underlying the work is the question of variation within cyclothems as compared with variation between. Probably the most instructive way of judging this is to compare the transition matrices and path diagrams for the 18 cyclothems separately with the matrix and diagram for a real or notional vertical sequence of profiles. This sequence could be either (1) the 231 cyclothems analysed (Allen 1970 table 5) from the Devonian of Britain and North America, or (2) the 68 cyclothems from the Red Marls of Pembrokeshire included amongst the 231 (Allen 1970 table 10), or (3) a notional one involving all profiles in the 18 cyclothems here described. Intuitively, the third choice appeals most, because no new data and circumstances of data collection are involved.

Figure 4 is the path diagram based on all of the profiles in the 18 cyclothems. In preparing the parent matrix, the cyclothems were treated in the sequence *1, 2, 3* etc. and the profiles in each assessed in the order *A, B, C* etc. The diagram has zero unoccupied transitions, 4·17 paths per state, and 2·50 steps per state. Table 3 shows that only one individual cyclothem (*12*) has unoccupied transitions of this order, and that two only (*7, 12*) have closely comparable values of the paths and steps per state. Cyclothem *12* actually has more paths per state than the path diagram for the combined profiles. Three cyclothems (*2, 3, 15*) have 1·80 steps per state, a value which is 72 per cent of the index for the combined profiles. Some cyclothems (e.g. *14, 4, 18*) have numbers of paths and steps per state that are small or very small compared with the values for the combined profiles, and numbers of unoccupied transitions which are very large. Other cyclothems, perhaps the majority, present intermediate values of these indices of variation.

This comparison, though empirical, demonstrates a continuum stretching between an individual variability which is small compared with the variation between cyclothems, to an extreme of variation that is comparable with the total variability. Some cyclothems, it must be concluded, exhibit real differences. Comparison with the 231 cyclothems gives a similar result. This sample has 4·39 representatives per profile, as compared with 5·14 for the 18 cyclothems, a difference that may reflect the less satisfactory overall conditions of exposure of the 231 cyclothems. The larger sample gave a transition matrix with 1 unoccupied state, 3·33 paths per state, and 2·50 steps per state.

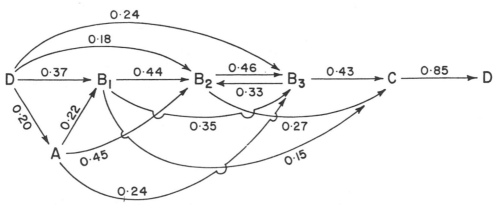

Fig. 4. Path diagram ($p \geqslant 0.15$) based on the upwards facies transition probability matrix constructed from the totality of profiles available in the set of 18 cyclothems.

It may be objected that cyclothem variability is largely set by the sampling condition of the geometrical relation between the row of profiles and the palaeocurrent direction, on the grounds that more variation might be expected perpendicular to than parallel with flow. The data of Figure 1 and Table 3 oppose the objection. Relatively variable as well as relatively uniform cyclothems are represented by rows of sections steeply related to palaeoccurrents (e.g. *4, 12*). Hence the palaeocurrent-outcrop relation appears to exert little or no control on variability.

A more serious objection to the comparison is that the rows of profiles are in no case extensive enough to allow a proper estimate to be made of the variability of a cyclothem. In order conclusively to refute this objection, it is necessary to have sampled over the full three-dimensional extent of a representative number of cyclothems. Such samples are not yet available, either from Pembrokeshire or any other region. Intuitively, however, it is felt that lines of profiles 50–100 times as long as the profile thickness (see Table 1) are sufficient to form an acceptable estimate of the degree of variation. It may perhaps never be possible conclusively to test this objection.

8. Sedimentological interpretation

As a broad interpretation, the alluvial origin of fining-upwards cyclothems in the Old Red Sandstone and related rocks is now well established, and need not be recapitulated in detail. Essentially, the coarse member (facies A, B_1, B_2, B_3) is interpreted as a channel deposit, with the intraformational conglomerates representing channel-floor conditions, whereas the fine member (facies C, D) is assigned a flood-plain origin in environments ranging between uppermost point bar, through levee and crevasse splay, to floodbasin. However, workers differ about the precise character and behaviour of the streams that gave rise to the cyclothems (e.g. Moody-Stuart 1966; Allen 1970).

The following general points arising from the preceding analysis and discussion of variation are prompted by two considerations. Firstly, most of the cyclothems (*5–18*) occur within the narrow stratigraphical interval of approximately 200–275 m. The second concerns time scales. Taking the radiometric limits on the Devonian, the thicknesses of cyclical formations in the Old Red Sandstone, and the thicknesses of the cyclothems, it is plausible that the cyclothems have a characteristic recurrence interval at any station comparable with 3×10^4 years. That is, subsidence combining with other factors caused the conditions initiating a cyclothem to recur at a station at an interval comparable with this value. Noting that each cyclothem has erosional upper and lower bounds, the *deposits* that make up a given cyclothem may however have formed in a much shorter period. Data on migration rates of modern stream channels (Wolman and Leopold 1957 table 4; Handy 1972) indicate that, where the coarse member of a cyclothem is known to be a lateral deposit, a single vertical profile in that member could have been completed in a matter of 10^2 years and often less. Extending the argument, a row of profiles of the order of 100 profile thicknesses might have been produced in as short a time as 10^3 years. Many more cyclothems are likely to include lateral deposits than can be proved in the field to have them (Allen 1970 p. 312), so that these arguments seem to apply to more than cyclothems *7*, *11* and *16*. The accumulation rate of the fine members is less clear. However, noting that many include one or more zones of limestone concretions interpreted as calcretes (e.g. Allen 1965), it may be suggested from knowledge of the radiocarbon ages of Quaternary pedogenic carbonates (Gile *et al.* 1966; Williams and Polach 1971) that there might have been *no* detrital deposition for periods comparable with 10^4 years. Hence the clastic deposits now found in each cyclothem may plausibly have formed in a period approaching an order of magnitude smaller than the estimated recurrence interval.

The results of the study of cyclothem variation, when viewed against these considerations (particularly of time scale), suggest as the chief sedimentological conclusions that (1) the variation within cyclothems is explicable largely in terms of Beerbower's autocyclic mechanisms, whereas (2) the variation between them can be satisfactorily accounted for by his allocyclic factors. Hence the differences of thickness, facies states, and facies sequence from profile to profile can be ascribed to one or a combination of (1) seasonal and comparably short-term variations of discharge, (2) continuous changes of channel position and plan and cross-sectional shape, and (3) channel cut-off or limited diversion (sudden changes) but with a periodicity measurable on the short term. The variation between cyclothems is probably related to one or a combination of (1) eustatic changes of base level, (2) climatic shifts, or (3) tectonic movements within the stream drainage basin. Avulsions may also be treated as allocyclic, provided if they depend on particularly rare events, for example, a 3000-year flood. All of these factors, as is to some extent known from studies in the Quaternary and from experiments (e.g. Schumm 1968; Schumm and Khan 1972), are capable of altering the gross character of streams, by changing slope, fluid and sediment discharge, and calibre of load. One would therefore expect some amongst stratigraphically adjacent cyclothems to exhibit substantial real differences.

[Editor's Note: The conclusions have been omitted.]

References

ALLEN, J. R. L. 1962. Petrology, origin and deposition of the highest Lower Old Red Sandstone of Shropshire, England. *J. sedim. Petrol.* **32,** 657.
—— 1965. Sedimentation and palaeogeography of the Old Red Sandstone of Anglesey, North Wales. *Proc. Yorks. geol. Soc.* **35,** 139.
—— 1970. Studies in fluviatile sedimentation: a comparison of fining-upwards cyclothems with particular reference to coarse-member composition and interpretation. *J. sedim. Petrol.* **40,** 298.
—— and FRIEND, P. F. 1968. Deposition of the Catskill facies, Appalachian region: with notes on some other Old Red Sandstone basins. *Spec. Pap. geol. Soc. Am.* **106,** 21.
BEERBOWER, J. R. 1964. Cyclothems and cyclic depositional mechanisms in alluvial plain sedimentation. *Bull. geol. Surv. Kansas* **169,** 31.
CANTRILL, T. C., DIXON, E. E. L., THOMAS, H. H. and JONES, O. T. 1916. The Geology of the South Wales Coalfield. Part XII. The country around Milford. *Mem. geol. Surv. U.K.*
DIXON, E. E. L. 1921. The Geology of the South Wales Coalfield. Part XIII. The country around Pembroke and Tenby. *Mem. geol. Surv. U.K.* 220 pp.
FRIEND, P. F. 1965. Fluviatile sedimentary structures in the Wood Bay Series (Devonian) of Spitsbergen. *Sedimentology* **5,** 39.
—— and MOODY-STUART, M. 1972. Sedimentation of the Wood Bay Formation (Devonian) of Spitsbergen: regional analysis of a late orogenic basin. *Norsk Polarinst. Skr.* Nr. 157.
GILE, L. H., PETERSON, L. F. and GROSSMAN, R. B. 1966. Morphological and genetic sequences of carbonate accumulation in desert soils. *Soil Sci.* **101,** 347.
HANDY, R. L. 1972. Alluvial cut-off dating from subsequent growth of a meander. *Bull. geol. Soc. Am.* **83,** 475.
MOODY-STUART, M. 1966. High and low-sinuosity stream deposits, with examples from the Devonian of Spitsbergen. *J. sedim. Petrol.* **36,** 1102.
OWEN, D. B. 1962. *Handbook of Statistical Tables.* Addison-Wesley, Reading (Massachusetts), 580 pp.
POTTER, P. E. and BLAKELY, R. F. 1968. Random processes and lithologic transitions. *J. Geol.* **76,** 154.
READ, W. A. and MERRIAM, D. F. 1972. A simple quantiative technique for comparing cyclically deposited successions. In (D. F. Merriam, editor) *Mathematical Models of Sedimentary Processes,* Plenum, New York.
SCHUMM, S. A. 1968. River adjustment to altered hydrologic regime—Murrumbidgee River and paleochannels, Australia. *U.S. geol. Surv. Prof. Pap.* **598,** 65 pp.
—— and KAHN, H. R. 1972. Experimental study of channel patterns. *Bull. geol. Soc. Am.* **83,** 1755.
SIEGEL, S. 1956. *Nonparametric Statistics for the Behavioural Sciences,* McGraw-Hill, New York.
WILLIAMS, G. E. and POLACH, H. A. 1971. Radiocarbon dating of arid-zone calcareous paleosols. *Bull. geol. Soc. Am.* **82,** 3069.
WOLMAN, M. G. and LEOPOLD, L. B. 1957. River floodplains: some observations on their formation. *U.S. geol. Surv. Prof. Pap.* **282C,** 87.

Sedimentology Research Laboratory,
Department of Geology,
The University,
Reading RG6 2AB.

6

Reprinted from pp. 347, 348, 349–357, 362, 363–364 of *Sedimentology* **20**(2):347–364
(1973)

Markov chain analysis applied to an ancient alluvial plain succession

ANDREW D. MIALL

Institute of Sedimentary and Petroleum Geology, Calgary, Alberta, Canada

ABSTRACT

Markov chain analysis is a comparatively simple statistical technique for the detection of repetitive processes in space or time. Coal measure cyclothems or fluvial fining-upward cycles are good examples of sedimentary successions laid down under the control of Markovian processes.

Analyses of stratigraphic sections commence with a transition count matrix, a two-dimensional array in which all possible vertical lithologic transitions are tabulated. Various probability matrices may be derived from this raw data, and these are then subjected to chi-square tests to determine the presence or absence of the Markov property. This technique is applied to four types of stratigraphic succession which occur in the Devonian rocks of Prince of Wales Island, Arctic Canada.

(1) A conglomerate succession of alluvial fan origin. Markov analysis is of little or no assistance in the interpretation of these rocks, in which only two principal lithologies are present.

(2) A conglomerate–sandstone succession. Fluvial fining-upward cycles are detectable by visual examination of the sections and are strongly indicated by Markov analysis.

(3) A sandstone–carbonate succession, of marginal marine origin, and including both marine and non-marine strata. Cyclicity is weak in these rocks, but analysis suggests that regressions took place much more rapidly than transgressions during their period of deposition.

(4) A succession in which the relative proportions of the various lithologies vary markedly with age. The varying nature of the cyclic tendencies is emphasized in this case by dividing the succession into two subintervals, for the purpose of analysis.

[*Editor's Note:* Material has been omitted at this point.]

[*Editor's Note:* Material, including figure 1, has been omitted at this point.]

ANALYTICAL METHOD

The methods described in this section are based mainly on those of Harbaugh & Bonham-Carter (1970, Chap. 4) and Gingerich (1969).

The transition count matrix

A simple, or first-order, Markov chain depends only on single steps, that is, the relationship between a given bed and the next bed immediately succeeding it. Thus coal seams commonly follow a seat-earth and, in the fluvial environment, a channel lag-conglomerate is normally succeeded by point-bar sandstones. More complex Markov processes are possible in which the nature of the Markov dependency includes reference to still earlier beds, or to changes in the dependency relationship with time, but these will not be considered in detail in the present paper.

The starting point in Markov chain analysis is the transition count matrix. This is a two-dimensional array which tabulates the number of times that all possible vertical lithologic transitions occur in a given stratigraphic succession. The lower bed of each transition couplet is given by the row number of the matrix, and the upper bed by the column number, each lithofacies present being assigned a code number for the purpose of the analysis. An example is given in Table 1, in which the lithofacies codes are as follows: 1, conglomerate; 2, coarse, pebbly sandstone; 3, coarse to medium sandstone; 4, fine sandstone and silty sandstone.

Elements in the transition count matrix are hereafter referred to by the symbol f_{ij}, where i = row number and j = column number. It will be noted that where $i = j$ zeros are present in the matrix, i.e. transitions have only been recorded where the

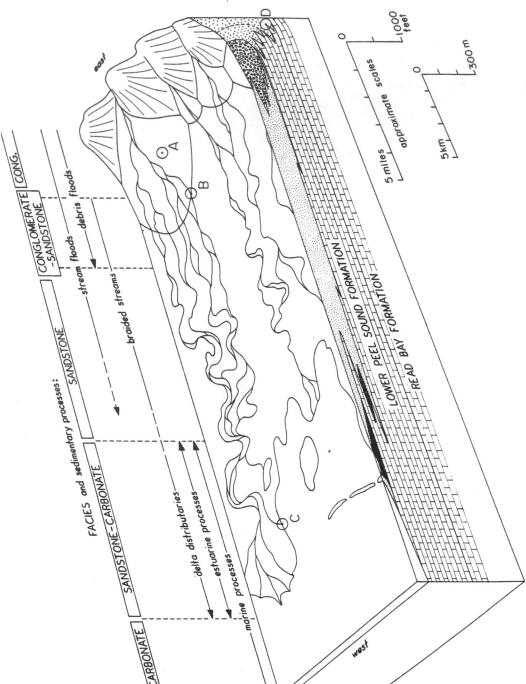

Fig. 2. Block diagram of a typical east–west segment of Prince of Wales Island during the Early Devonian showing principal depositional environments. Letters A to D refer to localities mentioned in the text. They are intended to provide a general indication of the fluctuating conditions that gave rise to the four types of vertical stratigraphic succession analysed in this paper.

Table 1. Transition count matrix, Conglomerate–Sandstone Facies

Lithofacies		1	2	3	4	Row sum
Conglomerate	1	0	3	11	1	15
Pebbly sandstone	2	5	0	1	8	14
Coarse to medium sandstone	3	4	0	0	15	19
Fine sandstone	4	5	13	6	0	24
Total						72

lithofacies shows an abrupt change in character, regardless of the thickness of the individual bed. This is referred to as a 'method 1' type of analysis throughout the rest of the present paper. It corresponds to the 'embedded Markov chain' of Krumbein & Dacey (1969). An alternative method of analysis, designated 'method 2', is to record the bed type at fixed sample intervals throughout the section. In this case the principal diagonal of the matrix will not necessarily contain zeros. The differences between these two methods of analysis will be discussed later.

Probability matrices

From the transition count matrix two probability matrices may be derived. The first is an independent trials probability matrix composed of r_{ij}, which represents the probability of the given transition occurring randomly. Given any state i the probability of this state being succeeded by any other state j is dependent only on the relative proportions of the various states present. Thus

$$r_{ij} = s_j/t$$

where t = total number of beds = $\sum_{ij}^{n} f_{ij}$, n = the rank of the matrix, i.e. the total number of rows or columns used, and s_j is the sum of the f_{ij} for the jth column of the f matrix. This formula is suitable only for method 2 analyses in which the upper and lower bed in a transition couplet may be the same. For the embedded chain method, $i = j$ transitions are not permitted, and the total range of possibilities must be set to exclude them. The above formula thus becomes:

$$r_{ij} = s_j/(t - s_i) \tag{1}$$

and the remaining probability values are thereby increased proportionally along each row of the r matrix, although remaining the same relative to one another.

The second matrix, containing elements p_{ij}, gives the actual probabilities of the given transition occurring in the given section:

$$p_{ij} = f_{ij}/s_i. \tag{2}$$

The values in the p matrix sum to unity along each row and they will necessarily reflect (although they cannot, by themselves prove) the presence of any Markovian dependency relationship. It is also useful to construct a difference matrix d_{ij}, derived thus:

$$d_{ij} = p_{ij} - r_{ij}. \tag{3}$$

Positive entries in the d matrix serve to emphasize the Markov property by indicating which transitions have occurred with greater than random frequency. Values in each row of the d matrix sum to zero. It is important to remember that for method 1 analyses no values may be calculated for any of the positions on the principal diagonals of the probability matrices.

The various probability matrices that can be calculated for the example given in Table 1 will be discussed in the section on results.

Tests of significance: within a stratigraphic succession

Although the differences between the p matrix and the r matrix may seem to be considerable, the differences may themselves be due to random chance, and thus it is important to apply tests of significance to the results. A chi-square test is suitable for the purpose. A formula for method 1 analyses is given by Billingsley (1961, p. 17) and Gingerich (1969, p. 331):

$$\chi^2 = \sum_{ij}^{n}(f_{ij} - s_i r_{ij})^2/s_i r_{ij}. \tag{4}$$

The number of degrees of freedom is given by the number of non-zero entries in the r matrix, minus the rank of the matrix, i.e. $n^2 - 2n$. Agterberg (personal communication, 1972) recommends using this equation only where each value of f_{ij} exceeds 5.

An alternative test is given by Anderson & Goodman (1957) and Harbaugh & Bonham-Carter (1970, p. 121):

$$\chi^2 = 2\sum_{ij}^{n} f_{ij} \cdot \log_e(p_{ij}/[s_j/\ t]). \tag{5}$$

This statistic has $(n-1)^2 - n$ degrees of freedom in method 1 analyses and $(n-1)^2$ degrees of freedom in method 2 analyses.

For both equations the null hypothesis is that the vertical succession of strata was derived by random variation in the depositional mechanisms.

It is important to ensure the size of the sample conforms to the normal statistical rules concerning sample validity and the repeatability of the results.

Simple computer programs may be written by the user (as in the present case) or obtained from published sources (Krumbein, 1967) to calculate the probability matrices and the chi-square statistics from the raw data.

Tests of significance: between stratigraphic sections

Although a given segment of a stratigraphic succession may exhibit the Markov property, examination of older and younger portions of the sequence (if this is possible) may produce different transition count distributions, showing that the nature of the cyclicity changed with time. Such a succession is said to exhibit 'non-stationary' Markovian dependency.

A test for stationarity is presented below, and forms an important statistic for comparing whole stratigraphic successions. The sample successions may be subdivisions of a single extended sequence or they may be lateral equivalents within the same sedimentary basin. The test statistic has the chi-square distribution; a formula for carrying out the calculation is given by Harbaugh & Bonham-Carter (1970, p. 124):

$$\chi^2 = 2\sum_{i}^{t}\sum_{jk}^{n} f_{ijk} \cdot \log_e(p_{ijk}/q_{jk}), \tag{6}$$

where $i = 1, 2 \ldots t$, giving the number of sections or subsections tested against each other, f_{ijk} and p_{ijk} are the transition count and probability matrix values for each subsection, and q_{jk} refers to the values calculated for an overall transition probability matrix.

The number of degrees of freedom $= (t - 1) \cdot [n(n - 1)]$.

The two sampling methods

As described earlier, there are two different methods available for sampling a stratigraphic section: (1) by counting only discrete lithofacies transitions, regardless of individual bed thickness and (2) by sampling at fixed vertical intervals.

The philosophy behind these two methods is essentially different. The first emphasizes the actual change, and the focus is therefore on the evolution of the depositional processes. The second method can give rise to a much more accurate measure of the relative frequencies of the lithotypes present, but at the expense of accuracy in measuring step-by-step depositional change. The choice of sample interval is crucial when applying the second method. Too large an interval will tend to bypass thin beds, whereas too small an interval will produce excessively large figures on the principal diagonal of the f matrix, overshadowing any Markovian tendencies present in the section, and giving rise to anomalously high values in the chi-square tests described above. These difficulties have been noted by several authors, e.g. Krumbein (1967) and Read (1969). The present author's experience is that a sample interval slightly less than the average bed thickness generally produces the most satisfactory results.

An attempt was made by Carr *et al.* (1966) to combine the virtues of both these methods by recognition of 'multistorey lithologies'. Thus, discrete lithologic changes are counted, as in the first method, but a change from, say, sandstone to sandstone is allowed also if the unit shows a sudden marked change in character. In the present author's opinion this is an unsatisfactory approach as it introduces the necessity for a subjective judgment as to what *is* a major change within a single lithology; grain size? colour? bedding characteristics? What, for example, should be done with a thinly laminated sandstone?

Recognition of changes *between* lithologies will not generally be difficult as it is necessary in any case to simplify stratigraphic sections down to an easily manageable number of gross lithotypes in order to prevent transition tendencies from being too diffused throughout the count matrix. Between four and six lithofacies appear to be ideal. It would be possible to set up a larger number of states by subdividing lithotypes on the basis of colour, presence or absence of crossbedding and other such criteria. This would largely circumvent the problem of multistorey lithologies, but would at the same time require a much greater quantity of observational data for statistically meaningful relationships to appear.

The first of the two principal methods, that of the embedded Markov chain, is used throughout most of this paper.

RESULTS AND DISCUSSION

General remarks

The Peel Sound Formation may be subdivided into five laterally equivalent facies zones as indicated in Fig. 2. Three of these are virtually dominated by single lithologies.

These are the Conglomerate, Sandstone and Carbonate Facies. There are also two facies that are transitional in nature, namely the Conglomerate–Sandstone Facies and the Sandstone–Carbonate Facies. Cyclicity in the single-lithology facies is limited to alternation between the dominant sediment type and one or two accessory lithologies. The application of Markov Analysis to such strata would provide very little useful geological information. However, a rapid examination of the Conglomerate Facies is presented below as it will serve to illustrate the differences between the two methods of Markovian analysis discussed above.

Conglomerate Facies

The environment under discussion in this section is indicated diagrammatically by the letter 'A' in Fig. 2.

Elsewhere (Miall, 1970b, p. 569) the conglomerates have been interpreted as the deposits of debris floods analogous to the mudflows of modern semi-arid desert regions, e.g. those described by Chawner (1935) and Sharp & Nobles (1953). The minor sandstone interbeds were deposited by the less violent waters at the tail end of a flood or by occasional non-catastrophic floods over alluvial fan surfaces that were probably devoid of surface run-off most of the time.

Two transition count matrices may be set up side by side, as shown in Table 2. On the left a method 1 analysis is used, on the right is a method 2 analysis, based on a sample interval of 30 cm. Conglomerate is Facies 1 and sandstone is Facies 2. One hundred transitions have been recorded in each case (the figures have been rounded off for convenience, but are based on field observations).

Table 2. Transition count matrices, Conglomerate Facies

		1	2		1	2
Conglomerate	1	0	50	1	97	1
Sandstone	2	50	0	2	1	1

The left hand matrix tells us only that the two lithologies alternate, whereas the right hand matrix gives us an indication of relative bed thickness. The proportion of sandstone in the Conglomerate Facies in fact averages 2%. Probability matrices for the two samples are given in Table 3. From a Markovian point of view the results are obvious, and there is clearly no geological reason for pursuing this type of analysis when only two lithotypes are present, as it provides us with no extra information with which to interpret the changes through time of the depositional environment.

Table 3. Transition probability matrices, Conglomerate Facies

		1	2		1	2
Conglomerate	1	0	1	1	0·99	0·01
Sandstone	2	1	0	2	0·50	0·50

Conglomerate–Sandstone Facies

The sedimentology of this facies has been discussed by Miall (1970a, pp. 128–131) and a full account will not be repeated here. As will be noted from Fig. 2 (on which the localities to be discussed are indicated by the letter 'B') the facies is situated at the distal edge of the alluvial fan zone, where conglomerates interfinger with fluvial sandstones.

Cyclic sedimentation has already been demonstrated for these deposits. It takes the form of an upward fining in grain size that is highly characteristic of fluvial deposits (Allen, 1965a, b, 1970; Visher, 1965, 1972) and recorded at numerous localities in ancient continental deposits (see Allen, 1970, p. 301, for references). Such cycles are particularly common in the Devonian, when continental deposition was unusually widespread. A statistical examination of the cyclic property is considered to be worthwhile for it does reveal several unexpected relationships between the bed types.

Two sections were measured in this facies near the north end of Prince of Wales Island. They total 62 and 78 m in thickness and include thirty-three and thirty-nine bed transitions, respectively. The two sections are approximately one third of a mile apart and owing to their similarity (see below for test results) and the rather low total number of beds their transition count matrices have been combined for the purpose of this analysis (Table 1).

The independent trials probability matrix, transition probability matrix and difference matrix for these data are presented in Tables 4, 5 and 6. Chi-square test results are given in Table 7, and the presence of the Markov property in these strata is clearly indicated.

Table 4. Independent trials probability matrix, Conglomerate–Sandstone Facies

		1	2	3	4	
Conglomerate	1	0·00	0·25	0·33	0·42	
Pebbly sandstone	2	0·26	0·00	0·33	0·41	$= r_{ij}$
Coarse to medium sandstone	3	0·28	0·26	0·00	0·45	
Fine sandstone	4	0·31	0·29	0·40	0·00	

Table 5. Transition probability matrix, Conglomerate–Sandstone Facies

		1	2	3	4	
Conglomerate	1	0·00	0·20	0·73	0·07	
Pebbly sandstone	2	0·36	0·00	0·07	0·57	$= p_{ij}$
Coarse to medium sandstone	3	0·21	0·00	0·00	0·79	
Fine sandstone	4	0·21	0·54	0·25	0·00	

Table 6. Difference matrix, Conglomerate–Sandstone Facies

		1	2	3	4	
Conglomerate	1	0·00	− 0·05	0·40	− 0·35	
Pebbly sandstone	2	0·10	0·00	− 0·26	0·16	$= d_{ij}$
Coarse to medium sandstone	3	− 0·07	− 0·26	− 0·00	0·34	
Fine sandstone	4	− 0·10	0·25	− 0·15	0·00	

Table 7. Tests of significance, Conglomerate–Sandstone Facies

Test equation	χ^2	d.f.	Limiting value*
4[+]	33·364	8	15·51
5	82·775	5	11·07

* From table of chi-square values with correct number of degrees of freedom, at 95% confidence level.

[+] Equation 4 from Billingsley (1961), equation 5 from Anderson & Goodman (1957).

105

The nature of the cyclic process may be derived by following through the highest values of the p matrix or the positive values in the d matrix (Fig. 3). Transitions not indicated on this diagram may be attributed to the occurrence of non-cyclic, 'random' changes in the nature of the depositional mechanisms, for example the change from facies 4 to facies 3 (fine sandstone to coarse sandstone), which was actually recorded six times.

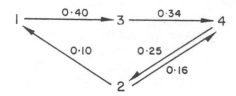

Fig. 3. Cyclic processes in the Conglomerate–Sandstone Facies. Lithofacies types (large figures) are numbered as in the text (1, conglomerate; 2, pebbly sandstone; 3, coarse to medium sandstone; 4, fine sandstone). The small figures represent the greater-than-random probability of occurrence of each transition (from the d matrix, Table 6).

Visual examination of the stratigraphic sections originally led the author to suggest that many, if not most of the fining-upward cycles were originated by debris floods sweeping across the pediment surface and creating new channels. Waning flow plus the process of lateral accretion acting during periods of more normal, quieter run-off, then combined to deposit beds of successively finer grain size, until the next flood altered the channel pattern once again (Miall, 1970a, pp. 130–131). It was also noted that several conglomerate beds could not be fitted into a fining-upward cycle, and these were interpreted as the deposits of debris floods that did not result in the establishment of new channel systems, so that once the flood had passed, the original fluvial pattern re-established itself.

In general the results of the Markov analysis support this interpretation, except that there appear, from Fig. 3, to be two types of fining-upward cycle.

(1) Conglomerate–coarse to medium sandstone–silty sandstone.

(2) Pebbly sandstone–silty sandstone.

Whether or not the two cyclic patterns are truly distinct remains a problem. For the purposes of the analysis it is necessary to assign nominal values to facies types, implying a rigid distinctiveness to each that may not always be justifiable. Gradational contacts between facies types cannot easily be included without using a much more detailed observational scale. The second of the two types of cycle shown above may simply be a condensed version of the first, Facies 3 (coarse sandstone) not appearing in the original section description because its thinness led it to be dismissed as a gradational contact between Facies 2 (pebbly sandstone) and 4 (fine sandstone). Pebbly sandstones may or may not be of different genetic origin to the conglomerates. The latter are debris flood deposits, whereas sandstones with scattered pebbles may represent non-violent run-off, the pebbles being carried as a traction load rather than in suspension. The presence of cross-bedding in some of the pebbly sandstones confirms this interpretation.

The importance of the facies transition from coarse sandstone to conglomerate is of interest, as it was not expected. It is interpreted as a result of debris flood flow interrupting quieter surface run-off processes, so that the two-member cycle was replaced temporarily by the three-member cycle. It might be expected that conglomerate would occasionally follow silty sandstone in the cyclic process. In fact this transition was recorded five times, but the analysis has shown that statistically this is not significant. The reason may lie in the erosive power of debris floods (described by Sharp & Nobles, 1953, p. 553), since the first action of a flood sweeping down from the mountains would be to remove the topmost layer of the alluvial fan surface. Facies 4, although more abundant than Facies 2, is characterized by a smaller average bed thickness (112 cm versus 216 cm) and would therefore be more susceptible to complete removal during a single, short-lived act of erosion.

It is possible to apply Allen's (1970) sedimentation model for fluvial cycles to the present study area in only a general sense owing to major differences between his field examples (located in Wales and the Catskill region) and the fluvial deposits of the Peel Sound Formation. The two most important differences are the much greater proportion of conglomerate in the sections described herein, and the paucity of sedimentary structures. Both these differences may be attributed to the much greater relative proximality of the Peel Sound Conglomerate–Sandstone Facies. The paucity of sedimentary structures in the latter suggests the predominance of upper regime, planebed flow (Harms & Fahnestock, 1965) during the deposition of the coarse to medium sandstone (Facies 3). Allen (1970, p. 318) has shown how this type of flow is characteristic of relatively high energy, low sinuosity streams, which are more likely to be found in the proximal part of a river system than near base level.

The two stratigraphic sections analysed herein give the same cyclic pattern shown in Fig. 3 when considered separately, but the chi-square test shows that the Markov property in each was barely significant at the 95% confidence level, probably owing to insufficient data. Application of the test for similarity between sections (equation 6) gives a value of 5·276 with 12 degrees of freedom. The limiting value at the 95% confidence level is 21·03, and thus the null hypothesis, that there is no significant difference between these sections, cannot be rejected.

[*Editor's Note:* Material has been omitted at this point.]

CONCLUDING REMARKS

The analysis presented herein has shown how the application of Markov chain analysis to several different types of stratigraphic succession has assisted in clarifying lithofacies relationships by defining these relationships statistically. The most obvious use of such an approach is to assist in the detection and definition of cyclic relationships, but even where these are weak or absent the method can still be useful in bringing out genetic relationships between two or three (or more) of the total number of facies present that might otherwise have been missed. Obviously this type of information can greatly assist in environmental interpretation.

An advantage of this type of analysis is its simplicity. Modern digital computers can calculate all the statistics necessary for a complete analysis in a matter of seconds. The author's own programs were designed to accept data in the form of a coded stratigraphic succession, one computer card per bed. Compilation of the data in this form is not particularly time consuming, and it is thus possible to use the Markov method as a standard analytical tool.

A few cautionary comments are necessary. Care must be taken in defining the facies used in the analysis so that they are fully representative. Too few or too many will obscure or distort the results. Secondly, it is important to ensure validity of the sample by making it as large as possible while still observing the third precaution, which is to bear in mind the possibility of the facies relationships being non-stationary in character throughout the section or sections measured.

REFERENCES

ALLEN, J.R.L. (1965a) Fining-upward cycles in alluvial successions. *Geol. J.* **4,** 229–246.

ALLEN, J.R.L. (1965b) A review of the origin and characteristics of recent alluvial sediments. *Sedimentology,* **5,** 89–191.

ALLEN, J.R.L. (1970) Studies in fluviatile sedimentation: a comparison of fining-upward cyclothems, with special reference to coarse-member composition and interpretation. *J. sedim. Petrol.* **40,** 298–323.

ANDERSON, T.W. & GOODMAN, L.A. (1957) Statistical inference about Markov Chains. *Ann. math. Statist.* **28,** 89–110.

BILLINGSLEY, P. (1961) Statistical methods in Markov Chains. *Ann. math. Statist.* **32,** 12–40.

BROAD, D.S. (1968) *Lower Devonian Heterostraci from the Peel Sound Formation, Prince of Wales Island, Northwest Territories.* Unpublished M.Sc. thesis, University of Ottawa, Canada.

BROWN, R.L., DALZIEL, I.W.D. & RUST, B.R. (1969) The structure, metamorphism and development of the Boothia Arch, Arctic Canada. *Can. J. Earth Sci.* **6,** 525–543.

CARR, D.D., HOROWITZ, A., HRABAR, S.V., RIDGE, K.F., ROONEY, R., STRAW, W.T., WEBB, W. & POTTER, P.E. (1966) Stratigraphic sections, bedding sequences, and random processes. *Science, N.Y.* **154,** 1162–1164.

CHAWNER, W.D. (1935) Alluvial fan flooding, the Montrose, California, flood of 1934. *Geogrl Rev.* **25,** 225–263.

DACEY, M.F. & KRUMBEIN, W.C. (1970) Markovian models in stratigraphic analysis. *J. int. Ass. mathl Geol.* **2,** 175–191.

DOVETON, J.H. (1971) An application of Markov Chain analysis to the Ayrshire Coal Measures succession. *Scott. J. Geol.* **7,** 11–27.

GINGERICH, P.D. (1969) Markov analysis of cyclic alluvial sediments. *J. sedim. Petrol.* **39**, 330–332.

GOULD, H.R. (1970) The Mississippi delta complex. In: *Deltaic Sedimentation, Modern andAncient. Spec. Publs Soc. econ. Paleont. Miner., Tulsa*, **15**, 3–30.

HARBAUGH, J.W. & BONHAM-CARTER, G. (1970) *Computer Simulation in Geology.* Wiley-Interscience, New York.

HARMS, J.C. & FAHNESTOCK, R.K. (1965) Stratification, bed forms and flow phenomena (with an example from the Rio Grande). In: *Primary Sedimentary Structures and their Hydrodynamic Interpretation. Spec. Publs Soc. econ. Paleont. Miner., Tulsa,* **12**, 84–115.

KERR, J.W. & CHRISTIE, R.L. (1965) Tectonic history of Boothia Uplift and Cornwallis Fold Belt, Arctic Canada. *Bull. Am. Ass. Petrol. Geol.* **49**, 905–926.

KRUMBEIN, W.C. (1967) Fortran IV computer programs for Markov Chain experiments in Geology. *Computer Contr. Geol. Surv. Kansas,* **13**, 38.

KRUMBEIN, W.C. (1968) Fortran IV computer program for simulation of transgression and regression with continuous time Markov models. *Computer Contr. Geol. Surv. Kansas,* **26**, 38 pp.

KRUMBEIN, W.C. & DACEY, M.F. (1969) Markov chains and embedded chains in geology. *J. int. Ass. mathl Geol.* **1**, 79–96.

MIALL, A.D. (1969) *The Sedimentary History of the Peel Sound Formation, Prince of Wales Island, Northwest Territories.* Unpublished Ph.D. thesis, University of Ottawa, Canada.

MIALL, A.D. (1970a) Continental–marine transition in the Devonian of Prince of Wales Island, Northwest Territories. *Can. J. Earth Sci.* **7**, 125–144.

MIALL, A.D. (1970b) Devonian alluvial fans, Prince of Wales Island, Arctic Canada. *J. sedim. Petrol.* **40**, 556–571.

READ, W.A. (1969) Analysis and simulation of Namurian sediments in central Scotland using a Markov-process model. *J. int. Ass. mathl Geol.* **1**, 199–219.

SCHWARZACHER, W. (1967) Some experiments to simulate the Pennsylvanian rock sequence of Kansas. *Computer Contr. Geol. Surv. Kansas,* **18**, 5–14.

SCRUTON, P.C. (1960) Delta building and the deltaic sequence. In: *Recent Sediments, Northwest Gulf of Mexico,* pp. 82–102. American Association of Petroleum Geologists, Tulsa, Oklahoma.

SHARP, R.P. & NOBLES, L.H. (1953) Mudflow in 1941 at Wrightwood, southern California. *Bull. geol. Soc. Am.* **64**, 547–560.

VISHER, G.S. (1965) Use of vertical profile in environmental reconstruction. *Bull. Am. Ass. Petrol. Geol.* **49**, 41–61.

VISHER, G.S. (1972) Physical characteristics of fluvial deposits. *Spec. Publs Soc. econ. Paleont. Miner., Tulsa,* **16**, 84–97.

VISTELIUS, A.B. (1949) On the question of the mechanism of formation of strata. *Dokl. Akad. Nauk SSSR,* **65**, 191–194.

(*Manuscript received* 1 *September* 1972; *revision received* 18 *December* 1972)

Editor's Comments
on Papers 7 Through 10

ALLUVIAL-PLAIN FACIES: PATTERNS OF ANCIENT
MEANDERING STREAMS

With increased understanding of the behavior of modern braided
and meandering streams, geologists are now able to analyze deposits of
some ancient ones. Friend and Moody-Stuart (1972) have presented a
useful review of the independent and dependent variables involved in
the development of river systems (pp. 53–78). Pertinent studies include a
comparison of modern point bars with Eocene and Pleistocene ones on
the Gulf coastal plain (McGowan and Garner 1970); description of
transitions from meandering- to braided-stream deposits in early
Cenozoic channel fill in Utah (Maxwell and Picard 1976) and of bedding geometry of a Pennsylvanian channel sandstone that accumulated
on the inside of a meander curve in Pennsylvania (Beutner et al. 1967);
calculation of the limits for bankful widths of paleochannels with data
from selected fining-upward cycles (Leeders 1973); and estimates of the
morphology and magnitude of channels in Carboniferous coal measures
in England (Elliott 1976). Schumm (1972) had summarized recent information about fluvial paleochannels.

This theme is illustrated by four short papers. From observations of well-exposed point-bar ridges and swales that record accretionary events in a late Carboniferous delta in southern Morocco, Padgett and Ehrlich (1976) have assembled paleohydraulic and paleogeographic information about the broad, shallow, very sinuous rivers (Paper 7). Their reconstruction includes distance from source to the sea, direction of flow, channel slope, and an estimated river discharge.

Cotter (1971) has reconstructed the flow characteristics of a Late Cretaceous river in Utah (Paper 8) by delimiting many of its physical parameters. In this analysis the geometry of sedimentary structures suggested the width and depth of flow (channel form), and sandstone textures recorded the type of load transported; but channel sinuosity, meander length, mean annual discharge, mean annual flood, channel slope, and flow velocity were estimated from empirically derived relationships of modern streams. These parameters also suggested an estimated area of the drainage basin and the length of the river from mouth to drainage divide.

Exceptionally good three-dimensional exposures of alluvial-plain deposits reveal significant details about ancient point bars and meander belts. Nami (1976) has described such a situation in fluvio-deltaic Jurassic strata in England (Paper 9). Measurement of the lateral accretion-surface dimensions of meander loops suggests the channel bankful depth and width as well as meander wavelengths, whereas mean annual discharge values were derived from empirical equations. Similarly, Puigdefabregas (1973) described a well-exposed, small-size Miocene point-bar deposit in northern Spain (Paper 10). Its internal structure consists of imbricated and concentrically arranged cross-stratified bundles that correspond to the accretionary topography of the point bar. These structures were easily related to the surface pattern of meander belts clearly visible on aerial photographs.

REFERENCES

Beutner, E. C.; Flueckinger, L. A.; and Gard, T. M. 1967. Bedding geometry in a Pennsylvanian channel sandstone. *Geol. Soc. America Bull.* **78:**911–916.

Elliott, T. 1976. The morphology, magnitude, and regime of a Carboniferous fluvial distributary channel. *Jour. Sed. Petrology* **46:**70–76.

Leeder, M. R. 1973. Fluviatile fining-upwards cycles and the magnitude of paleochannels. *Geol. Mag.* **110:**265–276.

Maxwell, T. A., and Picard, M. D. 1976. Small channel-fill sequences in the Duchesne River Formation near Vernal, Utah: Possible examples of transitions from meandering to braided stream deposits. *Utah Geology* **3:**61–66.

McGowan, J. H., and Garner, L. E. 1970. Physiographic features and stratigraphic types of coarse-grained point bars: Modern and ancient examples. *Sedimentology* **14:**77–111.

Schumm, S. A. 1972. Fluvial paleochannels. *Soc. Econ. Paleontologists and Mineralogists Spec. Pub. 16,* pp. 98–107.

Reprinted from *Geol. Soc. America Bull.* **87**(8):1101–1104 (1976)

Paleohydrologic analysis of a late Carboniferous fluvial system, southern Morocco

G. V. PADGETT*
ROBERT EHRLICH } *Department of Geology, University of South Carolina, Columbia, South Carolina 29208*

ABSTRACT

Differential weathering of paleosurfaces in upper Carboniferous sandstone of southern Morocco has exposed arcuate ridges and swales that record point-bar accretionary events. From the point-bar ridges and swales, paleohydraulic and paleogeographic data have been gathered that define an array of late Carboniferous rivers with moderate discharges calculated to be from 130 to 565 m³/sec. The broad sinuous channels flowed toward the east-southeast; the distance from the source terrain to the sea was about 100 km. Channel slope was approximately 0.16 m/km. *Key words: Carboniferous fluvial deposits, accretion scars, paleohydraulics, Morocco.*

INTRODUCTION

The present desert climate and gentle regional dips in southern Morocco combine to expose a Carboniferous point-bar accretion complex. The area comprises several hundred square kilometres on the north flank of the Tindouf Basin (Fig. 1) where Carboniferous strata contain Dinantian marine-shelf deposits overlain by Westphalian-Stephanian nonmarine red beds (Hollard and Jacquemont, 1956; Fabre and Greber, 1955; Vina y Villa and Cabezon, 1959). V. V. Cavaroc (1975, written commun.) described a fluvial-deltaic and a beach-barrier facies within the red beds (Fig. 2). The study reported here concerns the fluvial-deltaic facies, 300 m of deltaic deposits, and several thin fluvial sandstone units interbedded with interfluvial siltstone and shale.

Grain size within the sandstone units typically ranges from pebbly sand at the base to silty sand at the top. The lower parts of each sandstone unit are massive, grading upward into megaripples which, in turn, grade into small-ripple cross laminae. The basal pebbly zones contain plant fragments, and the uppermost silty zones are intensely root penetrated. These features are commonly associated with point-bar deposits and are described in Holocene deposits by Frazier and Osanik (1961) and in ancient strata by Visher (1965).

Sandy point-bar deposits crop out as cap rock, weathering having removed overlying siltstone and shale. Scarps with less than 1 m of relief produce point-bar ridges (Fig. 3) on these ancient surfaces. They are seen in aerial photographs as concentric arcs (Fig. 4) and look exactly like Holocene meander scars.

If we assume that the arcuate ridges represent the curvature of fossil channels, the radius of curvature of meanders of the paleochannels can be measured from the accretionary ridges. Thus, the empirical relations developed by Leopold and others (1964) relating the radius of curvature to meander length and channel width can be applied to these Carboniferous fluvial deposits. These empirically derived functions were used to estimate width, meander length, velocity, and mean annual discharge where meander scars were measurable from aerial photographs. The estimated morphologic properties (river width and meander wave length) at each sampling point were plotted to generate a paleogeographic map for the area. A plot of the derived discharge estimates resulted in a paleodraulic mosaic. This paper concerns only the dominant exposed channels; V. V. Cavaroc (1975, written commun.) reported many small, discontinuous sand units that he attributed to minor streams.

RADIUS OF CURVATURE AND CALCULATION OF CHANNEL DIMENSIONS

Radius of curvature was measured from aerial photographs by matching a transparent circle template to each arcuate point-

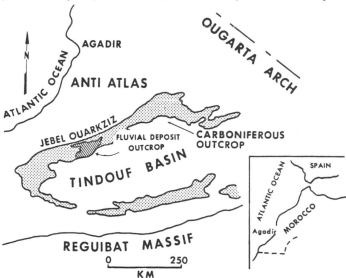

Figure 1. Location of Carboniferous point-bar deposits on north flank of the Tindouf Basin, southern Morocco.

* Present address: Exploration Division, Consolidation Coal Company, Bluefield, West Virginia 24605.

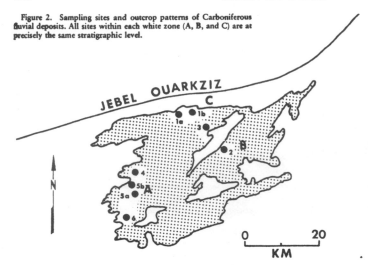

bar ridge. Distortion of the aerial photographs was lower than the precision of measurement (Lappman and Ray, 1965). Measurements were replicated and averaged; at least five distinct point-bar ridges were measured at each location, and each ridge was measured three times on three different occasions.

Channel width and meander length were calculated for each sampling point using formulas developed by Leopold and Wolman (1960). Two large channels were defined, each approximately 230 m wide, with meander wave lengths of 2,660 m. Three smaller courses were delineated: two were approximately 130 m wide and one was 120 m wide.

RIVER DISCHARGE AND VELOCITY

A high correlation exists between some physical parameters of Holocene meandering channels and their mean annual discharge (Schumm, 1968, 1972). The relationships are

$$(Q_m)^{0.34} = \frac{(L)(M^{0.74})}{1,890} \qquad (1)$$

$$(Q_m)^{0.38} = \frac{(W)(M^{0.37})}{37} \qquad (2)$$

$$M = \frac{(S_c)(W) + (S_b)(2D)}{W + 2D}, \qquad (3)$$

where Q_m is mean annual discharge (in cubic feet per second), L is meander wave length, S_c is the percentage of silt and clay in the channel alluvium, S_b is the percentage of silt and clay in the bank alluvium, B is width, and D is channel depth. The above formulas were developed from studies of stable alluvial channels in semiarid to semihumid regions in the United States and Australia; they have been applied to Pleistocene-Holocene paleochannels by Schumm (1968) and to Late Cretaceous paleochannels by Cotter (1971). Estimates of channel depth (D), percentages of silt and clay in the channel alluvium (S_c), and percentages of silt and clay in the bank alluvium (S_b) must be made in order to apply these formulas.

The proportion of silt and clay in the channel alluvium was estimated from samples and thin sections of the massively bedded part of a point-bar sequence, which represents channel alluvium; this portion is 5 percent. The portion of silt and clay in the basal alluvium is assumed to be nearly 100 percent, because it represents silt and clay in the interfluvial bank. Since point-bar sequences grade upward into silty sand and are invariably capped by silt and shale, it seems reasonable to assume that the interchannel plains are composed primarily of silt and shale; hence, S_b is 100 percent.

Allen (1965) showed that the vertical distance between the toe and crest of the point-bar accretionary unit is approximately the river depth at bankful stage. Therefore, good approximation of channel depth should be the thickness of the point-bar deposit.

CALCULATION OF RIVER DISCHARGE AND VELOCITY

Equations 1, 2, and 3 were solved to give discharge estimates for each sampling point. Ideally, the discharges determined from equation 1 should be equivalent to those calculated by equation 2, and the plot of Q in equation 2 (Q_2) versus Q in equation 1 (Q_1) should fall along a 45° line emanating from the origin. In fact, Q_2 is approximately two-thirds of Q_1. The discharge at each location was assumed to be the average of Q_2 and Q_1 (Q_{avg}). Discharges (Q_{avg}) at each of six sampling sites were calculated in this manner. The calculated discharges range from a high of 565 cm (19,900 cfs) to a low of 130 cm (4,600

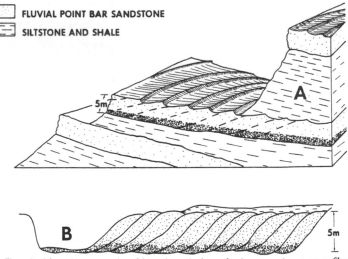

☐ **FLUVIAL POINT BAR SANDSTONE**

☰ **SILTSTONE AND SHALE**

Figure 3. Schematic representation of the ancient point-bar surface in outcrop. A, outcrop profile; B, cross section through point-bar sand.

cfs) (Fig. 5, Table 1). The greatest discharge then is approximately four times the minimum discharge.

Assuming Q_{avg} to be the discharge, velocities can be estimated using the following relationship:

$$Q = VA \qquad (4)$$

(Leopold and others, 1964), where Q is discharge, V is velocity, and A is the cross-sectional area of the channel. If the cross-sectional area is approximately rectangular, area equals depth times width. The calculated estimates of velocity range between 0.2 m/sec to 0.4 m/sec (Table 1).

ESTIMATING SLOPE

Slope can be estimated from a formula developed by Schumm (1968; equation 5) or from the Manning equation (from Barnes, 1967; equation 6):

$$S = 60(M^{-0.38})(Q_m^{-0.32}) \qquad (5)$$

$$V = \frac{1.49}{n} R^{2/3} S_c^{1/2}, \qquad (6)$$

where M is the sediment load parameter, Q_m is the mean discharge, S is the slope, V is velocity, and R is the hydraulic radius. The constant n in Manning's equation, estimated from the tables of Barnes (1967), is 0.06. The calculated slopes are gentle, approximately 0.0016.

STREAM ORDER

Samples were collected from three different stratigraphic time units. The samples from each time unit can be used to calculate channel dimensions of roughly contempor-

ary streams and thus permit reconstruction of the river system — or part of it — as it appeared at three different times. From this, a stream order to horizons A and B is proposed (Fig. 5). This assumption is tenuous,

APPROXIMATE LIMIT OF THE FLUVIAL DEPOSIT OUTCROP

Figure 5. River orientations, meander wave lengths, and discharge. Numbers represent discharge estimates calculated in cubic metres per second. Solid lines are river segments where radius of curvature was measured. Dashed lines are postulated river networks. A, B, and C are three stratigraphic horizons. Small rectangle represents location of aerial photograph of Figure 4.

and no unique solution exists for ordering the streams. Only one river course could be identified within cluster C, and it was oriented roughly east; its discharge is equivalent to that of the major stream in cluster A. Clusters A and B, about 40 km apart, are at about the same stratigraphic horizon. There is no way to determine whether these two channels were parts of one river or if they were in separate systems.

The channel orientations and direction of increasing discharge suggest southeasterly transport.

INDEPENDENT CHECKS OF THE METHOD

Two independent checks test the precision and accuracy of our results. The first is related to paleotransport data. Nearly 200 megaripple axial azimuths were determined throughout the fluvial sandstone facies (Fig. 5). The data agree extremely well with the paleoflow direction, as determined by meander wave length and discharge. An independent comparison can be made of the meander wave length calculated by the radius of curvature and the meander wave length measured from aerial photographs.

In two localities (1a and 1b, 5a and 5b in Fig. 2 and Table 1) there are two contemporary point bars on opposite banks of a paleochannel, and the point-bar ridges are

Figure 4. Aerial photograph of arcuate point-bar ridges (for location, see Fig. 5). Scale bar represents about 1 km.

TABLE 1. SUMMARY OF HYDROLOGIC AND GEOMORPHIC PARAMETERS
CALCULATED FOR EACH SAMPLING SITE

Sampling site	Meander wave length (m)	Channel width (m)	Average mean annual discharge (cm)	Water velocity (m/sec)	Average river slope (m/m)
1a	2,195	189	367	0.317	0.00016
1b	3,139	268	763	0.466	0.00019
Average (1a, 1b)	2,667	229	565	0.392	0.00018
2	2,438	113	453	0.354	0.00016
3	1,311	171	130	0.190	0.00016
4	1,987	213	298	0.285	0.00015
5a	2,499	174	474	0.362	0.00016
5b	2,012	229	309	0.291	0.00015
Average (5a, 5b)	2,256	202	425	0.340	0.00016
6	2,652	229	540	0.388	0.00017

arcuate in diametrically opposed directions. The meander wave lengths were thus obtained directly and could be compared with those calculated. In the first case, the measured wave length is 4,785 m, whereas the calculated mean wave lengths are 2,195 and 3,139 m, respectively. At the second sampling site, the measured wave length is 3,960 m, as compared to 2,499 and 2,012 m, respectively. In both cases calculated meander wave lengths and the measured wave length are within 60 percent of one another. Obviously the method used in this study must be applied with caution.

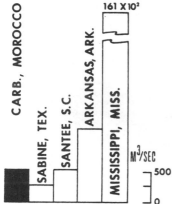

Figure 6. Composite discharges of selected United States east coast rivers and the postulated Stephanian river system.

SUMMARY

Rivers in late Carboniferous time (Stephanian) were broad, shallow, and highly sinuous. Calculated mean annual discharges of the rivers studied ranged between 140 and 850 m³/sec. Even if calculated estimates are off by a factor of 2, the discharge (Fig. 6) was similar to that of medium-sized rivers on the Atlantic and gulf coastlines of the United States. The channels were oriented west-northwest, with discharges increasing easterly, suggesting an easterly flow direction. This observation is corroborated by sedimentary structural data and regional paleogeographic studies by V. V. Cavaroc (1975, written commun.). The distance from the Atlas source uplands to the basin was about 100 km (V. V. Cavaroc, 1975, written commun.). The estimated discharges of the channels, assuming periodic flooding, were capable of carrying cobbles. The absence of pebbles and cobbles in the alluvial deposits, coupled with the nearness of the source terrain, suggests a source with subdued physiography.

ACKNOWLEDGMENTS

This research was done under the auspices of the University of South Carolina International Geology Programs funded by National Science Foundation Grant 325010X, which includes a grant from the Office of International Programs and a grant from the Earth Science Section of the Division of Environmental Sciences.

REFERENCES CITED

Allen, J.R.L., 1965, The sedimentation and paleogeography of the Old Red Sandstone of Anglesey, North Wales: Yorkshire Geol. Soc. Proc., v. 35, p. 139–185.

Barnes, H. H., 1967, Roughness characteristics of natural channels: U.S. Geol. Survey Water-Supply Paper 1849, 213 p.

Cotter, Edward, 1971, Paleo-flow characteristics of a Late Cretaceous river in Utah from analysis of sedimentary structures in the Ferron sandstone: Jour. Sed. Petrology, v. 41, p. 129–138.

Fabre, J., and Greber, C., 1955, Le carbonifère continental au norde de Tindouf: Publ. Service Carte Géologique de l'Algerie, Bull. 8, p. 9–23.

Frazier, D. E., and Osanik, A., 1961, Point bar deposits, Old River Locksite, Louisiana: Gulf Coast Assoc. Geol. Socs. Trans., v. 21, p. 121–370.

Hollard, H., and Jacquemont, P., 1956, Le gothlandian, le devonian, et le carbonifere des regions du Draa et du Zemoul: Maroc Service Géol. Notes, v. 15, p. 7–33.

Lappman, L., and Ray, R. 1965, Aerial photographs in field geology: New York, Holt, Rinehart and Winston, 230 p.

Leopold, L. B., and Wolman, M. G., 1960, River meanders: Geol. Soc. America Bull., v. 71, p. 769–794.

Leopold, L. B., Wolman, M. G., and Miller, J. P., 1964, Fluvial processes in geomorphology: San Francisco, W. H. Freeman, 522 p.

Schumm, S. A., 1968, River adjustment to altered hydrologic regimen — Murrumbidgee River and paleochannels, Australia: U.S. Geol. Survey Prof. Paper 598, 65 p.

——1972, Fluvial paleochannels, in Rigby, J. K., and Hamblin, W. K., eds., Recognition of ancient sedimentary environments: Soc. Econ. Paleontologists and Mineralogists Spec. Pub. 16, p. 98–107.

Vina y Villa, J., and Cabezon, C.S.N., 1959, Contribución al estudio del Carbonifero de la zona Sur del Oued Draa: Bol. Inst. Geol. y Minero de Espana, v. 70, p. 275–314.

Visher, G. S., 1965, Fluvial processes as interpreted from ancient and recent fluvial deposits, in Middleton, G. V., ed., Primary sedimentary structures and their hydrodynamic interpretation: Soc. Econ. Paleontologists and Mineralogists Spec. Pub. 12, p. 116–132.

MANUSCRIPT RECEIVED BY THE SOCIETY DECEMBER 18, 1974
REVISED MANUSCRIPT RECEIVED SEPTEMBER 17. 1975
MANUSCRIPT ACCEPTED OCTOBER 3, 1975

8

Reprinted from pp. 129, 131–138 of Jour. Sed. Petrology 41(1):129–138 (1971)

PALEOFLOW CHARACTERISTICS OF A LATE CRETACEOUS RIVER IN UTAH FROM ANALYSIS OF SEDIMENTARY STRUCTURES IN THE FERRON SANDSTONE[1]

EDWARD COTTER

Department of Geology and Geography, Bucknell University, Lewisburg, Pennsylvania 17837

ABSTRACT

Interpretation of sedimentary structures and detailed stratigraphic relations of the fluvial facies of the Upper Cretaceous Ferron Sandstone in the Castle Valley in east-central Utah permits reasonable reconstruction of many parameters of a Late Cretaceous alluvial system. After the width and depth of flow of the ancient Ferron river have been estimated from the geometry of preserved sedimentary structures, and the type of sediment transported by the river has been determined by study of the sandstone texture, various relationships of modern streams empirically derived by Schumm are used to estimate channel sinuosity, meander length, mean annual discharge, mean annual flood, channel slope, and flow velocity. Values derived for the slope and velocity are supported by other indirect methods based on the Manning equation. It appears likely that the Late Cretaceous Ferron river was about 300 feet wide and 25 feet deep, and that it was highly sinuous, with meander lengths of 2,500 to 4,100 feet. As the 200-mile-long river drained an area to the southwest of 6,000 to 8,000 square miles, it had a mean annual discharge of approximately 6,000 to 7,000 cubic feet per second and a mean annual flood of about 22,000 cubic feet per second. Although only 2 percent of the total river load was bedload, the flow velocity of between 2.0 and 4.6 feet per second in the upper part of the lower flow regime caused the fine- to medium-grained sand to be in a dune bed configuration.

Some caution is advised in using this approach because of unresolved questions of applying modern stream relations to pre-Quaternary deposits and because of uncertainties in determining the cross-sectional shape and sediment texture of the ancient Ferron river.

[1] Manuscript received August 26, 1970; revised October 1, 1970.

[Editor's Note: Material, including figures 1 and 2, has been omitted at this point.]

FIG. 3.—Laterally accumulated point bar deposit in the Ferron Sandstone. Located in Dry Wash, east of the village of Moore in Castle Valley, Utah. Thickness of point bar unit is 30 feet.

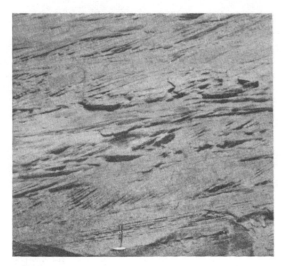

FIG. 4.—Closer view of epsilon cross-stratified point bar sandstone of figure 3. Sedimentary structure viewed here is seen to be trough cross lamination when examined on outcrop face normal to this (see figure 5).

$$M = \frac{Sc \times W + Sb \times 2D}{W + 2D} \qquad (1)$$

(where Sc is the percentage of silt and clay in the channel alluvium, Sb is the percentage of silt and clay in the bank alluvium, W is the channel width, and D is the channel depth) (Schumm, 1960b, p. 18).

Schumm found that channel morphology, as expressed by the width : depth ratio (F) and the sinuosity (P), is significantly controlled by the type of sediment load, reflected in the parameter M. The following regression equations indicate that quality of these relationships (r is the correlation coefficient, Se is the standard error of estimate) :

$$(2)$$
$$F = 255M^{-1.08} \qquad (r = 0.91, Se = 0.20) \text{ (Schumm, 1960b)}$$
$$(3)$$
$$P = 0.94M^{-0.25} \qquad (r = 0.91, Se = 0.06) \text{ (Schumm, 1963b)}$$
$$(4)$$
$$P = 3.5F^{-0.2} \qquad (r = 0.89, Se = 0.06) \text{ (Schumm, 1963b)}$$

The difficulty in applying the apparently simple relationship to the Ferron Sandstone fluvial facies is the ambiguity in defining the stream width and depth, because the sandstone intervals are broader and thicker than original stream channel cross sections. Estimates of the stream width and depth were made using characteristics of the laterally accumulated, sheet-like, epsilon cross-stratified, point bar deposits (fig. 3).

Estimation of stream depth.—The thickness of the intervals of epsilon cross stratified point bar sandstone represents the vertical distance from some elevation above mean river level to the depth of the talweg at a stream bend. Both Moody-Stuart (1966) and Allen (1965b, as reported by Moody-Stuart, 1966) suggest that the thickness of such cross-stratified units corresponds to the channel depth at bankfull stage. The laterally accumulated point bar sheet sandstones studied are fairly consistently about 30 feet thick. Thus, it is suggested that the bankfull stream depth at a meander bend was approximately 30 feet. However, because the sediment load parameter (M) was developed by Schumm (1960) using the depth measured in relatively straight reaches, and because the talweg depth in straight reaches is less than that of meander bends (Sundborg, 1956), it is probable that the depth to be used in equation (1) is somewhat less than 30 feet. In later equations in which a value for stream depth is inserted, 25 feet will be used.

Channel Form

The fluvial geomorphologist, S. A. Schumm, in a series of papers (1960a, 1960b, 1963a, 1963b, 1967a, 1968a, 1969) has convincingly developed the thesis that modern river and channel morphology is controlled principally by the type of sediment load and by the quantity of sediment and water moving through the channel (Schumm, 1969, p. 256). This conclusion is based on study of 36 stable alluvial rivers in subhumid to subarid parts of the United States and Australia. So important is the type of sediment load in determining fluvial morphology that Schumm has proposed classifying alluvial river channels on the basis of the ratio of suspended load to bedload. Thus the introduction of the phrases: bedload, mixed-load, and suspended-load channels (Schumm, 1963a; 1968a, p. 37–41).

Although the ratio of suspended load to bedload is not readily determined directly, the type of sediment being moved through the channel can be estimated. This ratio (suspended load : bedload) is directly related to the percentage of silt and clay (sediment finer than 0.074 mm.) in the perimeter of the channel (Schumm, 1963a, p. 4–7; 1968a, p. 37–41; 1969, p. 257). Sediment of the bed and banks can be sampled, and the type of sediment in the channel perimeter can be characterized by the parameter M:

Estimation of stream width.—The width of the Ferron river must also be approximated indirectly using the laterally inclined point bar deposits. The relationship between the width of a point bar and stream width has apparently not been investigated systematically. Information and maps in a number of sources, such as Sundborg (1956), Leliavsky (1955), and Leopold and Wolman (1960), indicate that the width of a point bar, from the stream bank down into the talweg, is generally three-fifths to four-fifths of the channel width. Allen (1966, p. 166) states that the width of a lateral bar in a low sinuosity stream is about two-thirds the channel width. In his attempt at paleoflow analysis of Devonian deposits, Moody-Stuart (1966) concluded that point bar width is about two-thirds the stream width. As an approximation in this present study, this two-thirds relationship is used. The width of the epsilon cross-stratification in the studied intervals in the Ferron Sandstone is about 200 feet (fig. 3). If this represents two-thirds of the channel width, the Ferron river was about 300 feet wide.

Estimation of sediment load parameter.—With values now available for stream width and depth, all that is needed for calculation of the sediment load parameter, M (equation 1), is the silt-clay percentage of the material of the stream bed and banks. On the north wall of Willow Springs Wash there is exposed a fluvial sandstone interval that is not very extensive laterally. At the margin of this interval exists the finer-grained former banks of the stream. By counting 800 points in each of three thin sections, it was determined that this sandstone had a silt-clay content of about 15 percent. This estimate was made through the haze created by some diagenetic calcitization, and it is also possible that some of the clay is diagenetic. Samples of the bank material marginal to the sandstone are composed of essentially all silt and clay. Thus, for the parameter M calculation (equation 1), $S_c = 15$ and $S_b = 100$. With these values added to those for stream width (300 feet) and depth (25 feet), M is calculated to be approximately 27. This sediment load parameter value is characteristic of a suspended-load channel in which the bedload forms about 2 percent of the total load (Schumm, 1968a).

Estimation of width/depth ratio and sinuosity.—Equations (2), (3), and (4) may now be

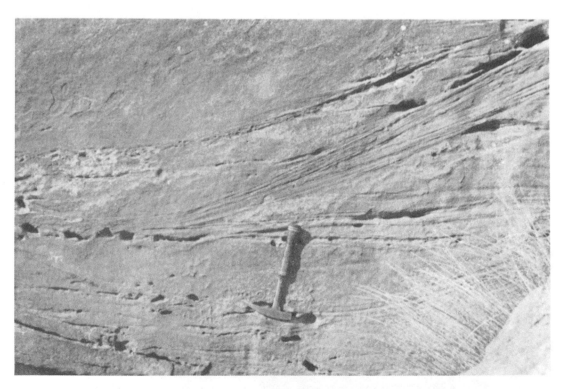

Fig. 5.—Segment of trough cross lamination in sandstone interval of figure 3, occurring on outcrop face normal to that of figure 4.

used to evaluate certain aspects of the channel form. It might seem superfluous to use equation (2) to calculate the width/depth ratio when it can come directly from the values just determined for width and depth. However, use of equation (2) and comparison of the width/depth ratio with the more directly determined value will offer a check on the values used for silt-clay percentages. The width/depth ratio calculated by equation (2) is 8, whereas the same ratio determined directly is 12, a reasonably close agreement.

The two width/depth ratio estimates would agree more closely if the value of the sediment load parameter were less, and the value of the sediment load parameter would be less if the estimate of the silt-clay percentage in the fluvial sandstone were lower. This might be a reason to reconsider the estimate of the silt-clay percentage, particularly in light of previous suspicion of some diagenetic clay. However, in my opinion, it is better to rely on the direct measurement of the silt-clay percentage, with awareness of the possibility of some error, than to adjust that percentage on the basis of such an involved chain of no more exact reasoning.

The two equations for estimation of sinuosity (equations 3 and 4) indicate that the ancient Ferron river was highly sinuous, with the distance along the channel about twice as long as direct-line distance. Channel sinuosity from equation (3), using the determined value of 27 for the sediment load parameter (M), is 2.1. Using equation (4), the sinuosity is calculated to be either 1.8 or 2.0, depending upon whether a value of 12 or 8 is used for the width/depth ratio.

Channel Dimensions

When data on discharge are added to those on sediment type, the dimensions of modern alluvial channels are better resolved (Schumm, 1968a, 1969). The following parts of this section contain multiple regression equations developed by Schumm in a series of studies of channel relationships. They were selected because of their applicability to interpretation of the fluvial deposits for the Ferron Sandstone. Most of the equations are summarized by Schumm (1968a, 1969); equation (5) came in personal correspondence with Dr. Schumm. Symbols used in the equation are: W, channel width; D, channel depth; L, meander length; M, sediment load parameter; Q_m, mean annual discharge, Q_{ma}, mean annual flood; r, correlation coefficient; and Se, standard error of estimate.

Mean annual flood.—This is defined as the flood

with a recurrence interval of 2.33 years (Leopold, Wolman, and Miller, 1964, p. 64). From Schumm (personal communication) comes the following equation:

$$\text{Log } Q_{ma} = 0.268 + 0.469 \log M + 1.378 \log W \quad (5)$$

Using previous estimates of M (27) and W (300 feet) in this equation the mean annual flood is calculated to be about 22,000 cubic feet per second.

Mean annual discharge.—Schumm (1968a, p. 49, 50) presents two equations which can be used to estimate the mean annual discharge.

$$W = 37 \frac{Q_m^{0.38}}{M^{0.37}} \quad (r = 0.93, Se = 0.14) \quad (6)$$

$$D = 0.6M^{0.34}Q_m^{0.29} \quad (r = 0.89, Se = 0.12) \quad (7)$$

These can be rearranged to give:

$$Q_m^{0.38} = \frac{WM^{0.37}}{37} \quad (8)$$

$$Q_m^{0.29} = \frac{D}{0.6M^{0.34}} \quad (9)$$

When previously determined values of W, D, and M are substituted, equation (8) indicates that the mean annual discharge was about 6,000 cubic feet per second, and equation (9) indicates about 7,000 cubic feet per second.

Meander length.—The agreement is not as close for two values of the meander length (L) estimated using two equations from Schumm (1968a, p. 47).

$$L = 234 \frac{Q_{ma}^{0.48}}{M^{0.74}} \quad (r = 0.92, Se = 0.19) \quad (10)$$

$$L = 1890 \frac{Q_m^{0.34}}{M^{0.74}} \quad (r = 0.96, Se - 0.16) \quad (11)$$

The meander length from equation (10) is 2500 feet, and from equation (11) it is 4,100 feet.

Channel Slope

One of the more important parameters used to describe the alluvial channel relationships is the channel slope, or gradient. If this parameter were known more exactly from some independent measurement for the Ferron fluvial sandstones, a number of other flow parameters could be calculated more accurately and a variety of other hydrodynamic relationships could be applied.

All the characteristics thus far adduced for the Ferron deposits indicate that the slope was relatively gentle. According to Schumm (1963, 1968a), suspended-load channels have gentle gradients. Harms and Fahnestock (1965, Pl. 1) have summarized that the large trough cross-lamination sets, so characteristic of the Ferron sandstones of fluvial origin, are most likely in low gradient, meandering, perennial streams. Meandering streams usually have lower gradients than braided streams of the same discharge (Leopold, Wolman, and Miller, 1964). And the physiographic setting of the Ferron river, an alluvial or deltaic plain, also suggests that the gradient was relatively gentle.

It is Schumm again who presents a relationship from which the slope of the Ferron river can be estimated (Schumm, 1968a, p. 51):

$$S_c = 60M^{-0.38} \, Q_m^{-0.32} \quad (r = 0.84, Se = 0.15) \quad (12)$$

The estimate of the sediment load parameter (M) is 27, but there are two values, closely in agreement, for the mean annual discharge, 6000 and 7000 cusecs. If these two values are averaged for use in equation (12), the slope can be calculated to be approximately 1 ft/mile (about 0.0002).

Additional estimates of the slope of the Upper Cretaceous Ferron river can be derived using the Manning equation for open channel flow:

$$V = \frac{1.49}{n} R^{2/3} S_c^{1/2} \quad (13)$$

where V is the average velocity in ft/sec, R is the hydraulic radius in feet, S_c is the channel slope, and n is the Manning roughness coefficient.

In the first use of this formula, it can be rearranged to:

$$S_c^{1/2} = \frac{n \, V}{1.49 \, R^{2/3}} \quad (14)$$

and estimated values substituted into the equation. Because the stream is relatively broad, the hydraulic radius, R, is approximately equal to the depth, estimated to be about 25 feet. As is shown in the next section, the average velocity of the Ferron river was probably about 3 ft/sec. And the roughness coefficient, n, for material of 0.25 mm average size and dune bed configuration can be estimated to be about 0.025 from the data of Guy, Simons, and Richardson (1966). With these values, the channel slope can be calculated by the above equation (14) to be 0.00029, or approximately 1.5 ft/mile.

Another approximation of the slope can be made if both sides of the Manning equation are multiplied by γ D S, where γ is the specific weight:

$$\gamma D S_c V = \frac{1.49}{n} R^{2/3} S_c^{1/2} \delta D S_c \quad (15)$$

Now, since γ D S is equal to τ (the shear stress), and R is approximately equal to D, this can be rearranged to:

$$S_c = \frac{(\tau V) \, n}{1.49 \, \gamma D^{5/3}} \quad (16)$$

In substituting values into this equation the only value that must be added to other previously determined is that for (τ V). Simons and Richardson (1963, p. 200) have presented a relationship between stream power (τ V), median fall diameter of bed material, and the form of bed roughness. For material about 0.25 mm in median diameter with bed form of dunes in the upper part of the lower flow regime, the stream power (τ V) is about 0.3. Calculations using this and the other values (n = 0.025, D = 25 ft, γ = 62.4) in the above equation (16) indicate that the channel slope of the Ferron river was 0.00019 (approximately 1 ft/mile).

Flow Velocity

In their unpublished manuscript, Nordin, Simons, and Richardson (1965) state that the dune bed configuration persists in the range between 1.5 V_c and 3.5 V_c, where V_c is the critical velocity for the initiation of particle motion. From Sundborg (1956, p. 177), the critical erosion velocity for quartz of 0.25 mm grain diameter is about 40 cm/sec (1.3 ft/sec) at a point between 1 and 10 meters above the bed. If this velocity value is placed in the relationship suggested by Nordin, *et al.*, the mean velocity is in the range 2.0 and 4.6 ft/sec.

The Manning equation (13) affords another assessment of the flow velocity.

$$V = \frac{1.49}{n} R^{2/3} S_c^{1/2}$$

By substituting values for n (0.025), R=D (25 feet), and S_c (1 ft/mile or 0.0002), the velocity is calculated to be 3.7 ft/sec. (The value of the channel slope is that obtained using equation 12.)

It is possible that this Manning equation velocity estimate is somewhat high. In the physiographic setting of the Ferron river it seems likely that there was some bank vegetation to increase the roughness coefficient above that due to the bed configuration alone. It is for this reason that in relationships in which a specific value of stream velocity must be used (such as in equation 14), the value of 3.0 ft/sec is used.

Additional Parameters

Drainage area.—A relationship between meander length and area of the drainage basin has been presented by Dury (1960, see Leopold, Wolman, and Miller, 1964, p. 312). In this study it was estimated that the Ferron river had a meander length of 2,500 to 4,100 feet. For such a range, Dury's relationship suggests that the drainage area was between about 2,000 and 8,000 square miles.

According to Eicher (1969), Leopold, Wolman, and Miller (1964, p. 251) suggest that in humid regions that area of the drainage basin (in square miles) is approximately equal to the mean annual discharge (in cubic feet per second). If this is true of the Ferron river system, the drainage area was about 6,000 to 7,000 square miles. This relationship was also used in an attempt to estimate paleoflow parameters of the streams that deposited the Cretaceous Lytle Formation in the Colorado Front Range (Eicher, 1969, p. 1081–2).

The overlap in these two approximations suggests that the drainage area of the Ferron River was about 6,000 to 8,000 square miles.

Stream Length.—The length of a stream, from mouth to drainage divide, is related to the area of the basin draining into that system. For a stream with a drainage area of about 6,000 to 8,000 square miles, the length is about 200 miles (Leopold, Wolman, and Miller, 1964, p. 144, fig. 5–6, B).

DISCUSSION OF THE APPROACH

This study has used a number of modern stream and flume relations to interpret paleoflow characteristics of a Late Cretaceous alluvial system, and it is quite in order to question the applicability of relations determined in one setting to a different geologic and climatic setting.

Great reliance has been placed on equations developed by Schumm from investigation of 36 stable alluvial channels in a semiarid to subhumid regions of the United States and Australia (Schumm, 1969, p. 256). The first demonstration of the use of these relations in ancient deposits of other climatic settings was by Schumm (1968a) in his study of Pleistocene and Holocene paleochannels in the Riverine Plain of Australia. Schumm found that this Australian study supported his hypothesis "that all aspects of river morphology are influenced by the type of sediment load moved through stream channels" (Schumm, 1968a, p. 1).

The climate of the Western Interior of the United States in Upper Cretaceous time was warmer and more humid than it is at present. This is evident from the facts that the Mesozoic Era essentially everywhere was a warm interval in the history of the earth (Schwarzbach, 1963, p. 162) and that the formation of Upper Cretaceous coal requires a humid climate (Krausel, 1964, p. 53). More detailed studies of plants (Dorf, 1942, p. 28, 103) and vertebrates (Ostrum, 1965, p. 40, 41) agree that in Late Creta-

ceous time the climate of the Rocky Mountain region was most likely warm temperate to subtropical. It is possible that there was a savannah-type climate, with a short dry season interrupting a longer period of hot, humid conditions (Moberly, 1960, p. 1172).

Because the modern river relations determined by Schumm have not yet been extended empirically into humid subtropical regions, it is possible that some of the equations used in this study have been extended beyond their range of applicability. It is not possible at this time to determine this. In my opinion, Schumm's equations will be shown to extend to more humid regions because of the general validity of his principle—that "the dimensions, shape, gradient, and pattern of stable alluvial rivers should be controlled by the quantity of water and quantity and type of sediment moved through their channels" (Schumm, 1969, p. 256).

Another circumstance that complicates interpretation of ancient paleohydrologic relations is the difference in vegetative cover of the earth in the geologic past. Such differences directly affect runoff volumes, flood peaks, and sediment yields (Schumm, 1967b, p. 189, 190). Late Cretaceous floras had a distinctly modern appearance that followed upon the rapid spread of the flowering plants in great numbers and variety (Darrah, 1960, p. 219), but it required the development of grasses in Late Cenozoic time to produce erosion and runoff conditions similar to those at present (but without the influence of man) (Schumm, 1968b, p. 1578). During the Late Cretaceous interfluve hillslopes became colonized, especially in the lower elevations, leading to stabilizing and reduction of sediment yield (Schumm, 1968b, p. 1578). Before hillslope sediment stabilization by modern vegetation, the hydrologic situation in ancient valleys resembled that in modern semiarid regions (Schumm, 1968b, p. 1584). Thus, it is possible that the paleoflow relations in humid regions in Upper Cretaceous time were relatively closer to those of semiarid and subhumid climatic areas today, and that Schumm's modern river equations can be applied directly.

While this somewhat conjectural situation awaits more data for resolution, the status of the appropriateness of using Schumm's modern river relations to interpret Upper Cretaceous fluvial deposits is well summarized by S. A. Schumm himself (personal communication, letter dated 11/14/68): "—considering that no grasses were present in the Cretaceous and that the river appears to have transported large quantities of sand, I believe that the relations are applicable. Nevertheless, you should state that there is some question about this in your paper."

An additional source of complication and potential error in applying modern empirical relations to ancient sandstones lies in the difficulties of transposing Ferron fluvial facies features into forms suitable for use with the modern relations. Examples of this on three different scales are: (1) the form of the stream cross section cannot be determined directly and must be approximated from the geometry of point bar deposits; (2) the former stream bed forms are not preserved as sedimentary structures without alteration; and (3) diagenetic alteration has clouded the details of the primary texture of the bed and bank materials, increasing the difficulty of measuring the silt-clay content.

And finally, a word to those who might assume that great accuracy of interpretation has been achieved when a single number has been derived as a solution to one of these equations. These results are merely reasonable approximations that fit the stratigraphic and geologic context in which the fluvial facies of the Ferron Sandstone was deposited. But even the less-than-exact information expands our knowledge of the paleogeography of east-central Utah in Late Cretaceous time in a way that cannot be obtained from other lines of geologic reasoning.

CONCLUSIONS

After analysis of the sedimentary structures and textures of the fluvial facies of the Ferron Sandstone in light of modern empirical stream relations, I conclude that:

1) In Late Cretaceous time in east-central Utah, a 200-mile-long river drained an area of 6,000 to 8,000 square miles as it flowed generally northward from part of the Sevier orogenic belt. Bedload formed only about 2 percent of the total load of this river, whose mean annual discharge was in the neighborhood of 6,000-7,000 cubic feet per second and whose mean annual flood was about 22,000 cubic feet per second. In keeping with the characteristics of such suspended-load stream, the channel pattern was highly sinuous, and the distance along the channel was about twice as great as the straight line distance.

As the river flowed on a slope of about 1 ft/mile through the alluvial plain toward the epeiric sea, it was about 300 feet wide, 25 feet deep (approaching 30 feet deep at meander bends), and had meander lengths of 2,500 to 4,100 feet. Flow characteristics in the upper part of the lower flow regime, and a flow velocity of between 2.0 and 4.6 feet/second caused the fine- to medium-grained sand to be in a dune configuration.

2) There are areas of knowledge and developments in hydrology and geomorphology that can help the sedimentary petrologist in his attempts to interpret paleogeographic parameters. In utilizing this knowledge, the sedimentary petrologist must be prepared to exercise a degree of judgment in order to place the stratigraphic relations in a form in which modern empirical relations can be used, and because the results obtained are only reasonable estimates that must fit the nature of the environment interpreted in other ways.

ACKNOWLEDGMENTS

Stimulation to attempt paleoflow estimates came from association with S. A. Schumm and D. B. Simons in the course of a National Science Foundation sponsored institute in Fluvial Geomorphology at Colorado State University. C. V. Campbell of Esso Production Research Company introduced me to the techniques of interpretation of depositional environments of sandstones. J. D. Howard of the University of Georgia Marine Institute suggested a study of th Ferron Sandstone. Discussions with Alan Donaldson have proved helpful. I am grateful to Bucknell University for generously providing funds for field expenses. This work formed the basis of a paper presented at the 1969 Meeting of the Society of Economic Paleontologists and Mineralogists in Dallas, Texas.

REFERENCES

ALLEN, J. R. L., 1963a, The classification of cross-stratified units, with notes on their origin: Sedimentology, v. 2, p. 93–114.
———, 1963b, Asymmetrical ripple marks and the origin of water-laid cosets of cross-strata: Liverpool and Manchester Geol. Jour., v. 3, p. 187–236.
———, 1965a, A review of the origin and characteristics of recent alluvial sediments: Sedimentology, v. 5, p. 91–191.
———, 1965b, The sedimentation and palaeogeography of the Old Red Sandstone of Anglesey, North Wales: Yorkshire Geol. Soc. Proc., v. 35, p. 139–185.
———, 1966, On bed forms and palaeocurrents: Sedimentology, v. 6, p. 153–190.
———, 1967, Depth indicators of clastic sequences: Marine Geology, v. 5, p. 429–446.
———, 1970, Studies in fluvatile sedimentation: a comparison of fining-upwards cyclothems, with special reference to coarse-member composition and interpretation: Jour. Sedimentary Petrology, v. 40, p. 298–323.
ARMSTRONG, R. L., 1968, Sevier orogenic belt in Nevada and Utah: Geol. Soc. America Bull., v. 79, p. 429–458.
BEUTNER, E. C., FLUECKINGER, L. A., AND GARD, T. M., 1967, Bedding geometry in a Pennsylvanian channel sandstone: Geol. Soc. America Bull., v. 78, p. 911–916.

DARRAH, W. C., 1960, Principles of paleobotany: second edition, Ronald Press Co., New York, 295 p.
DORF, ERLING, 1942, Upper Cretaceous floras of the Rocky Mountain region: Carnegie Inst. of Washington, Publication 508, 168 p.
EICHER, D. L., 1969, Paleobathymetry of Cretaceous Greenhorn sea in eastern Colorado: Am. Assoc. Petroleum Geologists Bull., v. 53, p. 1075–1090.
GUY, H. P., SIMONS, D. B., AND RICHARDSON, E. V., 1966, Summary of alluvial channel data from flume experiments, 1956–1961: U.S. Geol. Survey Prof., Paper 462–I, 96 p.
HALE, L. A., 1959, Intertonguing Upper Cretaceous sediments of northeastern Utah—northwestern Colorado, in Washakie, Sand Wash, Piceance Basins: Rocky Mtn. Assoc. Geologists Eleventh Field Conference Guidebook, p. 55–66.
HALE, L. A., AND VAN DE GRAAFF, F. R., 1964, Cretaceous stratigraphy and facies patterns—northeastern Utah and adjacent areas: Intermountain Assoc. Petroleum Geologists Guidebook, 13th Annual Field Conf., p. 115–138.
HARMS, J. C., AND FAHNESTOCK, R. K., 1965, Stratification, bed forms, and flow phenomena (with an example from the Rio Grande): p. 84–115 in G. V. Middleton, ed., Primary Sedimentary Structures and their Hydrodynamic Interpretation, Soc. Econ. Paleontologists and Mineralogists, Spec. Pub. No. 12, 265 p.
HARMS, J. C., MACKENZIE, D. B., AND MCCUBBIN, D. G., 1963, Stratification in modern sands of the Red River, Louisiana: Jour. Geology, v. 71, p. 566–580.
JOPLING, A. V., 1966, Some principles and techniques used in reconstructing the hydraulic parameters of a paleo-flow regime: Jour. Sed. Petrology, v. 36, p. 5–49.
KATICH, P. J., JR., 1953, Source direction of the Ferron sandstone in Utah: Am. Assoc. Petroleum Geologists Bull., v. 37, p. 858–861.
KRAUSEL, R., 1964, Introduction to the palaeoclimatic significance of coal: p. 53–56 in Nairn, A. E. M., editor, Problems in Palaeoclimatology: Interscience Publishers, 705 p.
LEOPOLD, L. B., AND WOLMAN, M. G., 1960, River meanders: Geol. Soc. America Bull., v. 71, p. 769–794.
LEOPOLD, L. B., WOLMAN, M. G., AND MILLER, J. P., 1964, Fluvial processes in geomorphology: Freeman, San Francisco, 522 p.
MCKEE, E. D., AND WEIR, G. W., 1953, Terminology for stratification and cross-stratification in sedimentary rocks: Geol. Soc. America Bull., v. 64, p. 381–390.
MIDDLETON, G. V., editor, 1965, Primary sedimentary structures and their hydrodynamic interpretation: Soc. Econ. Paleontologists and Mineralogists, Spec. Pub. No. 12, 265 p.
MOBERLY, RALPH, JR., 1960, Morrison, Cloverly, and Sykes Mountain Formations, northern Bighorn Basin, Wyoming and Montana: Geol. Soc. America Bull., v. 71, p. 1137–1176.
MOODY-STUART, M., 1966, High- and low-sinuosity stream deposits, with examples from the Devonian of Spitsbergen: Jour. Sed. Petrology, v. 36, p. 1102–1117.
NORDIN, C. F., SIMON, D. B., AND RICHARDSON, E. V., 1965, Interpreting depositional environments of sedimentary structures: unpublished manuscript of talk presented at the Southwest Regional Meeting, American Geophysical Union, Socorro, New Mexico, January 28–30, 1965.
OSTRUM, J. H., 1965, Cretaceous vertebrate faunas of Wyoming: Wyoming Geol. Assoc. Guidebook, 19th Annual Field Conf., p. 35–41.
SCHUMM, S. A., 1960a, The effect of sediment type on the shape and stratification of some modern fluvial deposits: Am. Jour. Sci., v. 258, p. 177–184.
———, 1960b, The shape of alluvial channels in relation to sediment type: U.S. Geol. Survey Prof. Paper 352–B, p. 17–30.
———, 1963a, A tentative classification of alluvial river channels: U.S. Geol. Survey Circ. 477, 10 p.
———, 1963b, Sinuosity of alluvial rivers on the Great Plains: Geol. Soc. America Bull., v. 74, p. 1089–1100.
———, 1967a, Meander wavelength of alluvial rivers: Science, v. 157, p. 1549–1550.
———, 1967b, Paleohydrology: application of modern hydrologic data to problems of the ancient past: Internat. Hydrology Symposium, Fort Collins, Colorado, Proc., v. 1, p. 185–193.
———, 1968a, River adjustment to altered hydrologic regimen-Murrumbidgee River and paleochannels, Australia: U.S. Geol. Survey Prof. Paper 598, 65 p.
———, 1968b, Speculations concerning paleohydrologic controls of terrestrial sedimentation: Geol. Soc. America Bull., v. 79, p. 1573–1588.
———, 1969, River metamorphosis: Am. Soc. Civil Engineers Proc. Jour. Hydraulics Div., v. 95, p. 255–273.
SCHWARZBACH, MARTIN, 1963, Climates of the past: D. Van Nostrand Co., Princeton, N.J., 328 p.
SUNDBORG, A., 1956, The river Klaralven, a study of fluvial processes: Geog. Annaler, v. 38, p. 127–316.
SIMON, D. B., AND RICHARDSON, E. V., 1963, A study of variables affecting flow characteristics and sediment transport in alluvial channels: Proceedings, Federal Inter-agency Sedimentation Conference, Agricultural Research Service Miscellaneous Publication No. 970, p. 193–207.
SPIEKER, E. M., 1946, Late Mesozoic and Early Cenozoic history of central Utah: U.S. Geol. Survey, Prof. Paper 205–D, 43 p.
———, 1949, Sedimentary facies and associated diastrophism in the Upper Cretaceous of central and eastern Utah: Geol. Soc. America, Mem. 39, p. 55–81.
VISHER, G. S., 1965, Fluvial processes as interpreted from ancient and recent fluvial deposits: p. 116–132 in G. V. Middleton, ed., Primary Sedimentary Structures and their Hydrodynamic Interpretation, Soc. Econ. Paleontologists and Mineralogists, Spec. Pub. No. 12, 265 p.
WEIMER, R. J., 1960, Upper Cretaceous stratigraphy, Rocky Mountain area: Am. Assoc. Petroleum Geologists Bull., v. 44, p. 1–20.
WRIGHT, M. D., 1959, The formation of cross-bedding by a meandering or braided stream: Jour. Sed. Petrology, v. 29, p. 610–615.
YOUNG, R. G., 1955, Sedimentary facies and intertonguing in the Upper Cretaceous of the Book Cliffs, Utah-Colorado: Geol. Soc. America Bull., v. 66, p. 177–202.

9

Reprinted from pp. 47, 49–52 of *Geol. Mag.* **113**(1):47–52 (1976)

An exhumed Jurassic meander belt from Yorkshire, England

M. NAMI

Summary. Three-dimensional coastal exposures of the Middle Jurassic (Bathonian) Scalby Formation in Yorkshire, England, exhibit a complex, exhumed meander belt. The lateral accretion surfaces (epsilon-cross stratification) of individual meander loops are described. Field and aerial photograph measurements of accretion-surface dimensions have enabled direct determination of channel bankful depths and widths and meander wavelengths. Values for mean annual discharge of the channel system are derived using existing empirical equations. Palaeocurrents measured in the meander belt sandstones show a wide dispersal due to many periods of meander migration and cut off.

1. Introduction

Recently, there has been much study of fluvial sedimentary deposits, with particular emphasis on comparisons between modern meandering streams and ancient fining-upwards cycles (e.g. Allen, 1962, 1965, 1970; Friend, 1961; Moody-Stuart, 1966; Friend & Moody-Stuart, 1972).

An important characteristic of a meandering stream channel is the accretion and migration of point bars. Material eroded from the outside of meander bends is deposited on downstream point bars located on the inside of the bends. The strata corresponding to the slope of successive depositional point bar slopes form a series of large-scale cross-strata with internal sedimentary structures indicating current flow parallel to the direction of the cross-strata strike. Allen (1963, 1965) defined this large-scale cross-stratification as epsilon-cross-stratification.

Much of the work in ancient point bar deposits consists of observations in relatively poorly exposed one- or two-dimensional outcrops, where epsilon-cross-stratification indicative of lateral accretion may be difficult to detect (Allen, 1970). Notable exceptions include the exhumed meander belt described by Puigdefabregas (1973) from the Miocene of the Ebro Basin, Northern Spain, and the vertical sections through lateral accretion units described by Allen (1965), Beutner, Flueckinger & Gard (1967) and Moody-Stuart (1966).

It is the purpose of this paper to describe and interpret a newly discovered exhumed meander belt magnificently exposed in three dimensions on the Yorkshire coast near Scarborough, England, and to propose suggestions concerning the magnitude of the fluvial system responsible.

[*Editor's Note:* Material, including figure 1, has been omitted at this point.]

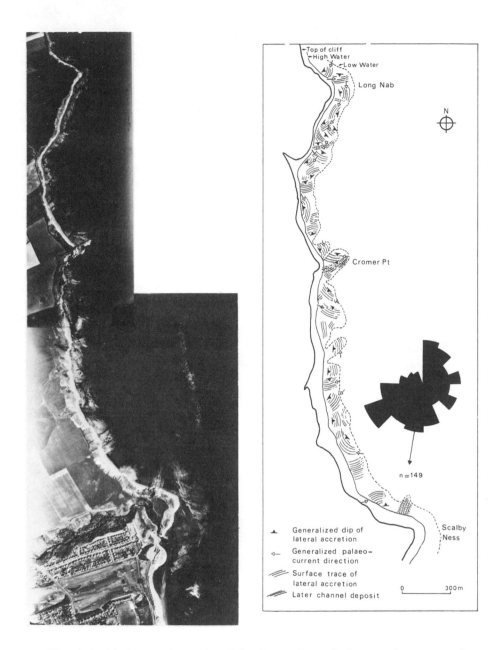

Plate 1. Aerial photograph mosaic and sketch map of meander loops on the coast, north of Scarborough. Note complex erosional relationships between individual meander loop deposits and large palaeocurrent variance. Arrow in rose diagram indicates the vector mean.

[*Editor's Note:* Plate 2 has been omitted.]

127

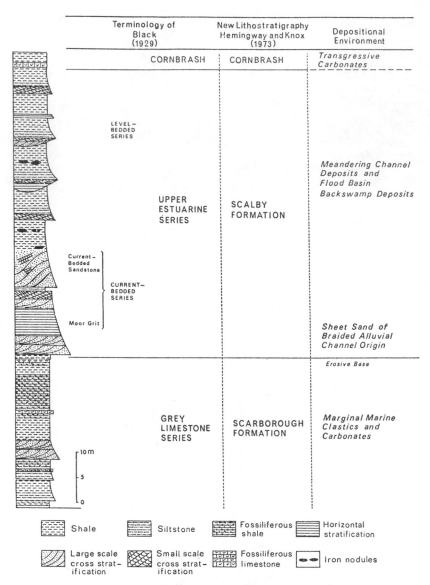

Figure 2. Generalized stratigraphic section and classification of upper part of Middle Jurassic in Yorkshire, showing the major depositional environments.

3. Exhumed meander belt

As the aerial photograph and sketch map show (Pl. 1), the point bar deposits are exposed along the foreshore, being covered by the sea during high water periods. Individual meander loops are well seen at low tide, being curvilinear in plan and displaying complex erosional relationships one to another (Pl. 2.*a*). The whole meander belt as now seen is thus complex, with many periods of meander cut-offs and migrations recorded. On the north side of Long Nab, where the Scalby Formation rises in the cliffs, good two-dimensional vertical exposures of individual meander loops may be seen.

In vertical and horizontal sections, the meander loop deposits consist of alternating sandstone members exhibiting large-scale sigmoidal cross-stratification interpreted as epsilon cross-sets. At the base of each set there is an erosion surface, frequently cut into the fine silty material at the top of the underlying set. The erosional surfaces are overlain by sands containing carbonized wood fragments and rarer clay intraclasts. Individual epsilon units consist of coarse to medium grained sandstone in the base with grain decreasing upwards, often to silt in the upper part. The silts and silty clays increase in thickness as individual epsilon sets are ascended and pass transitionally into overlying silty clay (Pl. 2.*b*), rich in plant fragments and containing sideritic iron nodules. The upper part of some point-bar deposits consists of convulated laminations. Compositionally, the sandstones are mature quartz arenites with only 2 % feldspar and 5 % rock fragments. Sedimentary structures in the base of units are small scale trough-cross-bedding or parallel laminations and climbing ripple cross-stratification passing into siltstone in the upper part. The cross-stratified sets show flow direction normal to the dip direction of the epsilon cross-stratification units. The total thickness of the epsilon units is 3–4 m, the thickness of individual epsilon members ranging from 30 to 70 cm. Palaeocurrent data from the whole meander belt outcrop indicate a general southerly flow with considerable variance due to the divergence of flow vectors in each abandoned meander loop.

The meander belt deposits described above in the Current Bedded Sandstone series are cut by erosion at Burnstone Wyke and Cromer Point by other younger point bar channel deposits in the level bedded series. They were originally described by Black (1928), who referred to them as washouts. Work continues on these deposits.

4. Palaeochannel magnitude

Since channel fills are not seen in the meander belt, deductions as to the magnitude of the palaeochannel involved must be made indirectly from the epsilon cross-stratification. The width and depth of the palaeochannels responsible for the meander loops were measured in vertical cliff sections, where the total width and height of the epsilon units is exposed. Following Allen (1965) the measured mean width of the epsilon units is taken as approximately two-thirds of bankfull width. The mean height of the epsilon units approximates to bankfull depth. Meander radius and meander wavelength values (the latter approximating to twice the radius) were measured directly from aerial photographs using 8 individual meander loops.

The calculated values of channel width, meander wavelength and mean annual discharge are based upon the equations, the method A discussed by Leeder (1973). The magnitude of the palaeochannels responsible is shown in Table 1.

Table 1. Palaeochannel magnitude

	$\bar{d}$	$\bar{w}$	w/d	$\bar{Lm}$	$\bar{Rm}$	$\bar{Q}$	Rm/w
Measured	4	21	5.25	130–180	77.5	—	3.1–4.3
Calculated	—	17–120	4.3–30	436–110	—	0.5–10.3	—
Units	m	m	—	m	m	$m^3\ s^{-1}$	—

$\bar{d}$ = Bankful depth; $\bar{w}$ = bankful width; $\bar{Lm}$ = meander wavelength; $\bar{Rm}$ = meander radius; $\bar{Q}$ = annual discharge.

5. Discussion and conclusions

Accretion of the meander belt point bars took place in highly sinuous streams varying in depth from 3 to 4 m and in width from 15 to 25 m. Meander wavelength was in the range 150 m and mean annual discharges around 0.5–10 $m^3\ s^{-1}$ were probably attained. Decreasing grain size upward and modified structures from small scale trough-cross-bedding to climbing ripple cross-stratification were possibly caused by fluctuating flow regime during falling stage flow. The silty material between the epsilon cross-sets was probably deposited during a period of sluggish flow.

The contorted and convoluted lamination of the upper parts of some point bars may have been caused by gravity sliding after deposition, possibly due to pore water expulsion at low stage or by differential stresses acting in the rapidly accumulating bed. After abandonment of the meander belt probably by channel avulsion, the area was occupied by flood-plain lakes in which clay material accumulated by flooding of adjacent rivers.

Acknowledgements. A grant from Ministry of Sciences and Higher Education of Iran is gratefully acknowledged. Dr M. R. Leeder is thanked for helpful suggestions.

References

Allen, J. R. L. 1962. Intraformational conglomerates and scoured surfaces in the Lower Old Red Sandstone of the Anglo-Welsh curette. *Lpool Manchr Geol. J.* **3**, 1–20.

Allen, J. R. L. 1963. The classification of cross-stratified units, with notes on their origin. *Sedimentology* **2**, 93–114.

Allen, J. R. L. 1965. The sedimentation and paleogeography of the Old Red Sandstone of Angelsey, North Wales. *Proc. Yorks. geol. Soc.* **35**, 139–85.

Allen, J. R. L. 1970. *Physical Process of Sedimentation.* Allen & Unwin, London, 248 pp.

Black, M. 1928. Washouts in the Estuarine Series of Yorkshire. *Geol. Mag.* **65**, 301–7.

Black, M. 1929. Drifted plant-beds of the Upper Estuarine Series of Yorkshire. *Q. Jl geol. Soc. Lond.* **85**, 389–439.

Beutner, E. C., Flueckinger, L. A. & Gard, T. M. 1967. Bedding geometry in a Pennsylvanian channel sandstone. *Bull. geol. Soc. Am.* **78**, 911–916.

Friend, P. F. 1961. The Devonian stratigraphy of north and central Vestpitsbergen. *Proc. Yorks. geol. Soc.* **5**, 39–68.

Friend, P. F. & Moody-Stuart, M. 1972. Sedimentation of the Wood Bay Formation of Spitsbergen: Regional analysis of a late orogenic basin. *Skr. norsk Polarinst.* **157**, 1.

Hemingway, J. E. 1974. Jurassic. *In* Rayner, D. H. & Hemingway, J. E. (Eds): *The Geology and Mineral Resources of Yorkshire*, pp. 161–223. Yorks. Geol. Soc.

Hemingway, J. E. & Knox, R. W. O'B. 1973. Lithostratigraphical nomenclature of the Middle Jurassic strata of the Yorkshire Basin of north east England. *Proc. Yorks. geol. Soc.* **39**, 527–35.

Leeder, M. R. 1973. Fluviatile fining-upwards cycles and the magnitude of palaeo-channels. *Geol. Mag.* **110**, 265–76.

Moody-Stuart, M. 1966. High and low sinuosity deposits with examples from Devonian of Spitsbergen. *J. sedim. petrol.* **36**, 1102–17.

Puigdefabregas, C. 1973. Miocene point-bar deposits in the Ebro Basin, Northern Spain. *Sedimentology* **20**, 133–44.

Department of Earth Sciences
University of Leeds
Leeds LS2 9JT

10

Reprinted from pp. 133, 135–136, 137–143, 144 of *Sedimentology* **20**(1):133–144
(1973)

Miocene point-bar deposits in the Ebro Basin, Northern Spain*

CAYO PUIGDEFABREGAS

Instituto de Estudios Pirenaicos, Jaca, Spain

ABSTRACT

Pyrenean debris dispersed in relation to the Miocene tectonic phase filled the northern part of the Ebro basin. In the studied area (50 km south of Pamplona) the Miocene sediments are represented by fluviatile sequences consisting of flood basin and point-bar deposits forming fining-upward cycles.

The point-bar character of part of these sediments is not only evidenced by the internal sedimentary structures, but also by the fact that the corresponding meander bends, about 200 m in radius, are clearly visible on aerial photographs. In this way a unique opportunity is given to compare directly the vertical profile and the surface pattern of ancient small size point bars.

The internal structure consists of many imbricated and concentrically arranged bundles corresponding to the accretional topography of the point bar. The separation planes between the bundles are erosional and dip toward the channel axis. The internal structure of each bundle is festoon mega-cross-stratification whose direction points toward the convex bank, away from the channel axis. The resulting structure resembles the Epsilon cross-stratification of Allen (1965a) and its origin is the same as postulated by Allen for this kind of cross bedding.

*Presented at the VIIIth International Sedimentological Congress in Heidelberg (1971).

[*Editor's Note:* Material, including figure 1, has been omitted at this point.]

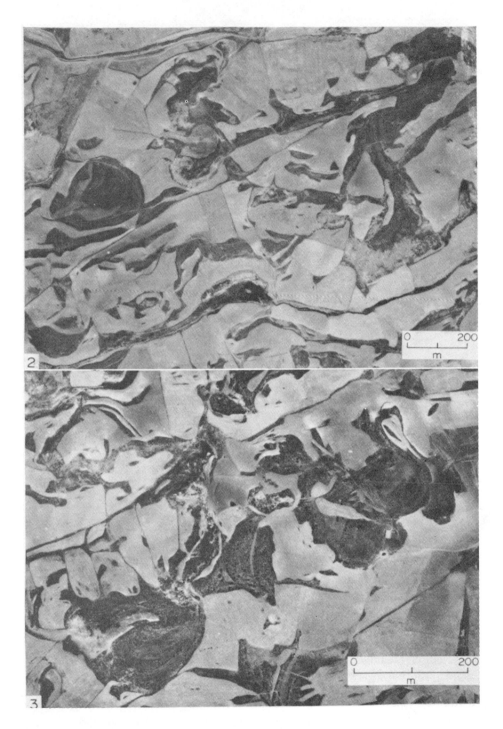

Figs 2 and 3. Aerial view of some meander bends. Dark parts are channel deposits and the flood basin deposits are the light parts, mainly cultivated. Accretion topography of the point bars can be distinguished.

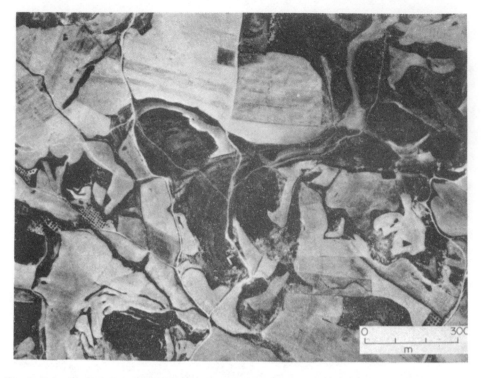

Fig. 4. Point bar near Murillo el Fruto. Because of its exceptional exposure, it is possible to compare the horizontal pattern with the vertical section.

GEOMORPHOLOGY AND OUTCROP SITUATION

The uniqueness of these point-bar deposits is their three-dimensional exposure, The outcrop is, in fact, delimited upwards by a plane of erosion which coincides with the highest terrace of the River Aragón. The resulting relief is of a type already described by Riba, Villena & Quirantes (1967) in other areas of the Ebro basin: the point bars and their channels are more consolidated and resistant and project above the finer alluvial sediments, so that each point bar forms a small mesa.

Similar morphological features occur along a belt about 50 km long between Pitillas and Sádaba, but the clearest examples are found in the Santacara and Murillo area.

The aerial photographs permit recognition of a large part of the channels and it is even possible to see the accretion topography of each point bar (Figs 2, 3 and 4). Unfortunately most details of the deposits do not show up very well in the field and only the example of Murillo is described here because of its superior exposure.

[*Editor's Note:* Material has been omitted at this point.]

THE MURILLO POINT BAR

The point bar exposed near Murillo el Fruto is shown in aerial view in Fig. 4 and sketched in Fig. 5. Apparently the pattern is that of a cut-off, but in fact there are two major bars (A and B) with the corresponding channels next to them. Channel B obviously cuts unit A. Section C–D shows the morphological outline seen on figure. The dip of each bar towards its channel is in the order of 1%.

The dimensions of this small-scale point bar are as in Table 1.

Table 1.

Channels	Point bar	Accretion units
wide: 3–5 m	radius: 200 m	wide: 1–2 m
depth: 1–2 m	thickness: 1–2 m	relief: 25 cm

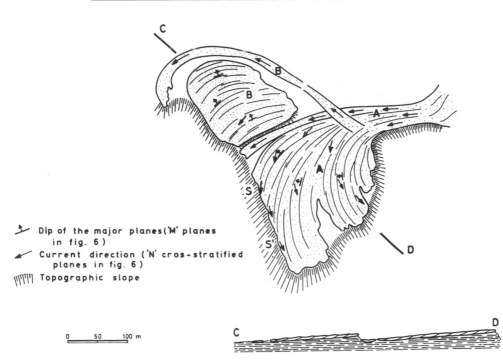

Dip of the major planes('M' planes in fig. 6)

Current direction ('N' cros-stratified planes in fig. 6)

Topographic slope

0   50   100 m

Fig. 5. Sketch of the Murillo point bar showing the principal features described in the text. A profile (C–D) is represented.

The petrographical composition of the sandstones includes a large percentage of rock fragments (80%) of which some 80% are calcareous. Feldspars are almost absent and where some occur, they come from disintegrated sandstone grains. The grain size is up to 1 mm in diameter and sorting is poor. There is some petrographic difference between these sandstones and the finer grained ones of the overbank deposits, in which 30% of quartz and 50% of carbonate grains are found while sandstone fragments are lacking. This increase in quartz content is clearly the result of the disintegration of small sandstone fragments.

The channels surrounding the point bar are composed of very low angle cross-stratified units exhibiting a weak down current dip, though rather irregularly and easily adapted to the channel form.

In aerial view the point bar accretion topography shows a succession of ridges and lows concentrically arranged (Fig. 5). Every ridge has a very marked surface ('M') which dips towards the channel axis of the meander. In contrast, in the lows one can see other surfaces ('N') with a dip oblique to the previous one and towards the convex bank, away from the channel axis.

A vertical section normal to the ridges makes an interpretation of this situation possible. Fig. 6 shows some imbricated units such that the interfaces are the principal planes ('M' in Fig. 5) dipping towards the channel (accretion planes). All of these units exhibit mega-cross-stratification in opposite direction ('N' surfaces in Fig. 5). Planes 'M' are erosional and are the cause of the accretional topography of the point bar. The simplified model, mainly from Allen (1968, 1965a) is shown in Fig. 7. This

Fig. 6. Photographs of the profile of the Murillo point bar (S–S′ in Fig. 5). The channel axis is always at the left. 'M' and 'N' surfaces are the same as in Fig. 5.

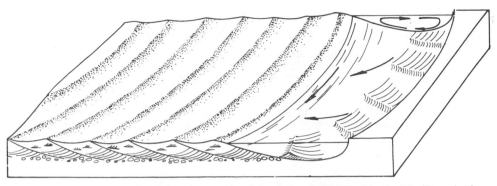

Fig. 7. Simplified model explaining the genesis of the structure described in the Murillo point bar. Mainly after Allen (1965a, 1968).

structure has a resemblance to the one described by Allen (1965a) from fluviatile Devonian deposits in Anglesey (Wales) as Epsilon Cross-Stratification (Allen, 1963). In this respect I would like to refer also to van Straaten's *lateral sedimentation* (1951) and Reineck's *longitudinale schrägschichtung* (1958) which concern similar phenomena, but observed in the intertidal Wadden environment. Latter field experience in diffcrent outcrops demonstrated that 'N' surfaces may vary in direction from parallel to flow to those shown in Fig. 7.

THE VERTICAL SEQUENCE (MURILLO SECTION)

Unfortunately, in the previous example, the vertical sequence comprising an alternation of point bar and floodbasin deposits cannot be examined. For that purpose another section was chosen between Murillo and Santacara (Fig. 1) in which no comparison with the horizontal pattern is possible. The following features are worth mentioning (Fig. 8):

(1) In a 50 m thick sequence there are only three channels with a total of 5 m of channel deposits. The ratio of channel deposits to overbank deposits is 1:10.

(2) Fining upward cycles are clearly visible.

(3) The sediments between the channel deposits are muds and silts with frequent intercalations of fine sand that exhibit small scale current ripples and very often climbing ripples. Plant remains and roots are not significant, but some plant debris occurs in the channel deposits which are often mineralized (copper). Polygonal sand dikes filling shrinkage cracks (Oomkens, 1966) often occur (Fig. 9).

(4) Thick beds are present (up to 2 m) made up of current rippled sand, with all types of climbing ripples (Jopling & Walker, 1968; Walker, 1969) in arbitrary superposition.

(5) Thin graded beds are found (5–10 cm) parallel laminated or unlaminated. On the fine grained top, current ripples may occur. Structurally they resemble turbidites and they may be the product of abnormal overflows.

Fig. 10 shows the lateral extension of channel 'B' in the section of Fig. 8. The structure resembles the point bar of Murillo, but here the upper point bar with small scale ripples is present. The channel migrates from left to right and a structure similar to epsilon cross-stratification occurs. The resulting vertical arrangement is analogous

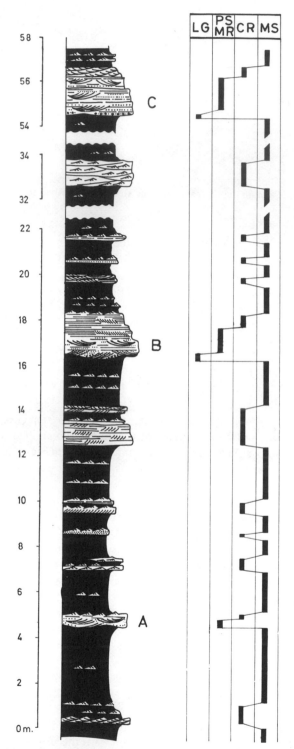

Fig. 8. Section of fining upwards cycles west of Murillo (see Fig. 1).
LG, lag deposits; PS–MR, mega-ripple and parallel stratification; CR, small scale current ripples and climbing ripples; MS, mud and silt with some thin sandstone layers.

Fig. 9. Polygonal sand dikes found in the section west of Murillo. (a) In the top of a fallen boulder; (b) in the correct stratigraphic position.

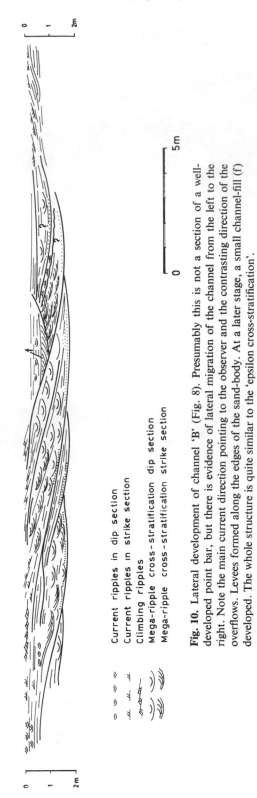

Current ripples in dip section
Current ripples in strike section
Climbing ripples
Mega-ripple cross-stratification dip section
Mega-ripple cross-stratification strike section

Fig. 10. Lateral development of channel 'B' (Fig. 8). Presumably this is not a section of a well-developed point bar, but there is evidence of lateral migration of the channel from the left to the right. Note the main current direction pointing to the observer and the contrasting direction of the overflows. Levees formed along the edges of the sand-body. At a later stage, a small channel-fill (f) developed. The whole structure is quite similar to the 'epsilon cross-stratification'.

to the structure and succession characterizing point bars of large modern meandering rivers. On the far sides of the sand bodies one finds levees weakly inclined to the channel. Parallel-stratification and small channel-fills occur, and also current ripples that climb the levee towards the floodbasin. This suggests that a real overflow took place.

CONCLUSIONS

In summary the characteristics of these sediments are as follows:

(1) Deposition took place in very small streams (generally 1 m in depth and 3–5 m wide).

(2) These were high-sinuosity streams (Moody-Stuart, 1966) with point-bar deposits.

(3) Levee deposits are present and overbank deposition took place.

(4) Epsilon cross-stratification is the characteristic structure of these small point bars.

(5) This kind of structure shows pronounced discontinuity planes which points to distinctly episodic deposition.

(6) The particular type of '*micro point bars*' described is the result of repeated current action with small, variable discharge and high load (important overbank deposits). The point bars were quickly built, but also soon abandoned due to constant change in drainage pattern.

[*Editor's Note:* Material has been omitted at this point.]

REFERENCES

ALLEN, J.R.L. (1965a) The sedimentation and Paleogeography of the Old Red Sandstone of Anglesey, North Wales. *Proc. Yorks. geol. Soc.* 35 (8), 139–185.

ALLEN, J.R.L. (1968) *Current Ripples*. North Holland, Amsterdam.

CRUSAFONT, M., TRUYOLS, J. & RIBA, O. (1966) Contribución al conocimiento de la estratigrafía del Terciario continental de Navarra y Rioja. *Notas Comun. Inst. geol. min. Esp.* 90, 53–76.

CRUSAFONT, M. & PONS, J.M. (1969) Nuevos datos sobre el Aquitaniense del Norte de la Provincia de Huesca. *Acta geol. Hisp.* 4, 124–125.

JOPLING, A.V. & WALKER, R.G. (1968) Morphology and origin of Ripple-Drift Cross-Lamination, with examples from the Pleistocene of Massachusetts. *J. sedim. Petrol.* 38, 971–984.

MOODY-STUART, M. (1966) High and low sinuosity deposits with examples from the Devonian of Spitsbergen. *J. sedim. Petrol.* 36, 1102–1117.

OOMKENS, E. (1966) Environmental significance of Sand Dikes. *Sedimentology,* 7, 145–148.

REINECK, H.E. (1958) Longitudinale Schrägschichtung im Watt. *Geol. Rdsch.* 47, 73–82.

RIBA, O., VILLENA, J. & QUIRANTES, J. (1967) Nota preliminar sobre la sedimentación en paleocanales terciarios de la zona de Caspe-Chipirana (Provincia de Zaragoza). *An. Edafol. Agrobiol.* 26.

SOLER, M. & PUIGDEFABREGAS, C. (1970) Lineas generales de la geología del Alto Aragón Occidental. *Pirineos* 96, 5–20.

VAN-STRAATEN, L.M.J.U. (1951) Texture and Genesis of Dutch Wadden Sea Sediments. *Proc. Third Int. Congr. Sediment.* Gröningen-Wageningen. Netherland. 5–12 July 1951.

WALKER, R.G. (1969) Geometrical analysis of Ripple-Drift cross-lamination. *Can. J. Earth Sci.* 6, 383.

Editor's Comments
on Papers 11 and 12

11 VAN HOUTEN
Excerpts from *Cyclic Lacustrine Sedimentation, Upper Triassic Lockatong Formation, Central New Jersey and Adjacent Pennsylvania*

12 RYDER, FOUCH, and ELISON
Excerpts from *Early Tertiary Sedimentation in the Western Uinta Basin, Utah*

ALLUVIAL-PLAIN FACIES: LACUSTRINE DEPOSITS

Detrital accumulation in an open lake with through-flowing drainage is essentially an intermediate stage between more active fluvial sedimentation on an alluvial plain and chemical sedimentation in a closed lake. Ancient deposits of this sort are rather rare, however, even though the largest of them developed in structural basins where the delicately adjusted control prevailed for as much as several million years, producing sequences more than a thousand meters thick. Commonly the detritus accumulated in small deltas and associated bottomset facies, but in some deeper lakes the sediment was transported to the basin center by density currents (Sanders 1968; Negendank 1972). Picard and High (1972a) have summarized criteria for recognizing ancient lacustrine deposits.

The few up-to-date descriptions of detrital lacustrine sequences include studies of the Devonian Middle Old Red Sandstone in northeastern Scotland (Donovan and Foster 1972), late Paleozoic sediments in maritime Canada (Belt 1968) and in Germany (Negandank 1972), early Mesozoic deposits in Connecticut (Sanders 1968; Hubert et al. 1976) and maritime Canada (Klein 1962), early Cenozoic sediments in the Rocky Mountain basins (Picard and High 1972b), and mid-Cenozoic deposits in southern England (Daley 1973).

Papers 11 and 12 present an interesting contrast in style of description and analysis, dictated in part by the nature of the outcrops and the data available. Late Triassic lacustrine deposits in New Jersey and adjacent Pennsylvania (Paper 11) occur in asymmetrical detrital and chemical cycles that resulted from expansion and waning of the lake (Van Houten 1969). Calcareous muddy sediments deposited by through-flowing drainage accumulated in the coarsening-upward detrital cycles. Commonly these sediments were disrupted by shrinkage cracks and mottled by small burrows. Some cycles contain lenses of cross-laminated siltstone and very fine-grained sandstone that apparently accumulated in prograding lobes of small deltas. Ryder et al. (1976) have reconstructed early Cenozoic open lacustrine, marginal lacustrine, and alluvial facies in northern Utah (Paper 12), based in large part on detailed stratigraphic analysis. Like the Triassic lacustrine sequence in New Jersey, the Eocene one in Utah records fluctuations of the alluvial-plain and lacustrine facies as well as of detrital and chemical sedimentation in the lake.

REFERENCES

Belt, E. S. 1968. Carboniferous continental sedimentation, Atlantic provinces, Canada. *Geol. Soc. America Spec. Paper 106*, pp. 127–176.

Daley, B. 1973. Fluvio-lacustrine cyclothems from the Oligocene of Hampshire. *Geol. Mag.* **110:**235–242.

Donovan, R. N., and Foster, R. T. 1972. Subaqueous shrinkage cracks from the Caithness Flagstone Series (Middle Devonian) of northeast Scotland. *Jour. Sed. Petrology* **42:**309–317.

Hubert, J. F.; Reed, A. A.; and Carey, P. J. 1976. Paleogeography of the East Berlin Formation, Newark Group, Connecticut Valley. *Am. Jour. Sci.* **276:**1183–1207.

Klein, G. deV. 1962. Triassic sedimentation, Maritime provinces, Canada. *Geol. Soc. America Bull.* **73:**1127–1146.

Negendank, J. F. W. 1972. Turbidite aus dem Unterrotliegenden des SaarNahe-Gebietes. *Neues Jahrb. Geologie u. Paläontologie Monatsh.* **9:**561–572.

Picard, M. D., and High, L. R. 1972a. Criteria for recognizing lacustrine rocks. *Soc. Econ. Paleontologists and Mineralogists Spec. Pub. 16*, pp. 108–145.

—— 1972b. Paleoenvironmental reconstruction in an area of rapid facies change, Parachute Creek Member of Green River Formation (Eocene), Uinta Basin, Utah. *Geol. Soc. America Bull.* **83:**2689–2708.

Sanders, J. E. 1968. Stratigraphy and primary sedimentary structures of fine-grained, well-bedded strata, inferred lake deposits, Upper Triassic, central and southern Connecticut. *Geol. Soc. America Spec. Paper 106*, pp. 265–305.

11

Reprinted from pp. 497, 501–511, 530–531 of *Kansas Geol. Survey Bull.* **169:**497–531
(1964)

Cyclic Lacustrine Sedimentation, Upper Triassic Lockatong Formation, Central New Jersey and Adjacent Pennsylvania

FRANKLYN B. VAN HOUTEN   *Princeton University, Princeton, New Jersey*

ABSTRACT

Upper Triassic Lockatong lacustrine deposits in central New Jersey and adjacent Pennsylvania are arranged in short asymmetrical "detrital" and "chemical" cycles that resulted from expansion and waning of the lake, presumably due to cyclic variation in climate.

Detrital short cycles, averaging 14 to 20 feet thick, comprise several feet of black shale succeeded by platy dark-gray, carbonate-rich mudstone in the lower part and gray, tough, massive calcareous silty mudstone in the upper. The massive mudstone has a small-scale contorted fabric produced largely by crumpled shrinkage cracks and burrows. Thicker, coarser grained detrital cycles contain 2 to 5-foot layers and lenses of thin-bedded, commonly cross-stratified siltstone and very fine grained sandstone, locally with small-scale convoluted bedding.

More common chemical short cycles average 8 to 13 feet thick. Lower beds are alternating dark-gray to black, platy dolomite-rich mudstone and marlstone 1 to 8 cm thick, extensively broken by crumpled shrinkage cracks. Locally initial deposits are crystalline pyrite or calcite as much as 2 cm thick. In the middle, several feet of dark-gray mudstone encloses 2 to 8 cm-thick layers of gray marlstone disrupted by syneresis. The upper part is gray, tough, massive analcime- and carbonate-rich mudstone containing as much as 7 percent soda and as little as 47 percent silica, and a maximum of about 35 to 40 percent analcime. The mudstone is brecciated on a micro-scopic scale, probably the product of syneresis. Much of the mudstone is also disrupted by slender crumpled shrinkage cracks irregularly filled with crystalline dolomite and analcime.

Some thinner chemical cycles are grayish red, especially in the uppermost part of the formation. These contain layers of greenish-gray weathering mudstone variously disrupted by shrinkage cracking. Thinner beds are broken into mosaic intraformational breccia; thicker ones are disrupted by long intricately crumpled cracks. Small lozenge-shaped pseudomorphs of dolomite and analcime after gypsum? or glauberite? are common in the grayish-red mudstone. Analcime and dolomite are also concentrated in shrinkage cracks and between mosaic flakes in dark dusky-red mudstone.

Varve-counts of black mudstone suggest that short cycles resulted from 21,000-year precession cycles. Bundles of detrital and of chemical short cycles occur in intermediate cycles 70 to 90 feet thick; these in turn occur in long cycles 325 to 350 feet thick. The patterns apparently resulted from alternating wetter and drier phases of intermediate and long climatic cycles, producing through-flowing drainage and a bundle of detrital cycles or a closed lake and a bundle of chemical cycles.

[*Editor's Note:* Material, including figures 1 and 2, has been omitted at this point.]

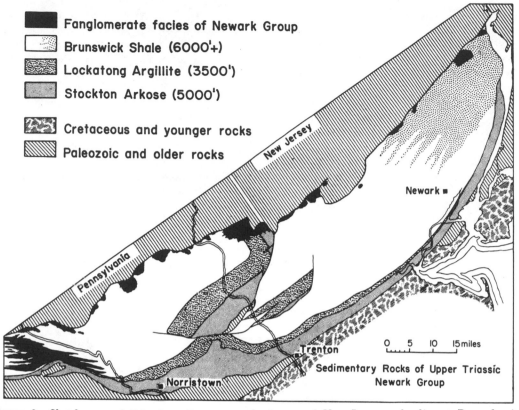

FIGURE 3.—Sketch map of Triassic sedimentary rocks in central New Jersey and adjacent Pennsylvania, showing relation of Lockatong Formation to distribution of fanglomerate facies and conglomeratic deposits of Brunswick Formation.

block consist principally of dark grayish-red and greenish-gray tough, micaceous mudstone, siltstone, and minor very fine grained sandstone. These deposits are characterized by small-scale irregular bedding and ripple- and cross-lamination. Moreover, the bedding is extensively disturbed by mudcracks and animal burrows, but bedding planes are seldom distinctly ripple marked.

(2) The characteristic rock type of the formation is tough, massive, homogeneous, very fine grained to aphanitic mudstone (argillite as a field term), most of which is medium to dark gray. Some of it is reddish brown to grayish red. This rock is used for building blocks and crushed stone, and is the type seen most commonly in outcrops.

It comprises two varieties which occur in different kinds of short cycles. Both exhibit extensively disturbed or disrupted fabric on a small scale.

(*a*) One kind which breaks with a hackly

ROCK TYPES

(1) Rocks in the southwestern extent of the formation and in the lower 400 to 500 feet of the thick section in the northwestern fault

145

or gnarly fracture is a detrital deposit composed of calcareous, feldspathic siltstone and silty mudstone with only a minor amount of quartz. It has indistinct irregular to wispy carbonaceous laminae that are extensively disturbed and generally deformed into a small-scale irregularly contorted fabric (Fig. 4C).

(b) The other kind of argillite is chiefly a colloidal-chemical deposit that is more brittle and breaks with a subconchoidal fracture. This rock type is rich in carbonate minerals of which dolomite is the more common. It contains as little as 47 percent silica and as much as 7 percent soda which is concentrated largely in analcime and albite. Analcime-rich argillite predominates in the upper part of the formation where some of it is grayish red. Because of its variation in color, colloidal-chemical argillite is the preferred building stone.

The groundmass generally has a unique brecciated aspect on a microscopic scale, and commonly is speckled with white patches of crystalline analcime and carbonates, producing a fabric similar to that of "birdseye" limestone.

(3) The rock type second in abundance in the Lockatong Formation is platy, very dark-gray to black, laminated, carbonate-rich mudstone and marlstone in which dolomite commonly predominates. Some of these rocks are varved (Van Houten, 1962, Pl. 2A), and many are extensively disrupted by shrinkage cracks (Fig. 15A, 15D, 15E).

(4) Thin-bedded, calcareous, well-sorted feldspathic siltstone with minor very fine grained sandstone occurs in 2 to 5 foot-thick layers and broad lenses in detrital argillite

(type 2a). Only very rarely is it associated with colloidal-chemical argillite (type 2b). This rock is characterized by distinct black micaceous laminae and irregular small-scale ripple-bedding, lenticular cross-bedding, convoluted bedding, and rare graded bedding (Fig. 5A, 5B, 5C). Microstylolites and isolated animal burrow-casts are common, and many of the thin black laminae are broken by slender, indistinct mudcracks and marked by small, indeterminate "tracks."

(5) Laminated black silty, calcareous shale, commonly with sublenticular laminae (Fig. 4A, 4B) constitutes a minor amount of the formation. Reddish-brown shale occurs in the uppermost part of the formation where it interfingers with the Brunswick Shale.

All of these rock types normally contain illite and some chlorite, albite, calcite or dolomite, and scattered pyrite. Quartz and potash feldspars are unusually scarce. Toward the southwest end of the basin, however, quartz is more common. The extent to which albite in the unmetamorphosed colloidal-chemical argillite is authigenic has not been determined.

SHORT CYCLES

GENERAL STATEMENT

An orderly asymmetrical repetition of the common rock types, reflected in weathered profiles, is accompanied by a regular vertical variation in color, composition, and sedimentary structures. Approximately 80 short cycles are present in a well-exposed 1,400-foot section of the middle part of the formation in a long road cut on New Jersey Highway 29,

━━━━▶

FIGURE 4.—Sedimentary features of detrital short cycles. *A*, Photomicrograph of lower black shale. Lower half composed of undisturbed, commonly discontinuous, black carbonaceous films in silty mudstone. Upper part massive, fine-grained siltstone with scattered pyrite, mottled by indistinct burrow-casts 0.1 to 0.2 mm in diameter. $\times$ 4.2. *B*, Photomicrograph of lower black shale with delicately disturbed bedding. Lower two-thirds composed of irregular films and laminae of black, dolomitic mudstone and sublenticular laminae of finely crystalline calcite; bedding apparently disturbed by organisms. Thicker bundles of black films recur about 0.4 to 1 mm apart. Upper third consists of several layers of black dolomitic mudstone with abundant scattered calcite crystals, disturbed by burrowers. Black lens with speckled center probably a coprolite. $\times$ 3.9. *C*, Photomicrograph of upper dark-gray, massive, silty mudstone with extensively disturbed fabric. Lighter, silty structure in lower left quarter may be burrow-cast. Black patch near center is pyrite. $\times$ 3.1. *D*, Photomicrograph of upper dark-gray, silty mudstone with siltier crumpled burrow-casts, and mottled or bioturbated fabric. Black spots are pyrite. $\times$ 3.7. *E*, Lower black shale with delicately disturbed films and laminae of black mudstone (white in directly reflected light) interbedded with very finely crystalline calcite (black in photograph). Thicker bundles of black films recur about 0.5 to 1.5 mm apart. Large black nodules are coprolites. $\times$ 1.7. *F*, Upper dark-gray, massive, silty mudstone extensively disrupted by siltier shrinkage crack casts offset by horizontal compaction shearing and by vertical zones of brecciation produced by upward-moving water and fluid mud. Upper part, medium-gray, siltier, and calcareous. White specks in lower part are pyrite, locally in shrinkage cracks and in thin zones of shearing. $\times$ 0.62.

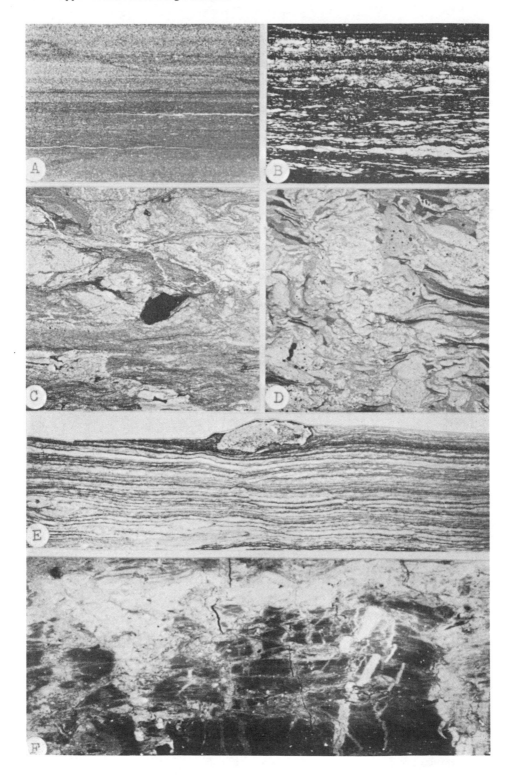

Figure 5.—Sedimentary features of detrital short cycles. *A*, Upper massive dark-gray silty mudstone with abundant slender crinkled shrinkage-crack casts; overlain by calcareous, cross-bedded well-sorted siltstone. × 0.54. *B*, Upper massive dark-gray silty mudstone mottled or bioturbated by organic burrowing. Overlain by calcareous ripple- and cross-bedded well-sorted siltstone with scouring and small load casts at base. × 0.54. *C*, Interbedded light-gray well-sorted siltstone and dark-gray mudstone of siltstone unit in upper part of cycle. White spots are pyrite mostly concentrated along base of beds. Contorted bedding in middle may be organic disturbance. × 0.58. *D*, Interbedded light-gray well-sorted siltstone and dark-gray mudstone of siltstone unit in upper part of cycle. Layers with distinct convoluted bedding and streaked-out "ripples." Load cast? in lower right corner and marked scouring of dark-gray mudstone below. × 0.58. *E*, Interbedded light-gray well-sorted siltstone and dark-gray mudstone of siltstone unit in upper part of cycle. Layers of cross-bedding and convoluted bedding with streaked-out "ripples." Several solitary burrow-casts parallel to bedding and a large cast at left rising vertically. × 0.58.

6 to 8 miles south of Frenchtown on the Delaware River (Fig. 2A, 19A). Short sequences of cycles are also well displayed in several large quarries in southeastern Pennsylvania (Fig. 2B, 2C, 2D; Van Houten, 1962, Pl. 1B).

In general form each short cycle consists of lower black shale and platy mudstone and marlstone, and of upper massive mudstone commonly mudcracked on top. There is no evidence of erosion between cycles but the most marked change in rock type occurs between the uppermost argillite and the basal shale or platy mudstone of the succeeding cycle. Observed in detail short cycles are of two rather distinct types, here informally referred to as detrital and chemical varieties (Fig. 6).

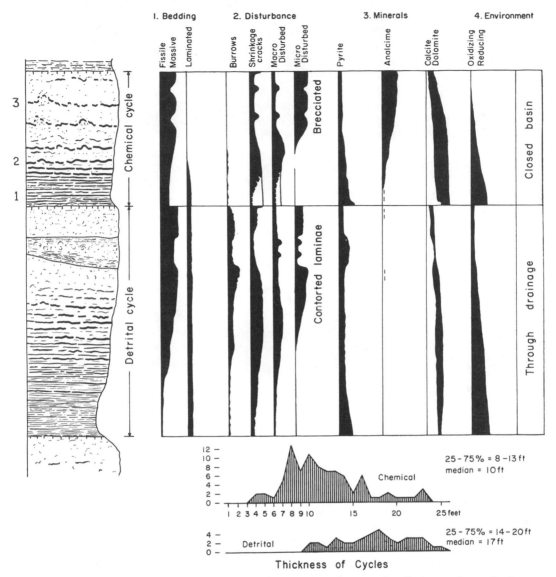

FIGURE 6.—Model of detrital and chemical short cycles, showing distribution and qualitative estimates of prevalence of sedimentary structures, selected minerals, and sedimentary environments; data on thickness of cycles. Numbers along left side of chemical cycle indicate units referred to in text.

DETRITAL VARIETY

The lower several feet of most detrital cycles consists of black shale succeeded by platy black to very dark-gray mudstone (laminae 0.5 to 2 mm thick) and interbedded tan-weathering disrupted, dark-gray, calcitic or dolomitic marlstone (laminae 0.1 to 2 mm thick) in alternating layers 1 to 6 cm thick. The fissile and platy beds normally are marked by rare short burrow-casts, but some thin beds of better-sorted siltstone are mottled by very small burrows (Fig. 4A). The upper part of a cycle is tough, massive dolomitic or calcitic silty mudstone and siltstone with indistinct irregular laminae. Only rarely are there distinct laminae 0.2 to 1 mm thick and some of these are graded. Instead, most of these rocks have been churned up on a small to microscopic scale to produce a contorted and disrupted fabric (Fig. 4C, 4D, 4F, 5A,

5B). In addition they commonly contain a 2 to 5-foot lens of rather well-sorted calcitic, feldspathic, laminated siltstone and rare very fine grained sandstone, much like deposits in the lower 500 feet of the formation (Fig. 5C, 5D, 5E).

Minute specks and larger crystals of pyrite are widespread in detrital cycles. Deep impressions of cubes of pyrite as much as 2 mm wide on surfaces of black laminae both above and below suggest growth of the crystals before complete induration of the mud. Coarsely crystalline patches of pyrite as much as 2 cm long occur in siltstone lenses, presumably as diagenetic concentrations in the more porous deposits.

Detrital cycles generally range from 14 to 20 feet in thickness, comprising the thicker short cycles; but they constitute only about one-fourth of the short cycles in the thickest section of the formation in northwestern New Jersey (Fig. 2A). In contrast, throughout the western third of the formation, west of Montgomeryville, Pennsylvania, only detrital cycles are present and these contain more quartz than occurs in cycles in the central part of the basin.

CHEMICAL VARIETY

A second, more common type of short cycle differs principally in containing less detrital and more colloidal-chemical sediment. These cycles generally range from 8 to 13 feet in thickness.

The lower part consists of interbedded, commonly disrupted black to very dark-gray, platy, laminated, dolomitic and calcitic mudstone and marlstone (Van Houten, 1962, Pl. 1C) and rare layers and lenses of crystalline carbonate (Fig. 7). The upper part is massive analcime- and carbonate- (predominately dolomite) rich mudstone brecciated on a microscopic scale. Presumably this nonlaminated mudstone resulted from flocculation of a colloidal sediment in the presence of a concentration of cations which produced random orientation of the clay minerals (White, 1961, p. 561). The analcime in these cycles was derived largely from saline water of a closed lake in which the mud itself was laid down. It may have been more nearly diagenetic than syngenetic, having been formed below the surface of the bottom mud, but it was introduced before the rock was lithified or deeply buried.

Many of the platy mudstone layers in the lower part of chemical cycles are composed of regularly alternating black and light-gray, carbonate-rich laminae. These couplets, which range from 0.05 to 0.2 mm in thickness, resemble carbonate- and organic-rich varves in the Eocene Green River lacustrine deposits (Bradley, 1930, p. 95-96). Basal beds of platy gray mudstone in some cycles consist of graded laminae of silt and clay ranging in thickness from 0.2 to 2 mm.

FOSSILS

Fossils found in the Lockatong Formation record a fauna conspicuously devoid of large aquatic bottom dwellers. It comprises, instead, mostly remains of elasmobranch, coelocanth, and palaeoniscid fishes, estheriids and ostracodes, plant fragments (cycads, equisitales, ferns, and possibly dasycladacean algae), and indeterminate microscopic spines or setae, as well as coprolites as much as 6 cm long and insect larvae faecal pellets about 2 mm long. In addition, there are traces of several kinds of small reptiles, phytosaurs, and large squat amphibians. There are also rare indistinct large footprints, presumably of large phytosaurs and tracks of small dinosaurs, rare small tetrapod tracks and minute indeterminate "tracks and trails."

Most of the fossils were recovered from the shaly and platy beds in the lower part of cycles, but estheriids occur in abundance in the upper part of detrital cycles as well. Plant remains are much more rare than in most lacustrine deposits, and efforts to recover spores or pollen from the mudstone have yielded very little.

Isolated silty casts of animal burrows 1 mm or less in diameter are scattered through the lower shale and platy mudstone. Rarely, irregular laminae of siltstone 1 to 2 mm thick in the black shale consist almost entirely of minute crumpled casts. In detrital cycles crumpled burrow-casts 1 to 3 mm in diameter are abundant in the upper gray muddy siltstone (Fig. 4D; Van Houten, 1962, Pl. 2C) which commonly is indistinctly mottled (*see,* Moore and Scruton, 1957, p. 2725-2731). In associated better-sorted coarser siltstone and very fine grained sandstone burrow-casts are rare, and bioturbation by bottom crawlers is preserved locally. Beds of reddish-brown to grayish-red mudstone and siltstone in the uppermost part of the Stockton Formation, in the lower 500 feet of the Lockatong Formation, and in the Brunswick Formation contain abundant burrow-casts as much as 1 cm in diameter. Locally these strata are mottled.

The fact that the fish faunas from the three formations of the Newark Group differ only in detail (Bock, 1959, p. 130-132) implies that the Lockatong deposits, like the Stockton and Brunswick Formations, are of nonmarine origin (Schaeffer, 1952, p. 58).

Most of the fish, amphibians, and reptiles found in the Lockatong Formation were recovered from its more marginal southwestern and northeastern parts, or from its lowest part near the middle. Although this distribution may have been controlled largely by available outcrops very few vertebrate remains have been found in quarries in the upper part of the formation in western New Jersey. According to these data, fossils occur mainly in detrital cycles and there apparently was a paucity of vertebrate life in the central part of the Lockatong lake.

INTERPRETATION

Of all the distinctive sedimentary features of the Lockatong Formation, crumpled shrinkage-crack casts and bedding disturbed by burrowing or by hydroplastic disruption are the most ubiquitous and varied, and generally point to slow accumulation of mud. In contrast, ripple-marked strata are rare. A greater

———————————————— ➤➤

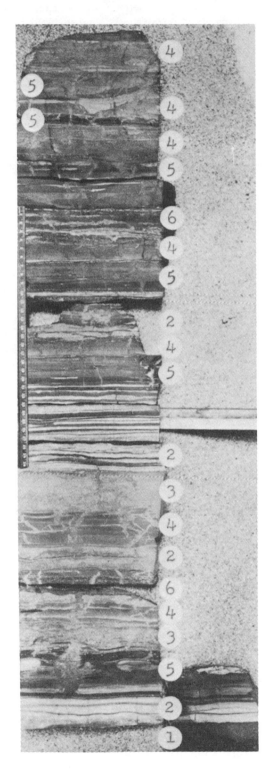

FIGURE 7.—Lower platy black mudstone and gray marlstone of gray chemical short cycle, showing various types of disruption by shrinkage. (1) Very dark-gray analcime-rich mudstone at top of preceding cycle. (2) Very light gray crystalline calcite with layers of dark-gray dolomitic mudstone. (3) Light-gray, extensively disrupted dolomitic marlstone with scattered small patches of calcite. (4) Medium-gray brecciated dolomitic marlstone with shrinkage-crack casts of gray calcitic and dolomitic or of black, dolomitic mudstone, offset laterally by compactional shearing. (5) Very dark gray to black dolomitic mudstone with medium-gray, dolomitic crumpled shrinkage-crack casts. (6) Black dolomitic mudstone with densely scattered crystals of calcite and medium-gray, dolomitic shrinkage-crack casts. Vertical zone of brecciation by upward-moving fluid in lower block. Specimen is 81 cm high.

thickness of most detrital cycles compared with chemical ones reflects a more rapid rate of detrital deposition.

In both lithic features and fauna the Lockatong Formation shares many characters with the mid-Devonian Caithness Flagstone of Scotland (Crampton, 1914), Mississippian Albert Shale of New Brunswick (Greiner, 1962), Permo-Pennsylvanian Dunkard Group in Pennsylvania, West Virginia, and Ohio (Beerbower, 1961), Triassic Blomiden Formation of Nova Scotia and New Brunswick (Klein, 1962a), Triassic Keuper rocks (Bosworth, 1912, p. 51-116; Elliot, 1961), and Eocene Green River Formation (Bradley, 1931), each largely of lacustrine origin.

Detailed comparison of the Lockatong lacustrine deposits with these formations emphasizes the uniqueness of the chemical cycles and the features they share with the distinctive 5 to 10-foot cycles in the Green River Formation. Lockatong detrital cycles, on the other hand, are more like lacustrine and alluvial Dunkard cyclothems and cyclic deposits in the Papigoe beds and Thurso flagstone of the Caithness Flagstone. In addition, detrital cycles share significant features with some deposits of marine tidal lagoons (Van Straaten, 1954). Nevertheless, the lagoonal deposits differ in possessing more numerous small channel and gully lenses pointing to some current action, abundant load casts, wave ripple marks, and ripple bedding (but only minor current ripple marks), a marine fauna, and shell beds, and abundant drifted plant fragments.

Members of the detrital and chemical short cycles record an asymmetrical sequence of events resulting from expansion and waning that recurred relentlessly during the life of the Lockatong lake. Each cycle began with the slow filling of the basin to form swamps and marshy ponds, and then extensive shallow lakes. During this initial stage mud accumulated in a reducing environment noxious to large bottom dwellers and burrowers. Gradually the inflow increased and the redox potential decreased. Then inflow waned and some cycles ended with exposure of the lake bottom; yet no erosion ensued, thus suggesting but little hiatus between cycles.

Reconstruction of an extensive shallow lake

in which only slight differences in elevation of the basin floor could cause widespread differences in conditions of deposition implies that members of a cycle may vary from place to place. Moreover, the paucity of bottom scour in these shallow-water sediments requires that there was no single large lake where wind-driven waves could easily have been generated, or that some deterrent, such as aquatic vegetation, restricted wave action.

In their study of two Pennsylvanian black shales Zangerl and Richardson (1963, p. 117-122) have recently explored the problem of accounting for an absence of disturbance of bottom mud below shallow water. As a solution they have elaborated the role of an algal flotant, a mat of floating vegetation as much as 3 feet thick which protected the mud from wave action. The flotant may also have prevented spores and leaves from reaching the bottom (p. 218-219) while it showered abundant microscopic plant debris on the mud. A flotant may have had a similar role in the Lockatong lake, but there is no direct evidence of its presence. In fact, Lockatong mudstone contains no such abundance of decomposition products of plants as occurs in the Pennsylvanian shales (p. 105-106). Moreover, a flotant could not have been present during oxidizing episodes that produced grayish-red chemical cycles.

The present attempt to account for the short cycles assumes that both detrital and chemical varieties had a common control, that the base of each is correctly identified, and that they accumulated in a lake. In its general setting the thick Newark Group accumulated in a continuously sinking basin supplied by a rising source area to the east or southeast as well as to the northwest. Secular fluctuations imposed on this framework produced the distinctive Lockatong cycles. But no obvious interpretation follows. Not only are there several possible causes of cyclic sedimentation but authors disagree as to which one may have been the fundamental control of a particular cyclic deposit (Weller, 1956; Goodlet, 1959; Wells, 1960; Beerbower, 1961).

Cyclic sedimentation has been attributed to continuous subsidence accompanied by *periodic compaction* (Van der Heide, 1950), or by *shifting of the pattern of sedimentation*

(Goodlet, 1959; Duff and Walton, 1962), as well as to a more regional tectonic control, either *periodic subsidence* (Trueman, 1948; Zangerl and Richardson, 1963) or *repeated uplift and submergence* as envisaged by Weller (1956). In contrast, theories of worldwide control have called upon *eustatic rise and fall* of sea level (Ham, 1960) or *cyclic variation in climate* accompanied by compaction in a continuously sinking basin (Brough, 1928; Beerbower, 1961).

My prejudice in favor of climatic control, rather than periodic sinking, stems largely from an assumption that alternations in climate are apt to be more regular and persistent than intermittent subsidence (*see*, Gilbert, 1895, p. 123, 127). Moreover, they account more reasonably for the sequence of members in the Lockatong short cycles. Admittedly, detrital short cycles alone could be attributed to tectonic control, but the chemical ones point, instead, to a climatic control. Significantly, climatic cycles have recently been advocated by Beerbower (1961) as the basic control of alluvial and lacustrine Dunkard cyclothems of Permo-Pennsylvanian age, Wanless (1963) has favored climatic oscillations and glacial episodes as basic causes of late Paleozoic cyclothems, and Elliot (1961) has ascribed cycles in the Triassic Keuper Series to a climatic control.

A tentative estimate of the duration of a short cycle, based on varves in the lower part, gives an average of 22,000 years, which agrees well with the 21,000-year precession cycle. Support for this interpretation is afforded by similar cycles in the Eocene Green River lacustrine deposits of Wyoming and Colorado. Varve-counts of these 5 to 10-foot cycles indicate an average duration of 21,630 years, suggesting control by the precession cycle (Bradley, 1930, p. 105-106).

VARIATIONS IN SHORT CYCLES

The characteristic features of detrital and chemical short cycles vary considerably in detail. Analysis of these variations yields basic data for a paleogeographic reconstruction (*see,* Duff and Walton, 1962).

DETRITAL CYCLES

Thickness

The detrital cycles measured range in thickness from 10 to 25 feet (Fig. 6). The 25 to 75 percent class is 14 to 20 feet thick; the median is 17 feet.

Stratification and grain size

Lamination of the lower fissile, black, pyritic mudstone varies considerably. Some of the mudstone with indistinct black films occurs in beds 0.5 to 3 mm thick (Fig. 4A). Very rarely there are minor scours 5 mm deep with tiny streaked-out "ripples" at the bottom. Distinct lamination 0.02 to 1 mm thick varies from a succession of delicately wrinkled black laminae with scattered silt and minute calcite crystals to wrinkled and delicately disturbed films and bundles of films of black mud with sublenticular laminae of silt or very finely crystalline calcite (Fig. 4B, 4E). Thicker bundles of black films commonly recur about 0.4 to 1.5 mm apart. In some beds the calcite and silt occur as sharply defined layers, stringers, and lenses 0.05 to 0.5 mm thick. Only very locally are laminae of black mud distinctly graded.

The pyritic, black shale in the lower part of detrital cycles apparently accumulated in marshy ponds and lakes as did the thin coaly shale (humulite) overlying the coal of some Pennsylvanian cyclothems of Illinois (Zangerl and Richardson, 1963, p. 69-74, 225-226). In contrast, the humulite contains abundant shells and microscopic flaky plant debris, and it is overlain by a transgressive marine facies of gray to black shale that was deposited in marginal lagoons along a deltaic coastal plain (p. 24, 226-227).

Thicker detrital cycles are consistently siltier in the upper part which contains a distinct 2 to 5-foot unit of gray, laminated (laminae commonly 0.2 to 1 mm thick, rarely 3 mm thick), well-sorted siltstone and fine-grained sandstone. Within each unit some of the siltstone and sandstone has irregularly paralled lamination, some has small-scale, ripple- and cross-lamination (Fig. 5C, 5D, 5F) and rarely some of the coarsest sandstone has crudely graded beds 1 to 2 mm thick. In some cycles the siltstone unit is scoured broadly a few feet into the darker mudstone below and contains very dark-gray intraformational mud clasts 1 to 10 mm long.

These coarse units mark the episode of maximum inflow, producing repeated scour (Fig. 5A, 5B) and spread of relatively well-sorted silt and very fine grained sand. Normally the coarse unit in the upper silty mudstone does not extend to the top of a cycle, suggesting a decrease in transportation energy toward the end of a cycle.

Thinner detrital cycles commonly begin with platy mudcracked mudstone and dolomitic marlstone, and have no distinct upper siltstone unit; the thinnest ones consist only of mudstone in the upper part, approaching chemical cycles in their sedimentary features. In fact, some cycles about 10 feet thick are intermediate in character between detrital and chemical varieties and contain traces of analcime.

Disturbed bedding

Inorganic.—Platy marlstone and mudstone in the lower part of detrital cycles commonly are conspicuously disrupted by shrinkage cracks of several different patterns like those in the platy lower part of chemical cycles.

Silty mudstone in the upper part of detrital cycles is marked by abundant long, slender, crumpled shrinkage-crack casts (Fig. 5A) and very irregular vertical brecciated zones 2 to 5 mm wide that apparently were paths of upward moving water and fluid mud (Fig. 4F). Disrupted bedding surfaces appear "shattered" by a complex of randomly oriented, slender, straight crack-casts rather than by a simple polygonal pattern. Locally, specks of pyrite are more abundant in crack-fillings and vaguely outline a mosaic type of intraformational breccia with all the fragments parallel to bedding. Much of this small-scale shrinkage disruption may have occurred under shallow water (*see*, Van Straaten, 1954, p. 38).

In profile, crumpled casts of silt vary from rare to abundant, and from obvious crack-fillings to obscure structures (Fig. 4C, 4F, 5A). Some may be crumpled siltstone dikes

(Shelton, 1962). Some may fill pockets and passageways formed by escaping gas (Cloud, 1960) or water released during compaction. Without adequate evidence, however, they cannot be distinguished from obscure shrinkage-crack casts or from animal burrow-casts.

Commonly, crumpled structures have been offset laterally (in profile) a few millimeters by shearing parallel to bedding on planes 2 to 15 mm apart (Fig. 4F), similar to slip-layers in fine-grained detrital deposits in the Keuper Series (Elliot, 1961, p. 199-202). Presumably the shearing resulted from local lateral flowage during dewatering and compaction. Vertical displacement of offset crumpled casts points to further compaction after shearing.

In a few detrital cycles bedding surfaces in the upper silty mudstone are marked by shrinkage cracks as much as 5 cm wide that outline polygons as much as 35 cm wide, like those "suncracks" in mudstone of the Caithness Flagstone (Crampton, 1914, Pl. VI, 2).

Many of the siltstone and fine-grained sandstone units are marked by layers of conspicuous irregular bedding (Fig. 5C, 5D, 5E), and by deformed load casts of silt protruding into underlying silty mudstone (Fig. 5C). Minor scouring and slumping or flowage over a distance measured in millimeters are common, and delicately convoluted beds with folds about 2 to 5 mm high are associated with streak-out "ripples" of the underlying mudstone (Fig. 5E). These features probably resulted from frictional drag of silt- and sand-laden currents on soft but cohesive mud over which the sediment was spread (Sanders, 1960), and in part, perhaps, from vertical loading (McKee and others, 1962). As pointed out by Sanders (p. 414), convoluted laminae in other deposits, such as the Keuper Series (Elliot, 1961, p. 202-204), are characteristically limited to sediments of silt and very fine grained sand sizes.

Rarely, anastomosing sharp-crested, slightly asymmetrical current ripplemarks with secondary crests are preserved on the surface of 1 to 2 cm-thick beds of well-sorted siltstone that are marked by conspicuous ripple-bedding, intraformational mud-chips, and scouring at the base. The main ripples have wave lengths of about 3 to 3.5 cm and amplitudes of 3 to 4 mm (ripple index = 8 to 12). Secondary crests are displaced somewhat toward the downcurrent side of each ripple.

In a few of the siltstone and sandstone units in the upper part of detrital cycles, as well as in similar rock types in the lowest part of the formation, lamination is disrupted by irregular vertical zones of brecciation as much as 2 cm wide and 8 cm high. At their bases these structures join shrinkage-crack casts, and like smaller ones in the silty mudstone (Fig. 4F), probably resulted from dilatation of sand and silt by gas, water, and fluid mud moving upward along shrinkage cracks during compaction.

The assemblage of structures in the siltstone and sandstone units is remarkably like that in cyclothemic Carboniferous deposits in England (Greensmith, 1956) and the regressive Dakota deposits in North Dakota (Shelton, 1962).

Organic.—Although much of the crumpled fabric of the silty mudstone is the result of inorganic disturbance some irregularly contorted structures are animal burrow-casts 1 to 3 mm in diameter that seldom have septal laminae and can be traced only a few centimeters (Fig. 4C). In siltier mudstone, burrow-casts crumpled by compaction are more abundant (Fig. 4D, 5B), as they are, for example, in the siltier Baggy Beds of the Old Red Sandstone (Goldring, 1962, p. 242, 248). Locally, the siltstone is indistinctly mottled. In many cycles, muddy siltstone with abundant burrowing (bioturbation or mottling) is interbedded with units of parallel- and cross-laminated well-sorted siltstone and fine-grained

sandstone that slightly truncate the extensively burrowed beds (Fig. 5B). Apparently the spreading sand sheet killed off the abundant mud-dwelling burrowers, for the well-sorted deposits, in contrast, are penetrated only by solitary burrow-casts as much as 5 mm in diameter and commonly with septal laminae (Fig. 5E). These features record minor variations in scour and sedimentation during accumulation of the coarser units.

Locally, several inches of the laminated siltstone and very fine grained sandstone unit has an irregularly and intricately contorted fabric of very lobate and crumpled folds as much as 1.5 mm high. The complex pattern suggests disturbance by bottom crawlers rather than by an inorganic agent.

As observed in other studies (Van Straaten, 1954, p. 29; Greensmith, 1956, p. 352; Goldring, 1962, p. 248; Middlemiss, 1962) muddy silt to very fine grained sand is the favorable range of sediment size for abundant burrowers in lakes, lagoons, and tidal flats, and extensive bioturbation or mottling points to a relative slow rate of deposition.

Many of the prominent features of the upper part of detrital cycles, and especially those in the coarser siltstone and very fine grained sandstone units, resemble features in the Devonian Psammites du Condroz which Van Straaten (1954, p. 43-44) believes accumulated in a marine lagoon that was covered by water most of the time.

[*Editor's Note:* Material has been omitted at this point.]

REFERENCES

BARRELL, JOSEPH, 1917, Rhythms and the measurements of geologic time: Geol. Soc. America Bull., v. 28, p. 745-904.

BASCOM, FLORENCE, 1938, Triassic system, *in* Geology and mineral resources of the Honeybrook and Phoenixville quadrangles, Pennsylvania: U. S. Geol. Survey Bull. 891, p. 65-78.

BEERBOWER, J. R., 1961, Origin of cyclothems of the Dunkard Group (Upper Pennsylvanian-Lower Permian) in Pennsylvania, West Virginia, and Ohio: Geol. Soc. America Bull., v. 72, p. 1029-1050.

BOCH, W., 1959, New eastern American Triassic fishes and Triassic correlations: Geol. Cent. Res. Ser., v. 1, p. 1-184.

BOSWELL, P. G. H., 1961, Muddy Sediments: W. Heffer and Sons, Cambridge, 140 p.

BOSWORTH, T. O., 1912, The Keuper Marl around Charnwood: Geol. Soc. London Quart., v. 68, p. 281-294.

BRADLEY, W. H., 1930, The varves and climate of the Green River epoch: U. S. Geol. Survey Prof. Paper 158, p. 87-110.

———, 1931, Origin and microfossils of the oil shale of the Green River Formation of Colorado and Utah: U. S. Geol. Survey Prof. Paper 168, 58 p.

BROUGH, J., 1928, On rhythmic deposition in the Yoredale series: Univ. Durham Philos. Soc. Proc., v. 8, pt. 2, p. 116-126.

CLOUD, P. E., 1960, Gas as a sedimentary and diagenetic agent: Am. Jour. Sci., v. 258A, p. 35-45.

CONANT, L. C., and SWANSON, V. E., 1961, Chattanooga shale and related rocks of central Tennessee and nearby areas: U. S. Geol. Survey Prof. Paper 357, 62 p.

CRAMPTON, C. B., 1914, Lithology and conditions of deposit of the Caithness flagstone series, *in* The geology of Caithness: Geol. Survey Scotland, Mem. p. 80-103.

DUFF, P. M. D., and WALTON, E. K., 1962, Statistical basis for cyclothems: a quantitative study of the sedimentary succession in the East Pennine Coalfield: Sedimentology, v. 1, p. 235-255.

ELLIOTT, R. E., 1961, The stratigraphy of the Keuper series in southern Nottinghamshire: York Geol. Soc. Proc., v. 33, p. 197-234.

ENGELS, B., 1957, Über die Fazies des Hunsrückshiefers: Geol. Rund., v. 45, p. 143-150.

GILBERT, G. K.. 1895, Sedimentary measurement of Cretaceous time: Jour. Geology, v. 3, p. 121-127.

GOLDRING, R., 1962, The trace fossils of the Baggy Beds (Upper Devonian) of North Devon, England: Paläont. Z., v. 36, p. 234-251.

GOODLET, G. A., 1959, Mid-Carboniferous sedimentation in the Midland Valley of Scotland: Edin. Geol. Soc. Trans., v. 17, p. 217-240.

GREENSMITH, J. T., 1956, Sedimentary structures in the upper Carboniferous of north and central Derbyshire, England: Jour. Sed. Pet., v. 26, p. 343-355.

GREINER, H. R., 1962, Facies and sedimentary environments of Albert Shale, New Brunswick: Am. Assoc. Petroleum Geologists Bull., v. 46, p. 219-234.

HAM, W. E., 1960, Middle Permian evaporites in southwestern Oklahoma: 21st Intern. Geol. Cong., sec. 12 Proc., Norden, p. 138-151.

HAWKINS, A. C., 1914, Lockatong Formation of the Triassic of New Jersey and Pennsylvania: N. Y. Acad. Sci. Ann., v. 23, p. 145-176.

———, 1928, Halite and gluaberite cavities and included minerals from central New Jersey: Am. Mineralogist, v. 13, p. 238-239.

HEIDE, S. VAN DER, 1950, Compaction as a possible factor in Upper Carboniferous rhythmic sedimentation: 18th Internat. Geol. Cong. Rept., pt. 4, p. 38-45.

JENKINS, P. M., 1932, Reports on the Percy Sladen Expedition to some Rift Valley Lakes in Kenya in 1929-1.: Ann. and Mag. Nat. Hist., ser. 10, v. 9, p. 533-553.

KLEIN, G. DEV., 1962a, Triassic sedimentation, Maritime Provinces, Canada: Geol. Soc. America Bull., v. 73, p. 1127-1146.

———, 1962b, Sedimentary structures in the Keuper Marl (Upper Triassic): Geol. Mag., v. 99, p. 137-144.

LANGBEIN, W. B., 1961, Salinity and hydrology of closed lakes: U. S. Geol. Survey Prof. Paper 412, 20 p.

LONGWELL, C. R., 1928, Three common types of desert mud-cracks: Am. Jour. Sci., v. 15, p. 136-145.

McKEE, E. D., and others, 1962, Laboratory studies on deformation in unconsolidated sediments: U. S. Geol. Survey Prof. Paper 450-D, p. 151-155.

McLAUGHLIN, D. B., 1946, The Triassic rocks of the Hunterdon Plateau, New Jersey: Pennsylvania Acad. Sci. Proc., v. 20, p. 89-98.

————, 1959, Mesozoic rocks, *in* Geology and mineral resources of Bucks County, Pennsylvania: Pennsylvania Geol. Survey, 4th ser., Bull. C9, p. 55-114.

MIDDLEMISS, F. A., 1962, Vermiform burrows and rate of sedimentation in the Lower Greensand: Geol. Mag., v. 99, p. 33-40.

MOORE, D. G., and SCRUTON, P. C., 1957, Minor internal structures of some recent unconsolidated sediments: Am. Assoc. Petroleum Geologists Bull., v. 41, p. 2723-2751.

RICHTER, R., 1931, Tierwelt und Umwelt in Hunsrückschiefer; zur Entstehung eines schwarzen Schlammsteins: Senckenberg., Bd. 13, p. 299-342.

SANDERS, J. E., 1960, Origin of convoluted laminae: Geol. Mag., v. 97, p. 409-421.

————, 1963, Late Triassic tectonic history of northeastern United States: Am. Jour. Sci., v. 261, p. 501-524.

SCHAEFFER, BOBB, 1952, The Triassic coelocanth fish *Diplurus*, . . . : Am. Mus. Nat. Hist. Bull., v. 99, p. 31-78.

SCHALLER, W. T., 1932, The crystal cavities of the New Jersey zeolite region: U. S. Geol. Survey Bull. 832, 85 p.

SHELTON, J. W., 1962, Shale compaction in a section of Cretaceous Dakota sandstone, northwestern North Dakota; Jour. Sed. Pet., v. 32, p. 873-877.

STURM, EDWARD, 1956, Mineralogy and petrology of the Newark Group sediments of New Jersey: Unpub. doctoral dissertation, Rutgers Univ., 219 p.

TRUEMAN, A. E., 1948, The relation of rhythmic sedimentation to crustal movement: Sci. Prog. v. 36, p. 193-205.

VAN HOUTEN, F. B., 1962, Cyclic sedimentation and the origin of analcime-rich upper Triassic Lockatong Formation, west-central New Jersey and adjacent Pennsylvania: Am. Jour. Sci., v. 260, p. 561-576.

VAN STRAATEN, L. M. J. U., 1954, Sedimentology of recent tidal flat deposits and the Psammites du Condroz (Devonian): Geol. en Mijnb. 16e Jg., p. 25-47.

WANLESS, H. R., 1963, Origin of late Paleozoic cyclothems (abs.): Am. Assoc. Petroleum Geologists Bull., v. 47, p. 375.

WELLER, J. M., 1956, Argument for diastrophic control of late Paleozoic cyclothems: Am. Assoc. Petroleum Geologists Bull., v. 40, p. 17-50.

WELLS, J. A., 1960, Cyclic sedimentation: a review: Geol. Mag., v. 97, p. 389-403.

WHERRY, E. T., 1916, Glauberite crystal-cavities in the Triassic rocks of easern Pennsylvania: Am. Mineralogist, v. 1, p. 37-43.

WHITE, W. A., 1961, Colloid phenomena in sedimentation of argillaceous rocks: Jour. Sed. Pet., v. 31, p. 560-570.

ZANGERL, R., and RICHARDSON, E. S., 1963, The paleoecological history of two Pennsylvanian black shales: Fieldiana, Geol. Mem., v. 4, 352 p.

12

Reprinted from pp. 496, 500–507, 511–512 of *Geol. Soc. America Bull.* **87**(4):496–512
(1976)

Early Tertiary sedimentation in the western Uinta Basin, Utah

ROBERT T. RYDER*
THOMAS D. FOUCH* } *Shell Oil Company, Western Division Exploration, P.O. Box 831, Houston, Texas 77001*
JAMES H. ELISON   *3132 18th Street, Suite 6, Bakersfield, California 93301*

ABSTRACT

During latest Cretaceous through middle Eocene time, over 3,000 m of siliciclastic and carbonate sediment accumulated in the Lake Uinta depocenter in northeastern Utah. Detailed stratigraphic analysis of this extensive lacustrine system, both along the western and southern outcrops and in the subsurface of the Uinta Basin, indicates three major facies: (1) open lacustrine, (2) marginal lacustrine, and (3) alluvial.

The open-lacustrine facies consists primarily of mud-supported carbonate and claystone units with minor amounts of sandstone and siltstone. Kerogen and other organic compounds produce shades of gray and brown. The open-lacustrine rock assemblages were deposited away from terrigenous clastic influxes either near the center of the lake or in nearshore settings. Rocks originating in the latter environment contain abundant fossils (mollusks and ostraco .es) and scattered desiccation features.

The marginal-lacustrine facies is composed of gray-green calcareous claystone, channel-form sandstone, and grain- to mud-supported carbonate units. The dominant depositional environments are interpreted to be lake-margin carbonate flat, deltaic, and interdeltaic. Lake-margin carbonate-flat deposits consist of carbonate beds as thick as 30 m which grade lakeward from an ostracode and oolite grain-supported texture to a mixed mud- and grain-supported texture. Deltaic deposits consist of channel-form sandstone units as thick as 15 m which cut adjacent thin beds of sandstone, siltstone, and gray-green claystone. Ostracode- and oolite-bearing grainstone beds, as much as 4.5 m thick, of lake-margin carbonate-flat origin are commonly interbedded with the deltaic rocks. Interdeltaic rocks exhibit characteristics similar to the deltaic rocks but contain more carbonate beds of lake-margin

carbonate-flat origin and far fewer sandstone units.

The alluvial facies, representing alluvial-fan, lower deltaic-plain, and high mud-flat environments, occupied the most proximal setting within the depositional system. The lower deltaic-plain environment is typified by 15- to 30-m-thick channel-form sandstone units and associated thin-bedded sandstone, siltstone, and red, mud-cracked claystone. The alluvial-fan environment is characterized by thick conglomerate beds with crude horizontal stratification. Red claystone, minor isolated channel-form sandstone, and thin fossiliferous gray-green claystone units characterize the high mud-flat environment. Thin mud- and grain-supported carbonate units of lake-margin carbonate-flat origin are locally interbedded with rocks of the above environments.

During most of early Tertiary time, the Lake Uinta system exhibited a northeast-trending core of open-lacustrine facies surrounded by successive halos of marginal-lacustrine and alluvial facies. The width of the open-lacustrine core continually fluctuated depending upon climatic and tectonic conditions prevailing at the time. Southerly derived feldspathic sands dominated the sediments supplied to the south flank of the basin, whereas sand contributed to the north flank was quartzose and originated from rocks exposed in the Sevier orogenic belt to the west and the Uinta uplift to the north. The north flank of the basin had a steeper depositional slope than the south flank, as manifested by the presence of coarser grained siliciclastic sediments and the narrower band of the marginal-lacustrine facies. *Key words: stratigraphic and historical geology, lacustrine depositional environments, fluvial depositional environments, Green River Formation, Uinta Basin, carbonate rocks, siliciclastic rocks.*

INTRODUCTION

During latest Cretaceous through middle Eocene time, more than 3,000 m of silici-

clastic and carbonate sediments accumulated in central and northeastern Utah. The lacustrine depocenters, known as Lakes Flagstaff and Uinta, developed shortly after the eastward retreat of the Late Cretaceous seaway and appear to have been localized within regional tectonic depressions where subsidence rates exceeded sedimentation rates. Lake Uinta coincided approximately with the structural axis of the Uinta Basin, and Lake Flagstaff was situated in the structural basin between the Sevier orogenic belt and the San Rafael swell (Fig. 1). These lakes were probably separate entities during their formative stages, but they eventually merged in the vicinity of the western Uinta Basin to produce one extensive lacustrine system. It is unlikely that the Lake Uinta–Lake Flagstaff system resulted from two separate overlapping lake systems of differing age, because the strata within the two depocenters appear to be contiguous and they lack major unconformities.

The objectives of this paper are to (1) describe the characteristics and distribution of the major facies within the western part of the Lake Uinta depocenter, (2) interpret the depositional environments of these facies, and (3) delineate the major stages through which Lake Uinta evolved. To these ends, an outcrop study of the southwest flank of the Uinta Basin between the towns of Thistle and Woodside, Utah, was combined with a subsurface study of the western and central Uinta Basin which utilized core and well-cutting data described by us from at least 50 drill holes. The study area is outlined in Figure 1.

* Present address: (Ryder and Fouch) U.S. Geological Survey, Branch of Oil and Gas Resources, Denver Federal Center, Denver, Colorado 80225

[*Editor's Note:* Material, including figures 1 and 2, has been omitted at this point.]

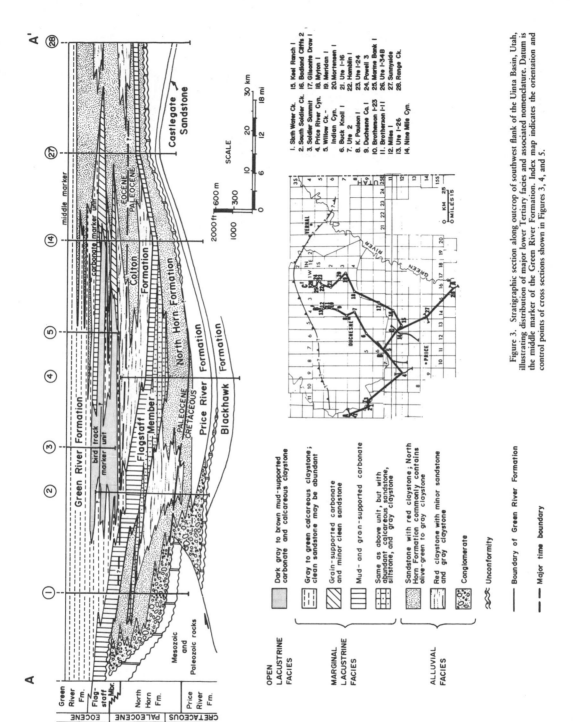

Figure 3. Stratigraphic section along outcrop of southwest flank of the Uinta Basin, Utah, illustrating distribution of major lower Tertiary facies and associated nomenclature. Datum is the middle marker of the Green River Formation. Index map indicates the orientation and control points of cross sections shown in Figures 3, 4, and 5.

SCALE

0   6   12   18 mi

0   10   20   30 km

0   300   600 m
1000   2000 ft

1. Sixth Water Ck.
2. South Soldier Ck.
3. Soldier Summit
4. Price River Cyn.
5. Willow Ck. – Indian Cyn.
6. Buck Knoll 1
7. Ute 2
8. K. Poulson 1
9. Duchesne Co. 1
10. Brotherson I-23
11. Brotherson I-11
12. Miles 1
13. Ute 1-26
14. Nine Mile Cyn.
15. Keel Ranch 1
16. Badland Cliffs 2
17. Gilsonite Drow 1
18. Myton 1
19. Meridan 1
20. Mortensen 1
21. Ute 1-16
22. Hamblin 1
23. Ute 1-24
24. Powell 3
25. Marine Bank 1
26. Ute I-34B
27. Sunnyside
28. Range Ck.

OPEN LACUSTRINE FACIES

Dark gray to brown mud-supported carbonate and calcareous claystone

MARGINAL LACUSTRINE FACIES

Gray to green calcareous claystone; clean sandstone may be abundant

Grain-supported carbonate and minor clean sandstone

Mud- and grain-supported carbonate

Same as above unit, but with abundant calcareous sandstone, siltstone, and gray claystone

ALLUVIAL FACIES

Sandstone with red claystone; North Horn Formation commonly contains olive-green to gray claystone

Red claystone with minor sandstone and gray claystone

Conglomerate

〰〰 Unconformity

——— Boundary of Green River Formation

– – – Major time boundary

159

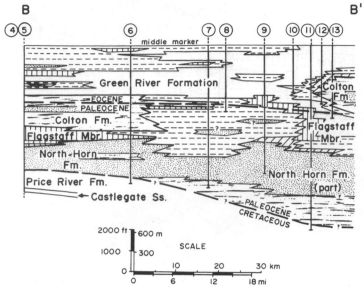

Figure 4. Stratigraphic section between the Willow Creek–Indian Canyon surface section and the Altamont oil field, illustrating distribution of major lower Tertiary facies and associated nomenclature. Datum is the middle marker of the Green River Formation. See Figure 3 for explanation.

CHARACTERISTICS OF LAKE UINTA DEPOSITIONAL ENVIRONMENTS

Lacustrine Depositional Model

Studies concerning the depositional environments of the lower Tertiary strata of the Uinta Basin are relatively common, but they generally are focused on only one aspect of the total depositional system. For example, Bradley (1929, 1931), Moussa (1965, 1970), Baer (1969), Picard and others (1973), and Williamson and Picard (1974) have described the carbonate-rich environments, whereas other investigators, including Jacob (1969), Picard and High (1970), and High and Picard (1971), were concerned mainly with the siliciclastic envi-

ronments. Prior to this paper, only in the Raven Ridge–Red Wash area (Koesoemadinata, 1970; Picard and High, 1972) has a Lake Uinta depositional model been offered to indicate the stratigraphic relations between siliciclastic and carbonate rock assemblages. These models resemble certain aspects of our model, but in part they were developed for a younger (above middle marker, see Fig. 2) and possibly more saline stage of the lake (Bradley, 1931; Ryder and Fouch, 1974). The lacustrine depositional model offered in this paper was first outlined by Fouch and others (1973).

Figure 6 illustrates the relative distribution and interpreted depositional settings of the three major facies within the Lake Uinta system soon after completion of a stage of marked regressive sedimentation during early Eocene time. In general, the lacustrine system at this time is composed of a core of open-lacustrine facies, flanked successively shoreward by the marginal-lacustrine and alluvial facies. A major feature of this suggested model is that the rock color changes systematically lakeward from red to green to dark gray and brown. Other aspects of the model include a general lakeward increase in carbonate content and organic richness of the units and a decrease in the volume of terrigenous sediment and grain-supported carbonate units. The following discussion is a detailed presentation of the components of the model.

The open-lacustrine facies occupies the most distal position of the three facies recognized in our lacustrine model. The thickness of this facies ranges from approximately 900 m near the center of the basin to 450 m along the exposed margin of the basin. In some cases the open-lacustrine rocks can be traced along the basin margin to an approximate zero edge where they are replaced by rocks of the marginal-lacustrine facies. The depositional axis of this rock sequence in the Uinta Basin is oriented east-northeast. The width of the open-lacustrine core ranges from a maximum of over 60 km to a minimum of 5 km.

Organic-rich, light brown to black, mud-supported carbonate and calcareous claystone are the chief constituents of the open-lacustrine facies (Fig. 7). The carbonate units consist primarily of calcite, but dolomite is a common constituent (Picard and others, 1973, Tables 7 and 8). Local interbeds of sandstone, siltstone, carbonate packstone, and bedded chert are also present. Thin horizontal laminations constitute the major bedding type in the carbonate and claystone units, but thick- and medium-sized horizontal beds are common. Convolute beds and associated penecontemporaneous microfaults are locally abundant. Minor sedimentary structures include loop bedding, curviplanar laminae

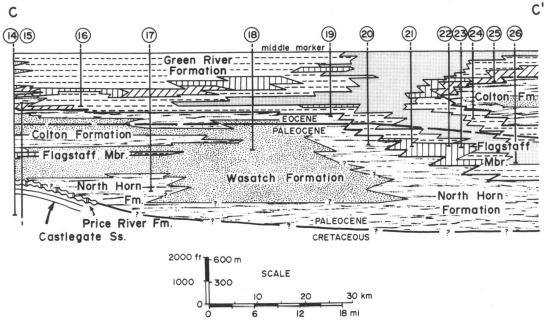

Figure 5. Stratigraphic section between the Nine Mile Canyon surface section and the Bluebell oil field, illustrating distribution of major lower Tertiary facies and associated nomenclature. Datum is the middle marker of the Green River Formation. See Figure 3 for explanation.

conforming to nodular chert, and desiccation cracks. The kerogen and other organic compounds associated with these rocks occur in three distinct modes: (1) very fine particulate matter dispersed throughout the thick- and medium-bedded carbonate mudstone and claystone; (2) thin laminae alternating with thin laminae of carbonate mudstone, calcite, and dolomite (this rock type is commonly referred to as oil shale when it yields oil in excess of approximately 15 gal/ton); and (3) algal coal beds as thick as 16 cm composed of *Botryococcus* sp. The sandstone and siltstone units exhibit small-scale cross-stratification, massive beds, and thin horizontal beds with minor linear bottom markings and burrow structures, whereas the carbonate packstone units consist of thin horizontal beds and local low-angle cross-stratification. Fossils are common in some units of the open-lacustrine facies, and they include ostracodes, charophytes, palynomorphs, algal filaments, gastropods (*Goniobasis* sp., *Physa* sp., *Viviparus* sp., *Valvata* sp., Planorbidae: indet., Sphaeriidae: indet., Acroloxidae: indet.), pelecypods (*Plesielliptio* sp.), and fish remains.

We suggest that most of the organic-rich strata was deposited as alternating layers of blue-green algal ooze and algae-generated low-Mg calcite in very quiet water ranging from approximately 5 to 30 m in depth.

Most likely the low-Mg calcite was precipitated during algal "blooms" when significant quantities of CO_2 were extracted from the lake water by the photosynthetic process (G. A. Desborough and J. K. Pitman, 1975, oral commun.). During photosynthesis, the blue-green algae also selectively removed Mg from the lake water and concentrated it in their sheaths. Thus, when the "blooms" abated and the dead algae settled to the bottom of the lake, their skeletons may have been enriched in Mg over Ca by a factor of three (G. A. Desborough and J. K. Pitman, 1975, oral commun.). The dolomite and high-Mg calcite were probably formed during burial as some of the Mg in the algae-rich laminae was transferred to the adjacent laminae of chemically precipitated low-Mg calcite (G. A. Desborough and J. K. Pitman, 1975, oral commun.). The remainder of the sediments are of terrigenous origin and were contributed to the open-lacustrine facies by fluvial and eolian processes.

Organic-rich sediments were also deposited in the nearshore open-lacustrine environment away from major terrigenous clastic influxes, but unlike their deeper water equivalents, they were replete with ostracodes and mollusks. The two stratigraphic sections illustrated in Figure 7 represent this type of depositional setting. During low stands of the lake, these nearshore

sediments were locally exposed to subaerial conditions, as manifested by local desiccation features and bird tracks. It should be emphasized that subaerial conditions were rarely experienced in the open-lacustrine setting, and when they occurred, only the outer margins of the open-lacustrine facies were exposed.

The playa-lake model proposed by Eugster and Surdam (1973) for the Green River Formation of Wyoming may explain the lithologic assemblage in the upper part of the Green River Formation (above the middle marker) in the western Uinta Basin, but it cannot be applied to the open-lacustrine facies that we have studied in the lower part of the Green River Formation. The evidence suggests that in the western Uinta Basin, Lake Uinta was an extensive, fresh to slightly saline perennial lake during the deposition of the lower part of the Green River Formation and the Flagstaff Member. Sodium-rich evaporites are rare to absent, and desiccation features are situated only along the perimeter of the open-lacustrine rocks. Moreover, we have observed no sedimentary features in open-lacustrine rocks — such as reworked ooliths and mudstone clasts — to indicate that the dolomite, interlaminated with the kerogen, was reworked and transported into the center of the lake from an adjoining playa flat, as suggested by Eugster and Surdam

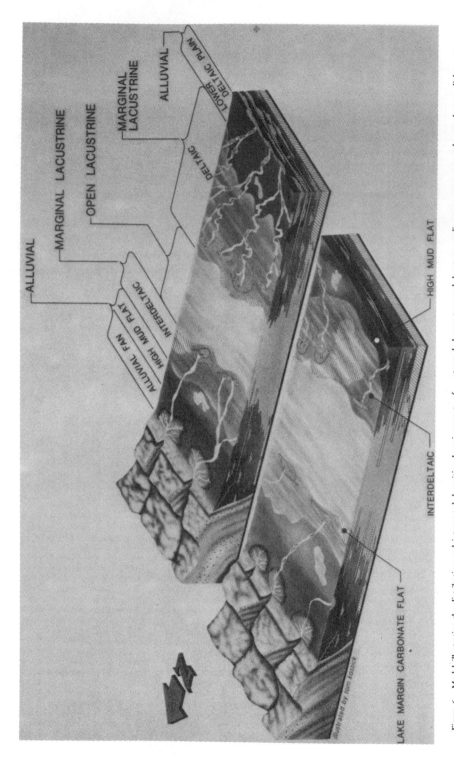

Figure 6. Model illustrating the distribution and interpreted depositional environments of open-lacustrine, marginal-lacustrine, and alluvial facies of western Lake Uinta, Utah: vertical stripes, mud-supported carbonate; diagonal stripes, grain-supported carbonate; heavy dots, sandstone; dark-gray tone, red claystone; medium-gray tone, gray and green claysone; light-gray tone, dark gray and brown claystone. Width of Lake Uinta in the diagram is approximately 40 km. Vertical exaggeration is between 15 and 20.

162

(1973). Anhydrite appears to be the only evaporite mineral associated with Lake Uinta at this time, and it is found primarily in the transition zone between the alluvial and marginal-lacustrine facies. In summary, we see no compelling evidence to suggest that early Lake Uinta was (1) highly saline, (2) ephemeral, or (3) had dolomite supplied to its organic-rich center from an adjacent playa flat.

Marginal-Lacustrine Facies

Lake-Margin Carbonate-Flat Environment. The rocks of the lake-margin carbonate-flat environment are peripheral to the open-lacustrine facies, and extensive intertonguing exists between them. The thickness of the carbonate units deposited in this setting ranges from less than 30 cm to 100 m, and width dimensions range from several kilometres to more than 50 km.

The lake-margin carbonate-flat units within the lower part of the Green River Formation and the Flagstaff Member have slightly different characteristics and thus will be discussed separately. Lake-margin carbonate-flat units within the Flagstaff Member exhibit medium to thick horizontal beds, and they grade progressively basinward from (1) interbedded sandstone, siltstone, gray calcareous claystone, kerogen beds (algal coal), and oncolite-bearing, gray grain-supported carbonate units to (2) highly fossiliferous, blue-gray, mud- and grain-supported carbonate units (Fig. 8) to (3) dark gray mud-supported carbonate beds of the open-lacustrine facies. The dominant mineral in these units appears to be calcite. Fossils include gastropods (*Goniobasis* sp., *Physa* sp., *Viviparus* sp.), pelecypods (*Elliptio* sp., *Lampsilis* sp.; La Rocque, 1960), ostracodes, palynomorphs, bone fragments, and blue-green algae in the form of stromatolites (laterally linked hemispheroid variety), charophytes, and oncolites. The sandstone and siltstone units present in the nearshore assemblage are highly calcareous, commonly channel-form, and contain small-scale cross-stratification and burrow structures.

The lake-margin carbonate-flat units in the lower part of the Green River Formation also vary systematically basinward. They grade from (1) yellow-orange, ostracode- and oolite-bearing grainstone beds (Fig. 8) with medium-sized horizontal beds and low-angle cross-stratification and minor sandstone units with thin horizontal beds to (2) light to dark gray, horizontally bedded, moderately fossiliferous, mud-supported carbonate units (Fig. 8) to (3) brown kerogenous ostracode-bearing mud-supported carbonate units of open-lacustrine origin. Commonly, the intermediate carbonate stage is absent, and the

ostracode and oolite grainstone units grade directly into the open-lacustrine rocks. Judging from the work of Picard and others (1973) and Williamson and Picard (1974), the lake-margin carbonate-flat units in the lower part of the Green River Formation have experienced varying degrees of dolomitization. Fossil assemblages are similar to those reported for the lake-margin carbonate-flat units of the Flagstaff Member.

Distinct depositional cycles, ranging from 2 to 5 m thick, are recorded in the ostracode-bearing grainstone units (Fig. 8). The cycle begins with alternating thin horizontal beds of gray claystone, kerogenous mudstone, and ostracode-bearing grainstone, which grade upward into ostracode-bearing grainstone units with medium-sized horizontal beds and low-angle, small-scale cross-stratification; the cycle ends with a several-centimetre-thick massive bed of carbonate-coated ostracode grains as large as pisolith size. In some localities the cycle is capped by a 2- to 30-cm-thick algal boundstone (laterally linked hemispheroid variety of stromatolite). The medium-bedded grainstone units near the upper part of the cycle locally contain large polygonal mud cracks and sedimentary breccia composed of bone and carbonate mudstone fragments.

The carbonate grainstone units containing the abundant oolites and ostracodes represent shoal and beach deposits that developed during local and regional transgressions of the shoreline. The water in this setting was generally agitated, as indicated by the current and oscillation ripple marks, low-angle cross-stratification, and lack of claystone interbeds. The individual cycles document conditions of upward shoaling. Judging from the abundance of ooliths and

pisoliths, thickness of the cycles, and observations of recent lacustrine stromatolites (Bradley, 1929; Eardley, 1938; Carozzi, 1962), the water depths probably ranged from several centimetres to 9 m. The large polygonal mud cracks indicate periodic subaerial exposure. This interpretation is similar to that offered by Williamson and Picard (1974), who recognized both shoal and beach-bar depositional environments within the grainstone-dominated carbonate units. Fluvial systems present during these high lake levels deposited most of their sediment load upstream, and consequently, only minor amounts of terrigenous clastic material reached the areas of carbonate deposition. Discharge from these streams, although relatively void of sediment, supplied nutrients to the growing ostracode communities.

Most of the oncolites within the Flagstaff Member probably also originated in the nearshore lake-margin carbonate-flat setting, except current activity was moderate and episodic. It is possible that the pisolith-size features we call oncolites were not formed by algae, but rather by inorganic precipitation (Estaban and Pray, 1975). If this is true, the pisoliths probably developed in bicarbonate-rich ponds adjacent to the main body of the lake. Either mode of origin is consistent with a nearshore origin for the Flagstaff carbonate units along the south flank of the basin.

The fossiliferous mud-supported carbonate rocks were generally deposited lakeward of the higher energy shoal environment in slightly deeper water where currents were evidently weak and intermittent. A muddy substrate and abundant nutrients in this setting allowed such mollusk genera as *Elliptio, Lampsilis, Goniobasis, Viviparus,* and *Physa* to thrive (La Rocque,

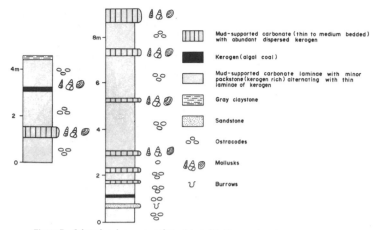

Figure 7.   Selected rock sequences that originated in the open-lacustrine environment.

Mud-supported carbonate (thin to medium bedded) with abundant dispersed kerogen

Kerogen (algal coal)

Mud-supported carbonate laminae with minor packstone (kerogen rich) alternating with thin laminae of kerogen

Gray claystone

Sandstone

Ostracodes

Mollusks

Burrows

1960). Our interpretation differs from that of Williamson and Picard (1974), who placed this rock assemblage in a lagoonal depositional environment. The lateral continuity of these fossiliferous mud-supported carbonate units, their consistent lakeward position relative to the grainstone units, and the lack of evidence for any type of associated barrier-island complex militate against the lagoonal interpretation.

The lake-margin carbonate-flat units were probably dolomitized by a process similar to the one proposed by Wolfbauer and Surdam (1974) to account for the origin of dolomite and dolomitic limestone units in the Green River Formation of Wyoming. According to their model, the dolomite formed on playalike flats adjacent to Lake Gosiute as a result of evaporative pumping of magnesium-rich brines through the pre-existing calcareous mud. More specifically, the evaporation that accompanied the subaerially exposed playalike flats created a hydraulic gradient between the relatively fresh water of the ground-water table below and the concentrated brines with high Mg^{+2}/Ca^{+2} ratios near the surface of the flats. It was in these near-surface areas that dolomite was precipitated and (or) replaced the pre-existing lime mud. The penecontemporaneous dolomite was concentrated near the lower part of the playalike flat, because at this locality the water table was closest to the playa surface and thus was more susceptible to the evaporation process. This model appears to explain the formation of dolomite in the lake-margin carbonate-flat environment in our study area; however, the term "playalike flat" is unacceptable because it implies that the adjacent lake was ephemeral and highly saline. As previously discussed, the rocks produced by early Lake Uinta exhibit little evidence to suggest that they were deposited under ephemeral and saline conditions. We have chosen the term "lake-margin carbonate flat" because it depicts an environment of extensive nearshore lacustrine carbonate sedimentation — part of which has probably experienced some subaerial exposure — without implying that the lake is ephemeral or highly saline. Another point of departure from the Wolfbauer and Surdam model concerns the dispersal patterns of the dolomite after it formed along the fringes of the lake. Wolfbauer and Surdam (1974) and Eugster and Surdam (1973) have suggested that large quantities of the dolomite mud were transported and redeposited into the central organic-rich part of Lake Gosiute. We maintain that most dolomitic sediments originating in the lake-margin carbonate flat bordering early Lake Uinta experienced only local transport.

Deltaic Environment. The deltaic environment occupied a marginal position within Lake Uinta similar to that of the lake-margin carbonate-flat environment. Salient lithologic types within the deltaic environment include numerous channel-form sandstone units, thin-bedded sandstone and siltstone, and gray-green calcareous claystone (Fig. 9). Individual deltaic assemblages range in thickness from 20 to 40 m, and they commonly are interbedded with and laterally equivalent to rocks of lake-margin carbonate-flat origin. Width dimensions of the deltaic assemblages along the south flank of the basin range from 20 to 45 km, whereas along the north flank of the basin they range from 15 to 25 km.

The gray-green calcareous claystone units exhibit thin to medium horizontal beds, thin laminae of siltstone, mud cracks, burrow structures, local red coloration, and minor faint remnants of root casts. Charophytes, ostracodes, algal filaments, mollusk fragments, and plant remains are locally abundant. Thin-bedded siltstone and very fine sandstone beds between 15 and 45 cm thick accompany the calcareous claystone units. The siltstone and sandstone units are generally structureless except for minor small-scale cross-stratification, horizontal laminae, tracks and trails, and burrow structures.

Individual and composite channel-form sandstone units are the most prominent features of the deltaic setting, and they commonly account for more than 50 percent of the assemblage (Fig. 9). Most of the channel-form sandstone units clearly truncate the adjacent claystone, siltstone, and very fine sandstone units, but in some cases, the channel shape is barely discernible because of a flat to slightly undulatory base.

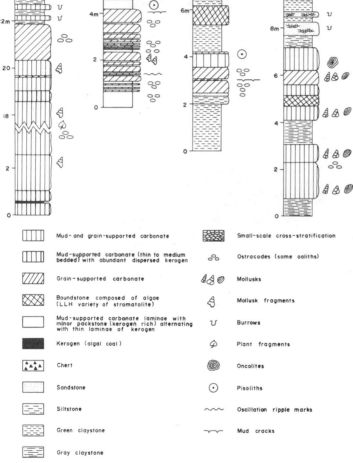

Figure 8.   Selected rock sequences that originated in the lake-margin carbonate-flat environment.

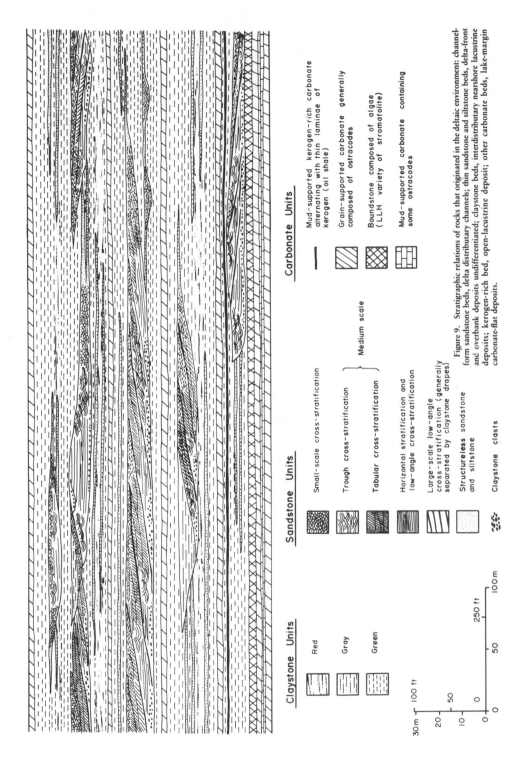

Claystone Units

	Red
	Gray
	Green

30m ─ 100 ft
20 ─
─ 50
10 ─
0 ─ 0

Sandstone Units

Small-scale cross-stratification

Trough cross-stratification

Tabular cross-stratification } Medium scale

Horizontal stratification and low-angle cross-stratification

Large-scale low-angle cross-stratification (generally separated by claystone drapes)

Structureless sandstone and siltstone

Claystone clasts

0 ─ 0
─ 50
10 ─ 250 ft
20 ─
30 ─ 100 m

Carbonate Units

Mud-supported kerogen-rich carbonate alternating with thin laminae of kerogen (oil shale)

Grain-supported carbonate generally composed of ostracodes

Boundstone composed of algae (LLH variety of stromatolite)

Mud-supported carbonate containing some ostracodes

Figure 9. Stratigraphic relations of rocks that originated in the deltaic environment: channel-form sandstone beds, delta distributary channels; thin sandstone and siltstone beds, delta-front and overbank deposits undifferentiated; claystone beds, interdistributary nearshore lacustrine deposits; kerogen-rich bed, open-lacustrine deposit; other carbonate beds, lake-margin carbonate-flat deposits.

165

At some localities, the edges of the channel units may be traced directly into the thin-bedded siltstone and sandstone units (Fig. 9). The individual channel-form sandstone units attain a thickness of 15 m and a width of 400 m. Sedimentary structures include large-scale, low-angle cross-stratification (with associated claystone drapes), uni-directional medium-scale (trough and tabular) and small-scale cross-stratification, horizontal stratification, low-angle cross-stratification with streaming lineations, and convolute bedding. The flat-based units are locally associated with large mound-shaped sandstone bodies containing horizontal beds (Fig. 9). Most of the sedimentary structures are arranged systematically such that the medium-scale cross-stratification is situated near the base of the channel and the small-scale cross-stratification appears near the top. The base of the channel sandstone units commonly contains lag deposits of claystone and carbonate mudstone clasts as large as 5 cm in diameter, *Ganoid* plates, and assorted bone fragments. Upward-fining grain-size sequences are prevalent and range from lower and upper fine sand at the base of the channel to upper fine and lower very fine sand at the top. Locally, the uppermost part of a given channel sequence may contain varying proportions of gray claystone, small-scale cross-stratified siltstone, and thin horizontally bedded carbonate grainstone units.

The entire lithologic assemblage is interpreted as a deltaic system where the channel-form sandstone units are delta distributary channels, the thin siltstone and very fine sandstone units are undifferentiated delta front and overbank deposits, and the gray-green calcareous claystone units represent interdistributary nearshore lacustrine deposits. The width of the deltaic system is greater along the south flank of the basin than the north flank because of the lower depositional slope associated with the former locality.

The delta front, overbank, and interdistributary deposits consist of terrigenous sediments supplied to the lake by the numerous delta distributary channels. In general, the undifferentiated delta front and overbank units were deposited in close proximity to the channels, whereas the finer grained interdistributary deposits occupied a more distal position relative to the channels. When the distributary channels were relatively dormant owing to (1) local or regional transgressions of the shoreline, (2) channel abandonment, or (3) a reduction in sediment supply from the source, thin but extensive lake-margin carbonate-flat and open-lacustrine deposits replaced the clay-rich sediments (Fig. 9). (Regarding the first condition, the Lake Uinta shoreline experienced nearly continual oscillations because it had no apparent drainage outlet and con-

sequently was susceptible to variations in rainfall and subsidence rates.) During low stands of the lake, these deposits were exposed for a short time along extensive mud flats so that red coloration, mud cracks, and clay clasts were locally produced in the clayey sediments.

We believe it is important to distinguish between fluvial-channel sandstone units deposited at or near the shoreline (fluvial-lacustrine channels) and those located in nonlacustrine alluvial-plain settings (fluvial-alluvial channels). As an example, delta distributary channels are included with the fluvial-lacustrine group of channels. They contain diagnostic fluvial features, are abundant and thus represent major areas of terrigenous sediment influx into the lake, and record frequent interactions with carbonate and fossiliferous green claystone units of lacustrine origin. In contrast, the thicker, wider, and more composite fluvial-channel units associated with carbonate-deficient, highly desiccated, red claystone and siltstone units in the lower deltaic-plain environment (see below) are classified with the fluvial-alluvial channels.

Ubiquitous large-scale, low-angle cross-stratification associated with the delta distributary channels suggests that these channel systems continually meandered and migrated and that they deposited the majority of their bed load on point bars (Fig. 9). The meandering stream–point bar origin of the distributary channel sandstone units is also supported by the common upward-fining grain-size sequences, the upward decrease in cross-stratification size, and the numerous sequences interpreted to represent channel abandonment. We see no evidence of natural levees and their accompanying dense vegetation. Apparently, the constant fluctuations of the shoreline impeded their development. Despite such compelling evidence for the meandering channel origin of these sandstone units, they contain some aspects of a braided-channel complex, as mentioned by Jacob (1969). Jacob reasoned that the flat-based channels resulted from aggrading streams incapable of deep scouring action. He interpreted the mound development within the channels as longitudinal or spool bars similar to those present in braided rivers of Iceland. We concur with Jacob's interpretation of the flat-based mounds and thus suggest that the distributary channel system had local braided segments between the more prevalent meander loops. These braided-meandering channels may have resembled the Cimarron River in north-central Oklahoma described by Shelton and Noble (1974).

Interdeltaic Environment. The interdeltaic environment is located up depositional dip from the open-lacustrine facies and along depositional strike from the deltaic

environment. Dominant lithologic types in the interdeltaic environment include gray-green calcareous claystone, siltstone, and channel-form sandstone units. Interdeltaic rock assemblages closely resemble the deltaic rocks, except that the claystone units and intercalated lake-margin carbonate-flat units are more abundant and channel-form sandstone units are smaller (4.5 m thick and 100 m wide) in the former environment. In addition, the 5- to 20-km width of the interdeltaic system is significantly narrower than width dimensions of the deltaic system.

This lithologic assemblage accumulated in a relatively quiet, shallow-water interdeltaic setting. The supply of terrigenous clay, silt, and fine sand was very low compared to the deltaic setting, but it was sufficient to keep it from becoming a permanent carbonate province. The narrow width of the interdeltaic system is probably related to the relatively minor quantity of terrigenous clastic sediments that it received and, secondly, to the clear and quiet water that permitted open-lacustrine sediments to be deposited close to the shoreline. The interdeltaic environment was intermittently exposed to subaerial conditions as indicated by mud cracks, clay clasts, and oxidized iron minerals (some red coloration).

[Editor's Note: Material has been omitted at this point.]

REFERENCES CITED

Abbott, W. O., 1957, Tertiary of the Uinta Basin, *in* Guidebook to the geology of the Uinta Basin: Intermtn. Assoc. Petroleum Geologists Guidebook 8th Ann. Field Conf., p. 102–109.

Baer, J. L., 1969, Paleoecology of cyclic sediments of the lower Green River Formation, central Utah: Brigham Young Univ. Geol. Studies, v. 16, pt. 1, p. 3–96.

Baker, D. A., and Lucas, P. T., 1972, Strat trap production may cover 280+ square miles: World Oil, v. 174, p. 65–68.

Born, S. M., 1970, Deltaic sedimentation at Pyramid Lake, Nevada [Ph.D. thesis]: Madison, Univ. Wisconsin, 234 p.

Bradley, W. H., 1929, Algae reefs and oolites of the Green River Formation: U.S. Geol. Survey Prof. Paper 154-G, p. 203–223.

——1931, Origin and microfossils of the oil

shale of the Green River Formation of Colorado and Utah: U.S. Geol. Survey Prof. Paper 168, 58 p.

Carozzi, A. V., 1962, Observations on algal biostromes in the Great Salt Lake, Utah: Jour. Geology, v. 70, p. 246–252.

Cashion, W. B., 1967, Geology and fuel resources of the Green River Formation, southeastern Uinta Basin, Utah and Colorado: U.S. Geol. Survey Prof. Paper 548, 48 p.

Cashion, W. B., and Donnell, J. R., 1974, Revision of the upper part of the Green River Formation, Piceance Creek basin, Colorado, and eastern Uinta Basin, Utah: U.S. Geol. Survey Bull. 1394-G, 9 p.

Dane, C. H., 1954, Stratigraphic and facies relationships of upper part of Green River Formation and lower part of Uinta Formation in Duchesne. Uintah. and Wasatch Counties, Utah: Am. Assoc. Petroleum Geologists Bull., v. 38, p. 405–425.

——1955, Stratigraphic and facies relationships of the upper part of the Green River Formation and the lower part of the Uinta Formation in Duchesne, Uintah, and Wasatch Counties, Utah: U.S. Geol. Survey Oil and Gas Inv. Chart OC-52.

Dunham, R. J., 1962, Classification of carbonate rocks according to depositional texture, in Ham, W. E., ed., Classification of carbonate rocks: Am. Assoc. Petroleum Geologists Mem. 1, p. 108–122.

Eardley, A. J., 1938, Sediments of Great Salt Lake, Utah: Am. Assoc. Petroleum Geologists Bull., v. 22, p. 1305–1411.

Estaban, M., and Pray, L. C., 1975, Subaqueous syndepositional growth of in-place pisolite, Capitan Reef Complex (Permian), Guadalupe Mountains, New Mexico and West Texas: Geol. Soc. America Abs. with Programs, v. 7, no. 7, p. 1068–1069.

Eugster, H. P., and Surdam, R. C., 1973, Depositional environment of the Green River Formation of Wyoming: A preliminary report: Geol. Soc. America Bull., v. 84, p. 115–120.

Fouch, T. D., 1975, Lithofacies and related hydrocarbon accumulations in Tertiary strata of the western and central Uinta Basin, Utah, in Bolyard, D. W., ed., Symposium on deep drilling frontiers in the central Rocky Mountains: Rocky Mtn. Assoc. Geologists, p. 163–173.

——1976, Revision of the lower part of the Tertiary System in the central and western Uinta Basin, Utah: U.S. Geol. Survey Bull. 1405-C (in press).

Fouch, T. D., Ryder, R. T., and Elison, J. H., 1973, Upper Cretaceous–lower Tertiary lacustrine depositional environments of the western Uinta Basin, Utah: Geol. Soc. America Abs. with Programs, v. 5, no. 6, p. 480–481.

Gilbert, G. K., 1885, The topographic features of lake shores: U.S. Geol. Survey Fifth Ann. Rept., p. 69–123.

Griesbach, F. R., and MacAlpine, S. R., 1973, Reconnaissance palynology and micropaleontology of the Late Cretaceous–early Tertiary North Horn Formation, central Utah: Geol. Soc. America Abs. with Programs v. 5, no. 6, p. 483.

High, L. R., Jr., and Picard, M. D., 1971, Nearshore facies relations, Eocene Lake Uinta,

Utah: Am. Assoc. Petroleum Geologists Bull., v. 55, p. 343.

Jacob, A. F., 1969, Delta facies of the Green River Formation (Eocene), Carbon and Duchesne Counties, Utah [Ph.D. thesis]: Boulder, Univ. Colorado, 182 p.

Koesoemadinata, R. P., 1970, Stratigraphy and petroleum occurrence, Green River Formation, Redwash Field, Utah: Colorado School Mines Quart., v. 65, 85 p.

La Rocque, A., 1960, Molluscan faunas of the Flagstaff Formation of central Utah: Geol. Soc. America Mem. 78, 100 p.

Lucas, P. T., and Drexler, J. M., 1975, Altamont-Bluebell: A major fractured and overpressured stratigraphic trap, Uinta Basin, Utah, in Bolyard, D. W., ed., Symposium on deep drilling frontiers in the central Rocky Mountains: Rocky Mtn. Assoc. Geologists, p. 265–273.

McBride, E. F., 1974, Significance of color in red, green, olive, brown, and gray beds of Difunta Group, northeastern Mexico: Jour. Sed. Petrology, v. 44, p. 760–773.

McCave, I. N., 1969, Correlation of marine and nonmarine strata with example from Devonian of New York State: Am. Assoc. Petroleum Geologists Bull., v. 53, p. 155–162.

Merrill, R. C., 1972, Geology of the Mill Fork area, Utah: Brigham Young Univ. Geol. Studies, v. 19, pt. 1, p. 65–88.

Moussa, M. T., 1965, Geology of the Soldier Summit quadrangle, Utah [Ph.D. thesis]: Salt Lake City, Univ. Utah, 129 p.

——1969, Green River Formation (Eocene) in the Soldier Summit area, Utah: Geol. Soc. America Bull., v. 80, p. 1737–1748.

——1970, Nematode fossil trails from the Green River Formation (Eocene) in the Uinta Basin, Utah: Jour. Paleontology, v. 44, p. 304–307.

Newman, Karl R., 1974, Palynomorph zones in early Tertiary formations of the Piceance Creek and Uinta basins, Colorado and Utah, in Guidebook to the energy resources of the Piceance Creek basin, Colorado: Rocky Mtn. Assoc. Geologists Guidebook 25th Ann. Field Conf., p. 47–55.

Picard, M. D., 1955, Subsurface stratigraphy and lithology of Green River Formation in Uinta Basin, Utah: Am. Assoc. Petroleum Geologists Bull., v. 39, p. 75–102.

——1959, Green River and lower Uinta Formations subsurface stratigraphy in western Uinta Basin, Utah: Intermtn. Assoc. Petroleum Geologists Guidebook 10th Ann. Field Conf., p. 139–149.

Picard, M. D., and High, L. R., Jr., 1970, Sedimentology of oil impregnated lacustrine and fluvial sandstone, P. R. Springs area, southeast Uinta Basin, Utah: Utah Geol. and Mineralog. Survey Spec. Studies 33, 32 p.

——1972, Paleoenvironmental reconstructions in an area of rapid facies change, Parachute Creek Member of Green River Formation (Eocene), Uinta Basin, Utah: Geol. Soc. America Bull., v. 83, p. 2689–2708.

Picard, M. D., Thompson, W. D., and Williamson, C. R., 1973, Petrology, geochemistry, and stratigraphy of Black shale facies of Green River Formation (Eocene), Uinta Basin, Utah: Utah Geol. and Mineralog. Survey Bull. 100, 52 p.

Pinnell, M. L., 1972, Geology of the Thistle quadrangle, Utah: Brigham Young Univ. Geol. Studies, v. 19, pt. 1, p. 89–130.

Russell, I. C., 1885, Geological history of Lake Lahontan: A Quaternary lake of northwestern Nevada: U.S. Geol. Survey Mon. 11, 288 p.

Ryder, R. T., and Fouch, T. D., 1974, Stratigraphic framework and nature of Uinta Basin, Utah oil shales and their implications for the evolution of early Lake Uinta: Geol. Soc. America Abs. with Programs, v. 6, no. 7, p. 935–936.

Shelton, J. W., and Noble, R. L., 1974, Depositional features of braided-meandering stream: Am. Assoc. Petroleum Geologists Bull., v. 58, p. 742–749.

Spieker, E. M., 1946, Late Mesozoic and early Cenozoic history of central Utah: U.S. Geol. Survey Prof. Paper 205-D, p. 117–161.

——1949, Transition between the Colorado Plateau and the Great Basin in central Utah: Utah Geol. Soc. Guidebook to the Geology of Utah, no. 4, 106 p.

Spieker, E. M., and Reeside, J. B., Jr., 1925, Cretaceous and Tertiary formations of the Wasatch Plateau, Utah: Geol. Soc. America Bull., v. 36, p. 435–454.

Thompson, A. M., 1970, Geochemistry of color genesis in red-bed sequence, Juniata and Bald Eagle Formations, Pennsylvania: Jour. Sed. Petrology, v. 40, p. 599–615.

Van De Graaff, F. R., 1972, Fluvial-deltaic facies of the Castlegate Sandstone (Cretaceous), east-central Utah: Jour. Sed. Petrology, v. 42, p. 558–571.

van de Kamp, P. C., 1973, Holocene continental sedimentation in the Salton Basin, California: A reconnaissance: Geol. Soc. America Bull., v. 84, p. 827–848.

Van Houten, F. B., 1968, Iron oxides in red beds: Geol. Soc. America Bull., v. 79, p. 399–416.

——1972, Iron and clay in tropical savanna alluvium, northern Colombia: A contribution to the origin of red beds: Geol. Soc. America Bull., v. 83, p. 2761–2772.

Walker, T. R., 1967a, Formation of red beds in modern and ancient deserts: Geol. Soc. America Bull., v. 78, p. 353–368.

——1967b, Color of recent sediments in tropical Mexico: A contribution to the origin of red beds: Geol. Soc. America Bull., v. 78, p. 917–920.

——1973, Bioherms in the Minturn Formation (Des Moinesian age), Vail-Minturn area, Eagle County, Colorado, in Paleozoic stratigraphy and structural evolution of Colorado: Colorado School Mines Quart., v. 67, p. 249–278.

Williamson, C. R., and Picard, M.D., 1974, Petrology of carbonate rocks of the Green River Formation (Eocene): Jour. Sed. Petrology, v. 44, p. 738–759.

Wolfbauer, C. A., and Surdam, R. C., 1974, Origin of nonmarine dolomites in Eocene Lake Gosiute, Green River Basin, Wyoming: Geol. Soc. America Bull., v. 85, p. 1733–1740.

MANUSCRIPT RECEIVED BY THE SOCIETY MAY 13, 1974
REVISED MANUSCRIPT RECEIVED JULY 9, 1975
MANUSCRIPT ACCEPTED JULY 11, 1975

Printed in U.S.A.

Editor's Comments
on Papers 13 Through 16

DUNE DEPOSITS AND PALEOSOLS

Ancient eolian sediments and soils, in different ways, record events that prevailed when normal nonmarine to paralic sedimentation was inactive. Correct identification of dune sandstones is seldom straightforward, however, because they share physical characteristics with other sandy sediments such as those in beaches and barrier bars. For example, McKee's (1962) evaluation of the Cretaceous Nubian Sandstone in northern Africa, commonly considered to be partly of eolian origin, showed that it is more reasonably a sandy paralic deposit. Inasmuch as large dune fields develop in deserts, the rather rare ancient deposits of windblown sand have received considerable attention as clues to paleoclimates (Bigarella 1973). Recent reconstruction of the dune facies of the Permian formations of northwestern Europe (Glennie 1972) and analysis of its silica diagenesis (Waugh 1970) emphasize the role of aridity. Bigarella (1972) has summarized the diagnostic features of ancient dune sand.

The two papers selected on dune deposits testify that environments during late Paleozoic time were conducive to the development and preservation of dunes. Steidtmann (1974) analyzed the significant features of the Pennsylvanian Casper Sandstone in southern Wyoming (Paper 13) and found that low-crested ripple marks oriented parallel to the dip of cross-strata, and evenly spaced lag grains on bedding surfaces are especially suggestive of an eolian origin. A cross-bedded Permian flagstone in central Colorado (Paper 14) has also been attributed to eolian processes by Walker and Harms (1972). Large-scale, steep sets of cross-strata deposited on accretion faces of dunes and major scour surfaces developed in interdune blowouts with lag grains equally spaced on some of the surfaces are cited as evidence of an ancient dune field.

Well-preserved soil profiles are seldom found beneath marine transgressions across weathered terrains. In contrast, they are relatively common beneath nonmarine deposits. For example, Williams (1968) has described a weathered mantle beneath late Precambrian nonmarine sediments that has chemical characteristics of a pedalfer and a pedocal, depending on its host rock. Freytet (1971) has demonstrated similar host-rock control of the composition of pre-Cretaceous residual paleosols in southern France.

More commonly, ancient soils that developed on slowly aggrading alluvial plains have been preserved by burial where succeeding channel scour was shallow, as in the fining-upward fluvial cycles of the early Cretaceous Wealdon Formation in the eastern part of the Paris Basin (Meyer 1976) and in late Cretaceous to early Eocene deposits of Languedoc, France (Freytet 1971). Many of the preserved paleosols in fluvial successions are characterized by accumulation of nodular calcium carbonate like that in Recent caliche of dry regions. Nodules and nodular layers of this sort were called *cornstone* by Scottish farmers. Examples occur in the Devonian Old Red Sandstone of Great Britain (Allen 1974a; 1974b, pp. 110–122), late Paleozoic rocks of northeastern Spain (Nagtegaal 1969), and in early Mesozoic deposits in the Connecticut Valley (Hubert 1977). Similar calcareous soils have been preserved on alluvial fans by periodic sedimentation (Gile and Hawley 1966). Leeder (1975) has constructed a quantitative model of arid-zone alluvial sedimentation rates based on development of pedogenic carbonates. Yaalon (1971) summarized many aspects of the study of paleosols, and Ortlam (1971, 1975) has demonstrated their role in reconstructing the Permian and Triassic stratigraphic record of central Europe. The current trend in paleosol analysis involves describing and interpreting them in terms of the micromorphology of Recent soils.

The two papers reprinted emphasize quite different aspects of paleopedology. Steel's (1974) description of a succession of cornstones in the Permo-Triassic New Red Sandstone of Scotland (Paper 15) provides a guide to the characteristics of calcareous soil profiles developed in slowly accreting or abandoned floodplain sediments and preserved beneath succeeding fining-upward cyclothems. The ancient caliche is characterized by laminated, pisolitic, and brecciated textures developed during subaerial diagenesis. Because of their regional persistence, cornstones are useful stratigraphic marker beds in alluvial sequences. Moreover, the cumulative thickness and maturity of multiple cornstone profiles better reflect the length of geological time involved than does the total thickness of the formation.

In contrast to Steel's focus on fabric of ancient caliche, Kalliokoski (1975) has analyzed the mineralogical and chemical composition of Precambrian (1 b.y. ago) paleosols and their granodiorite and peridotite bedrock (Paper 16). His comparisons reveal the development of a soil with caliche horizons like that found in present-day semiarid regions with relatively little vegetation. As in most other examples of paleosols, these very ancient ones were preserved by burial beneath a nonmarine deposit, the late Precambrian reddish-brown Jacobsville Sandstone.

REFERENCES

Allen, J. R. L. 1974a. Studies in fluviatile sedimentation: Implications of pedogenic carbonate units. Lower Old Red Sandstone, Anglo-Welch outcrops. *Geol. Jour.* **9:**181–208.

———— 1974b. Sedimentology of the Old Red Sandstone (Siluro-Devonian) in the Clee Hills area, Shropshire, England. *Sed. Geol.* **12:**73–167.

Bigarella, J. J. 1972. Eolian environments: Their characteristics, recognition, and importance. *Jour. Sed. Petrology Spec. Pub. 16,* pp. 12–62.

———— 1973. Paleocurrents and the problem of continental drift. *Geol. Rundschau* **62**(H. 2):447–477.

Freytet, P. 1971. Paléosols résiduels et paléosols alluviaux hydromorphes associés *aux d*épots fluviatiles dans le Cretacé superieur et l'Eocene basal du Languedoc. *Rev. Geographie Phys. et Geologie Dynam. (2)* **13**(F. 3):245–268.

Gile, L. H., and Hawley, J. W. 1966. Periodic sedimentation and soil formation on an alluvial-fan piedmont in southern New Mexico. *Soil Sci. Soc. America Proc.* **30:**261–268.

Glennie, K. W. 1972. Permian Rotliegendes of northwest Europe interpreted in light of modern desert sedimentation studies. *Am. Assoc. Petroleum Geologists Bull.* **59:**1048–1071.

Hubert, J. F. 1977. Paleosol caliche in the New Haven Arkose, Newark Group, Connecticut. *Paleogeography, Paleoclimatology, Paleoecology,* in Press.

Leeder, M. R. 1975. Pedogenic carbonates and flood sediment accretion rates: A quantitative model for alluvial arid-zone lithofacies. *Geol. Mag.* **112:**257–270.

McKee, E. D. 1962. Origin of the Nubian and similar sandstones. *Geol. Rundschau* **52**(H. 2):551–587.

Meyer, R. 1976. Continental sedimentation, soil genesis and marine transgression in the basal part of the Cretaceous in the east of the Paris Basin. *Sedimentology* **23**:235–253.

Nagtegaal, P. J. C. 1969. Microtextures in recent and fossil caliche. *Leidse Geol. Meded.* **42**:131–142.

Ortlam, D. 1971. Paleosols and their significance in stratigraphy and applied geology in the Permian and Triassic of southern Germany. In *Paleopedology; origin, nature and dating of paleosols,* ed. O. H. Yaalon. Jerusalem: Internat. Soc. Soil Sci., Israel Univ. Press, pp. 321–327.

———— 1974. Inhalt und bedeutung fossiler boden-komplexe in Perm und Trias von Mitteleuropa. *Geol. Rundschau* **63**(H. 3):850–884.

Waugh, B. 1970. Petrology, provenance and silica diagenesis of the Penrith Sandstone (Lower Permian) of northwest England. *Jour. Sed. Petrology* **40**:1226–1240.

Williams, G. E. 1968. Torridonian weathering, and its bearing on Torridonian paleoclimate and source. *Scottish Jour. Geol.* **4**:164–184.

Yaalon, D. H., ed. 1971. *Paleopedology, origin, nature and dating of paleosols.* Jerusalem: Internat. Soc. Soil Sci., Israel Univ. Press, pp. 1–350.

13

Reprinted from *Geol. Soc. America Bull.* **85**(12):1835–1842 (1974)

Evidence for Eolian Origin of Cross-Stratification in Sandstone of the Casper Formation, Southernmost Laramie Basin, Wyoming

JAMES R. STEIDTMANN, *Department of Geology, University of Wyoming, P.O. Box 3006, Laramie, Wyoming 82071*

ABSTRACT

The Casper Formation of Pennsylvanian and Permian age in the southernmost Laramie Basin contains large trough cross-stratification for which both eolian and subaqueous origins have been proposed. To determine which is the most reasonable interpretation of origin, information concerning depositional processes was obtained from a study of associated sedimentary structures and settling velocities of light and heavy minerals.

Two important characteristics of sandstone units in the Casper Formation strongly suggest an eolian origin: (1) The cross-stratified units typically contain low-crested, locally truncated ripple marks that nearly everywhere are oriented with their crests parallel to the dip of the cross-strata, and (2) there are evenly spaced lag grains along bedding surfaces.

Other characteristics of the sandstone units are compatible with an eolian environment, although their origin is more ambiguous. For example, low-dipping cross-strata, often cited as evidence against an eolian origin, are similar to the internal stratification of eolian dunes where moisture prohibits development of large avalanche faces. Moreover, large-scale contemporaneous deformation of sand, although commonly attributed to subaqueous gliding, probably requires more cohesion than would be present in saturated subaqueous sand. Moist eolian conditions may provide this cohesion. Finally, settling velocities of associated light and heavy minerals indicate deposition from eolian suspension. Therefore, I conclude that sandstone units in the Casper Formation of the southernmost Laramie Basin were deposited as a coastal dune field that bordered the ancestral Front Range uplift and migrated laterally in response to a fluctuating sea level. *Key words: sedimentary petrology, sedimentary structures, sedimentation, paleogeography.*

INTRODUCTION

The question of subaqueous or eolian origin for many ancient sandstone units containing large trough cross-stratification continues to perplex sedimentologists. Numerous studies concerning possible modern eolian counterparts (for example, Cooper, 1958; Finkel, 1959; Inman and others, 1966; McKee, 1966; Bigarella and others, 1969; Glennie, 1970) and possible modern shallow-marine counterparts (for example, Jordan, 1962; Off, 1963; Stride, 1963a,

1963b; Harvey, 1966; Houbolt, 1968; Terwindt, 1971) provide many details for comparison with ancient rocks. However, they also demonstrate the possible similarities in primary sedimentary features originating in environments with totally different sedimentary processes and attributes. As pointed out by Stanley and others (1971, p. 13), it can no longer be assumed that large festoon cross-strata prove an eolian dune origin because of the essentially identical form and scale of modern submarine dunes or sand waves.

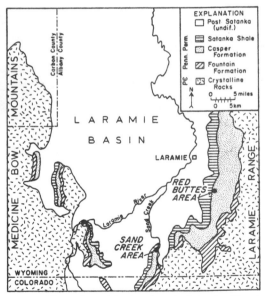

Figure 1. **Index map showing the location and general geologic setting of the study areas along Sand Creek and near Red Buttes.**

Figure 2. Longitudinal section through large trough cross-stratification in Casper sandstone near Sand Creek. For scale, man at center of photograph is 1.8 m tall.

In this regard, Pryor (1971a, 1971b) interpreted some cross-stratified Permian sandstone units in England and Germany, interpreted by others to be eolian, as shallow marine deposits. In doing this, he envisioned a shallow, sandy sea floor with sand ridges or banks similar to those described by Houbolt (1968). Other recent studies have also suggested subaqueous processes for cross-stratified sandstone previously thought to be mainly eolian. Marzolf (1969) and Visher (1971) presented analyses of sedimentary structures and textures that they interpreted as evidence for subaqueous processes in the Navajo Sandstone. On the other hand, some deposits that earlier writers interpreted to be subaqueous have recently been reinterpreted as eolian. A notable example is the study by Walker and Harms (1972) of the Lyons Sandstone in Colorado, originally interpreted as a beach deposit by Thompson (1949).

The problem of origin of the cross-stratification in the Casper Formation in the southernmost Laramie Basin (Fig. 1) is similar to that in the studies cited above. Knight (1929) first described these spectacular exposures of trough cross-stratification (Figs. 2 and 3) and proposed the term "festoon cross-lamination." Since then, these structures have been cited as a classical occurrence of large trough cross-stratification in a number of texts and articles (Dunbar and Rodgers, 1957; Frazier and Osanik, 1961; Harms and

Figure 3. Plan view of trough cross-stratification in Casper sandstone near Sand Creek. Man is 1.8 m tall.

others, 1963; Potter and Pettijohn, 1963). Knight (1929) postulated a submarine origin for this cross-stratification primarily because of associated large-scale deformational structures in which the sediment apparently moved as a cohesive mass and because marine limestone units are associated with the sandstone in the Casper Formation at Red Buttes (Fig. 1).

Since the original interpretation by Knight (1929), there have been numerous speculations concerning the origin of this cross-stratification. Many workers have disagreed with Knight's interpretation, and almost every conceivable depositional environment has been suggested; however, no interpretation has provided a completely satisfactory explanation for the origin of the cross-stratification and associated sedimentary features. In this study, to determine the origin of the large cross-stratification in the Casper Formation, two investigative approaches were used: (1) comparison of sedimentary structures associated with the cross-stratification with possible modern counterparts and with previously published data; and (2) analysis of settling velocities of light and heavy minerals as quantitative indicators of depositional processes.

The conclusions from both these approaches are quite definitive and show that sandstone units of the Casper Formation in the southernmost Laramie Basin were deposited in an eolian environment. These findings are significant in the reconstruction of Pennsylvanian and Permian paleogeography throughout Wyoming and adjacent areas.

GEOLOGIC SETTING

The stratigraphic relations and lithologic character of the Casper Formation in the Laramie Basin have been established in a number of local and regional studies of Pennsylvanian and Permian strata (Knight, 1929; Del Monte, 1949; Pederson, 1953a, 1953b; Thompson and Thomas, 1953; Thomas and others, 1953; Agatston, 1954; Berg, 1956; Maughan and Wilson, 1960; Tenny, 1966; Mallory, 1967; Kirn, 1972; Whitaker, 1972). In southeastern Wyoming, the Casper thins southwestward from 244 m on the east side of the Laramie Range just north of the state line and 183 m near the center of the Laramie Basin to about 41 m in the study area and a feather edge only 5 km farther south in Colorado. It is underlain by and intercalated with the Fountain Formation (Pennsylvanian), a partially contemporaneous coarse arkosic sandstone and conglomerate interpreted by Hubert (1960) as coalescing alluvial fan deposits. The Casper is overlain by red siltstone and shale of the Satanka Shale (Permian) interpreted by Hoyt (1963) as a marine deposit and by Pearson (1972) as a continental flood-plain deposit.

In general, the Casper Formation in the Laramie Basin consists of interbedded sandstone and limestone. The sandstone is a fine-grained, well-sorted, calcareous subarkose or orthoquartzite (Pettijohn, 1957, p. 291) that exhibits large trough cross-stratification. Permian fusulinids (Thompson and Thomas, 1953) and other marine fossils and fossil fragments, some reworked, are present in the sandstone at several localities (Kirn, 1972, p. 20). The interbedded limestone units are between 2.5 and 9.0 m thick and are only locally continuous where they are best exposed along the west flank of the Laramie Range. They pinch out toward the southwest and are not present at Sand Creek. The limestone is dense, sandy micrite containing fusulinids, nautiloid cephalopods, trilobites, brachiopods, crinoids, and bryozoa (Miller and Thomas, 1936).

Along Sand Creek (Fig. 1), the Casper Formation differs somewhat from its character throughout the remainder of the basin. There is no interbedded marine limestone, but thin, lenticular, nonmarine limestone lithosomes occur at several stratigraphic levels in the cross-stratified sandstone. In addition, only a few small, reworked fossil fragments are present in the sandstone (Kirn, 1972, p. 21), and large-scale contortion and brecciation of sandstone laminae (Figs. 4 and 5) occur throughout much of the area.

Although the Casper Formation in southeastern Wyoming ranges in age from Pennsylvanian (Desmoinesian) to Permian (Wolfcampian), the outcrops examined in this investigation are restricted to the uppermost Casper and are therefore probably all Wolfcampian.

SUMMARY OF PREVIOUS WORK

Of the many local and regional studies of Pennsylvanian and Permian strata, few have come to grips with the problem of origin of the large trough cross-stratification of the Casper Formation. Knight (1929, 1960) described the troughs as from 1.5 m (5 ft) wide, 7.5 m (25 ft) long, and 0.3 m (1 ft) deep to 305 m (1,000 ft) wide, several times as long, and at least 15 m (50 ft) deep. They are roughly symmetrical, both symmetrically and asymmetrically filled, and plunge in only one direction (generally southwest). Dips on the bounding surfaces of the cross-strata average between 10° and 15°, and laminae within the sets generally dip between 15° and 25° (Knight, 1929, p. 66). Knight ascribed the origin of these structures to marine processes. Wilson (1950) agreed with this interpretation, but his study dealt only with the Casper Formation at Red Buttes (Fig. 1) where marine limestone is interbedded with sandstone. Pederson (1953a, p. 83; 1953b, p. 24) suggested that both marine and eolian processes acted to form the cross-stratification, but this conclusion if not clearly substantiated by the evidence presented. Opdyke and Runcorn (1960) used evidence from the upper Casper in the southernmost Laramie Basin and elsewhere to determine late Paleozoic wind directions. In doing this, they interpreted the Casper cross-stratification as eolian primarily because of its similarity to Permian sandstone of Great Britain interpreted as eolian by Shotton (1937). Allen (1968, p. 118) and Conybeare and Crook (1968, p. 136) mentioned the sets of large trough cross-strata from the Casper Formation and suggested an eolian origin but did not substantiate this interpretation.

Hanley and others (1971) recently described trace fossils from the cross-stratified sandstone in the Casper Formation. Before this discovery, all interpretations concerning the origin of the cross-strata were based on physical evidence because no biogenic remains had been found (a fact cited by some observers as proof of eolian origin). On the basis of their discovery, Hanley and others (1971) concluded that sediment conditions must have been moist in order to preserve the burrows but that conditions could not have been entirely subaqueous because the organisms causing the markings did not inhabit a subaqueous environment. Finally, in their study of interbedded nonmarine limestone in the Casper, Hanley and Steidtmann (1973) cited mudcracks and desiccation cavities as evidence for subaerial exposure during the accumulation of carbonate mud. The limestone occurs along truncation surfaces of low relief similar to those that Stokes (1968) described as representing deflation surfaces controlled by ground-water level. According to Hanley and Steidtmann (1973), the limestone was probably deposited in interdunal ponds that formed intermittently because of rise in the water table and (or) surface runoff.

SEDIMENTARY STRUCTURES

Ripples

Low-amplitude ripples similar to those described by Harms (1969, p. 387) and Walker and Harms (1972, p. 282) occur on cross-stratification surfaces in the sandstone of the Casper Formation. The ripples are almost everywhere oriented parallel to the dip of the cross-strata. The ripple crests are relatively straight, parallel,

Figure 4. Large-scale deformation of Casper sandstone near Sand Creek. Man is 1.8 m tall.

slightly asymmetric, and spaced 5.5 to 9.0 cm apart. Their heights are from 1.5 to 3 mm; therefore, the ripple indices are 36 to 30. Because of the low crests, the ripples are difficult to observe; the best definition occurs at times of low-angle illumination.

As pointed out by Walker and Harms (1972, p. 282), ripples of this type are similar in regularity and ripple index to those developed on lee slopes of modern dunes where sand is transported across the face by lee-side eddies. If the ripples had been formed in shallow water by wave action, some sets should be oriented so that crests are not parallel to the dip of the slopes on which they were formed. Furthermore, if the ripples had been formed in water by waves acting on a sloping face, the spacing and height of the ripples would vary in response to depth changes down the slope (Walker and Harms, 1972, p. 282). Ripples in the Casper do not show such a variation.

Many of the ripples in the Casper display truncated crests that appear as lineations where the foreset bedding of the ripple intersects the bounding surface (Fig. 6). It is doubtful that this truncation is caused by recent erosion of complete ripples at the outcrop because it can be traced under overlying strata. Furthermore, the striking similarity between truncated ripples in the Casper (Fig. 6) and those in periodically moist modern eolian dunes (Fig. 7) suggests a similar origin. Truncation of modern ripples takes place

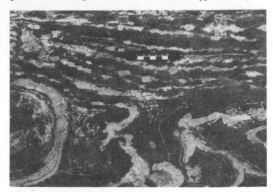

Figure 5. Small-scale folding and brecciation in gray- and white-banded Casper sandstone near Sand Creek.

Figure 6. Truncated ripples in Casper sandstone near Sand Creek.

Figure 7. Truncated ripples on modern eolian dune. Crests of wind ripples moistened by precipitation are dried and blown away exposing light-colored dry sand beneath. Troughs appear darker where surface is moist. Trowel is 28 cm long.

when dry sand is blown into them and subsequently becomes moist during periods of precipitation and (or) changes in ground-water level (Steidtmann, 1973). Greater exposure of certain parts of the crests to wind and sun causes drying and remobilization. Portions of the crests are blown away while other parts of the crests and almost all of the troughs remain wet and intact. If burial occurs before the rest of the ripple is remobilized, truncated ripples are preserved. Similar truncation could occur on subaqueously formed ripples during subsequent subaerial exposure; but as pointed out above, the nontruncated ripples in the Casper sandstone have all the characteristics of eolian ripples, and truncation by subaerial processes further substantiates the interpretation of an eolian origin.

Lag Grains

Uniformly distributed coarse grains (Fig. 8) occur along surfaces at the base of troughs that are otherwise filled with well-sorted, fine-grained sand. These lag grains are composed primarily of quartz and feldspar, are as much as 5 mm in diameter, and are spread 5 to 10 mm apart. Knight (1929, p. 61) noted these and interpreted them as lag deposits left by aqueous currents. As pointed out by Walker and Harms (1972, p. 284) however, uniform spacing of lag grains is apparently uniquely related to an eolian process described by Bagnold (1954, p. 157). Grains with diameters more than six times those of the grains in saltation are nearly immobile. Where these large grains are scattered among fine sand, as is the case in the Casper, they become concentrated as lag grains on deflation surfaces as they are progressively exposed. Subsequently, each large grain is undercut during removal of the fine sand by impact of saltating grains, and it rolls into the newly formed depression. A stable surface of uniformly spaced lag grains is formed when undercutting can no longer occur. This condition is a function of lag-grain size and the trajectory of saltating grains.

Dip of Cross Strata

As shown by Knight (1929, p. 66), approximately 75 percent of the cross-strata dips in the Casper at Sand Creek are between 10° and 25°, with an average of 18°. Few, if any, cross-strata dip at the maximum angle of repose for sand (34°). Inasmuch as angle-of-repose bedding is present in some modern eolian dune sands (for example, McKee, 1966; Bigarella and others, 1969), the lack of it is often cited as evidence contrary to an eolian interpretation for ancient sands.

Walker and Harms (1972, p. 280) attributed the absence of high dips in the Lyons Sandstone to postdepositional compaction and

the relative paucity of angle-of-repose slopes on some types of eolian dunes. The latter is probably the best explanation for dips as low as those cited above for the Casper. As pointed out by Walker and Harms (1972, p. 280), angle-of-repose slopes occur along the crests of some modern dunes while the lower slopes of such dunes accrete by sand driven laterally along the slope. The presence of ripples parallel to the dip of cross-strata and the lack of avalanche features in sandstone of the Casper Formation are indicative of this type of dune deposition. Because the lower part of the dune is most likely to be preserved, angle-of-repose cross-strata probably would not be present in many ancient eolian dune deposits.

Observations that I have made on modern eolian dunes support this explanation of low-dipping eolian strata (Steidtmann, 1973). Figure 9 shows the surface form and internal structure of a transverse dune in southwestern Wyoming. Only the uppermost part of the dune crest is composed of dry, mobile sand with a well-developed slip face. The lower nine-tenths of the dune, which is moist to within a few inches of the surface because of capillary action and local climatic and ground-water conditions, contains strata that dip 14° to 18° downwind and that are more or less parallel to the lower lee face where sand is transported laterally. The cohesiveness of the moist sand prohibits the development of

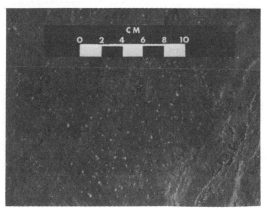

Figure 8. Uniformly distributed lag grains on bedding plane in Casper sandstone near Sand Creek.

avalanche faces, and therefore low-angle bedding dominates the internal structure of the dune. Furthermore, the immobility of the moist sand stabilizes the dune to some degree, thereby causing a relatively slow migration rate (Ahlbrandt, 1973, p. 55). The low-angle stratification in the lower part of the dune is likely to be preserved in the resulting deposit.

The moisture content of the sand is a controlling factor in determining the nature of sand movement and therefore the internal structure and external shape in the perennially moist dunes of southwestern Wyoming. This suggests that low-angle eolian cross-stratification may indicate a moisture-controlled eolian environment where only small or temporary slip faces form, whereas angle-of-repose bedding may indicate conditions which were temporarily or dominantly dry, thus permitting the formation of slip faces.

Local Deformation

Deformation of cross-strata in sandstone of the Casper Formation at Sand Creek (Figs. 4 and 5) ranges in scale from complex folds with amplitudes of as much as 7.5 m with diverse orientations to minute crinkles and brecciations. Deformation occurs at several stratigraphic levels and in discontinuous zones at the same level. In general, it is more common in the upper part of sets of cross-strata just beneath the superjacent truncation surface, but this is not true in every case. Relations between deformed strata and overlying undeformed strata or overlying truncation surfaces show that the deformation was contemporaneous or penecontemporaneous with sand deposition. In no instances are the truncation surfaces deformed.

The cause of deformation in the Casper Formation is not known. Knight (1929, p. 74) attributed the deformation to oversteepening and slumping during filling of the troughs. Furthermore, he noted the similarity between deformation in the Casper and that attributed to subaqueous gliding of marl and lime mud, thus maintaining consistency with his theory of a subaqueous origin for the cross-stratification. Later studies indicate that slumping because of gravity can also produce both small-scale folding and rupture of lee-side laminae in eolian dunes (McKee and others, 1971), and current drag of sediment-laden waters flowing along upper bounding surfaces of cross-stratification has been invoked to explain overturned cross-strata (McKee, 1962, p. 569; McKee and others, 1962b; Allen and Banks, 1972). It is doubtful, however, that these mechanisms produced deformation in the Casper because lee-side slumping on such a large scale has not been reported in modern dunes. Furthermore, widespread brecciation, common in the Casper, has not been described with overturned cross-strata; and preferred orientation of overturning, related to direction of current flow, is not present in the Casper. Loading, related to the encroachment of coastal eolian dunes, has been identified as the cause for large-scale deformation in moist muddy sediment (Brown, 1968; Parker, 1973). However, there is no shale related to Casper deformation, and the displacement of moist sand by eolian dunes is not documented. Furthermore, it is difficult to imagine how this mechanism would leave the truncation surfaces undeformed.

It is obvious that avalanching of dry, noncohesive grains commonly observed in modern eolian dunes cannot produce contortion and brecciation of laminae similar to that in sandstone of the Casper Formation. However, experimental work by Rettger (1935) and McKee and others (1962a, 1962b) on soft-sediment deformation and investigations of modern eolian sands (McKee and others, 1971; McKee and Bigarella, 1972) have demonstrated that a number of styles and scales of deformation can occur in eolian sand. As pointed out by Rettger (1935, p. 286) and McKee and others (1971, p. 373), moisture content of sand is a controlling factor in soft-sediment deformation. Wetted sand (for example, eolian dunes in areas of high rainfall or permeated by ground water) is much more cohesive than saturated sand (for example, sand in a standing body of water) and can be deformed by faulting as well as folding. It is not surprising then that eolian dune sand is commonly deformed. In fact, because water saturation provides less cohesion between grains than does moistening, it is more likely that folding and rupture would occur in moist eolian dunes than in saturated subaqueous dunes. The causative mechanism remains a question, however, and deserves further investigation.

SETTLING VELOCITIES OF LIGHT AND HEAVY MINERALS

Theory

The theory behind this investigative approach was explicitly described by Hand (1967). He explained that two grains of differing densities falling with the same terminal velocity in air will not settle together in water. The reduction in effective density due to the buoyancy of water is relatively greater for the less-dense grain, thus giving it a slower settling velocity. This relation can readily be seen in the theoretical velocity-in-water distributions for quartz and several heavy minerals having equal fall velocities in air (Hand, 1967, Fig. 1). It should be stressed, however, that this relation holds only for grains deposited by settling. For grains deposited from traction transport, the relations of the velocity-in-water curves for light and heavy grains are controlled by processes related to entrainment; in this case, the velocity-in-water curves would show the heavy grains somewhat slower than the associated light grains (Hand, 1967, p. 516).

White and Williams (1967) used these relations to distinguish between pure suspension deposits and combination suspension-traction deposits in subaqueous Holocene cross-stratification, and similar relations should hold for eolian cross-stratification. Grains deposited primarily by settling from suspension plumes form laminae at varying positions on the lee slope depending on the scale and geometry of the dune, grain size, and wind velocity. Velocity-in-water curves for light and heavy grains from these laminae should show the curve for light grain offset toward slower velocities, as predicted by Hand (1967, p. 517). In this study, the primary objective was to examine velocity-in-water distributions of light and heavy grains from sandstone of the Casper Formation in order to determine whether the suspension deposits are the result of aqueous or eolian processes.

Procedure

Samples of Casper sandstone were collected down the dip of individual laminae in sets of cross-strata in the Sand Creek and Red Buttes areas. Velocity-in-water curves for quartz and tourmaline were determined in a 140-cm settling tube by counting an average of 2,000 grains of each. Tourmaline was used as the heavy mineral because it is the most abundant, easily separable heavy mineral in Casper sandstone. Samples that showed significant calcite replacement of quartz in thin section were not used. Details of procedure and a discussion of the effects of availability are given by Haywood (1973).

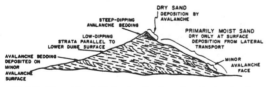

Figure 9. Schematic longitudinal section of modern moist eolian dune that shows the relations between mode of surface deposition and dip of internal stratification.

Results

Velocity-in-water curves for quartz and tourmaline from cross-strata at Sand Creek and Red Buttes are shown in Figure 10. Quartz curves from samples taken in both areas are offset toward slower velocities relative to the curves for tourmaline. As pointed out earlier, this relation indicates deposition from eolian suspension. This interpretation is substantiated by the fact that the velocity-in-water curves for tourmaline and quartz become essentially coincident when converted to velocities in air using the method described by Hand (1967, p. 515).

Discussion

The presence of eolian suspension deposits in the cross-stratified Casper sandstone near Sand Creek corroborates the eolian interpretation of other sedimentary structures from the same area. Consequently, I conclude that, for the most part, early Permian deposition of sand in this area took place on eolian dunes. On the other hand, eolian deposits at Red Buttes are somewhat surprising inasmuch as Casper sandstone units at that location are interbedded with marine limestone. In this area eolian sand accumulation apparently alternated with deposition of marine carbonate—a situation not too difficult to imagine because the many well-documented cyclic deposits of this age indicate relatively large and rapid fluctuations in the position of the strand line.

PALEOGEOGRAPHIC IMPLICATIONS

Several paleogeographic implications emerge from the study (Fig. 11). The first and most obvious implication is that during transgressive episodes of upper Casper deposition in the southernmost Laramie Basin, the shoreline lay between the Sand Creek and Red Buttes localities. It trended approximately northwest, paralleling the ancestral Front Range uplift to the west. A coastal dune complex separated areas of shallow marine sand and carbonate deposition to the east from the coarse alluvial accumulations of the Fountain arkose fringing the uplift to the west. During regressive phases, the shoreline migrated northeastward; this permitted the accumulation of eolian sand at least as far east as the present location of Red Buttes. However, because of the limited extent of this study, the easternmost position of the shoreline cannot be defined.

The second and more general implication concerns the depositional environment of the Casper Formation and the partially equivalent Tensleep Formation of central Wyoming. Casual observations on these formations at various places in Wyoming suggest

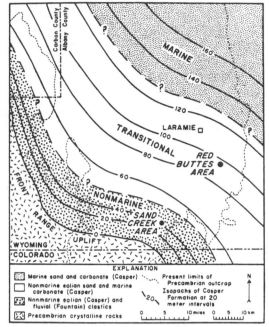

Figure 11. Generalized paleoenvironments and lithotopes in southernmost Laramie Basin at the beginning of Permian time.

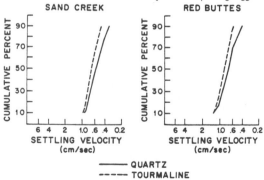

Figure 10. Velocity-in-water curves for quartz and tourmaline from cross-stratified Casper sandstone at Sand Creek and Red Buttes areas.

that the lower portions are dominantly marine. At several localities, however, the upper parts of these units contain cross-stratification and associated sedimentary structures (including large-scale deformation) similar in every respect to those observed in the Casper at Sand Creek. These structures are especially common in Fremont Canyon near Casper, Wyoming, where lower Casper sandstone is burrowed and interlayered with limestone containing a normal marine fauna; whereas the upper part of formation is dominantly highly deformed sandstone with large cross-strata and contains only one thin, nonfossiliferous carbonate unit (John C. Harms, 1973, oral commun.).

If these observations are indicative of the regional picture, then speculation suggests that throughout the region in Desmoinesian time, sand that would form the lower parts of the Casper and Tensleep Formations accumulated in a shallow marine environment. Later, in Missourian to Wolfcampian time, positive regional tectonism in central Wyoming (that is, the Big Horn–Wind River platform of Mallory, 1967, p. G23) resulted in an unconformity at the top of the Tensleep (Mallory, 1967, Fig. 3), caused general regressive conditions, and eventually provided sites for marginal marine eolian deposition. The location of these coastal dune complexes was controlled by the distribution of the landmasses, and their lateral extent was, at least in part, a function of fluctuating sea level.

SUMMARY OF CONCLUSIONS

1. Ripples, truncated ripples, and lag deposits in the Casper sandstone of the southernmost Laramie Basin are similar to those formed by modern eolian processes.

2. Low-dipping cross-strata (10° to 25°) and large-scale deformation, both of which have been cited as evidence against eolian deposition of the Casper sandstone, are in fact consistent with such an interpretation. Modern, moist eolian dunes in western Wyoming consist of low-dipping cross-strata, and published experimental work suggests that the cohesion necessary for large-scale deformation is more likely in moist eolian sand than in saturated subaqueous sand.

3. The presence of eolian suspension deposits in Casper sandstone, determined from settling velocities of light and heavy minerals, provides independent quantitative evidence that corroborates the eolian interpretations of sedimentary structures.

4. Apparently, the Casper sandstone in the southernmost Laramie Basin was deposited as a coastal dune field that bordered the ancestral Front Range uplift and that migrated laterally in response to a fluctuating sea level.

ACKNOWLEDGMENTS

I thank John H. Hanley, Harry C. Haywood, Douglas J. Kirn, William H. Partin, and Richard M. Whitaker for field and laboratory help and for providing many hours of enlightening debate. Discussions with John C. Harms were particularly instructive. Theodore R. Walker, Roger G. Walker, and Darwin R. Spearing critically read the manuscript, and the American Philosophical Society provided financial assistance.

REFERENCES CITED

Agatston, R. S., 1954, Pennsylvanian and Lower Permian of northern and eastern Wyoming: Am. Assoc. Petroleum Geologists Bull., v. 38, p. 508–583.

Ahlbrandt, T. S., 1973, Sand dunes, geomorphology and geology, Killpecker Creek area, northern Sweetwater County, Wyoming [Ph.D. thesis]: Laramie, Wyoming Univ., 174 p.

Allen, J.R.L., 1968, Current ripples: Their relation to patterns of water and sediment motion: Amsterdam, North Holland Publishing Co., 433 p.

Allen, J.R.L., and Banks, N. L., 1972, An interpretation and analysis of recumbent-folded deformed cross-bedding: Sedimentology, v. 19, p. 257–284.

Bagnold, R. A., 1954, The physics of blown sand and desert dunes: London, Methuen and Co., 265 p.

Berg, R. R., 1956, Subsurface stratigraphy of the pre-Niobrara formations in the Shirley and Laramie basins, Wyoming: Wyoming Geol. Assoc., p. 77–84.

Bigarella, J. J., Becker, R. D., and Duarte, G. M., 1969, Coastal dune structures from Parana (Brazil): Marine Geology, v. 7, p. 5–55.

Brown, R. G., 1968, Modern deformational structures in sediments of the Coorong Lagoon, South Australia: Geol. Soc. Australia Spec. Pub. 2, p. 237–242.

Conybeare, C.E.B., and Crook, K.A.W., 1968, Manual of sedimentary structures: Australia Bur. Mineral Resources, Geology and Geophysics Bull. 102, 327 p.

Cooper, W. S., 1958, Coastal dunes of Oregon and Washington: Geol. Soc. America Mem. 72, 169 p.

Del Monte, L., 1949, Correlation of Pennsylvanian and Permian rocks, southeastern Wyoming [M.S. thesis]: Laramie, Wyoming Univ., 63 p.

Dunbar, C. O., and Rodgers, J., 1957, Principles of stratigraphy: New York, John Wiley & Sons, Inc., 356 p.

Finkel, H. J., 1959, The barchans of southern Peru: Jour. Geology, v. 67, p. 614–647.

Frazier, D. E., and Osanik, A., 1961, Point-bar deposits, Old River Locksite, Louisiana: Gulf Coast Assoc. Geol. Socs. Trans., v. 11, p. 127–137.

Glennie, K. W., 1970, Desert sedimentary environments, in Developments in sedimentology, No. 14: New York, Elsevier Pub. Co., 222 p.

Hand, B. M., 1967, Differentiation of beach and dune sands using settling velocities of light and heavy minerals: Jour. Sed. Petrology, v. 37, p. 514–520.

Hanley, J. H., and Steidtmann, J. R., 1973, Petrology of limestone lenses in the Casper Formation, southernmost Laramie Basin, Wyoming and Colorado: Jour. Sed. Petrology, v. 43, p. 428–434.

Hanley, J. H., Steidtmann, J. R., and Toots, H., 1971, Trace fossils from the Casper sandstone (Permian) southern Laramie Basin, Wyoming and Colorado: Jour. Sed. Petrology, v. 41, p. 1065–1068.

Harms, J. C., 1969, Hydraulic significance of some sand ripples: Geol. Soc. America Bull., v. 80, p. 363–396.

Harms, J. C., MacKenzie, D. B., and McCubbin, D. G., 1963, Stratification in modern sands of the Red River, Louisiana: Jour. Geology, v. 71, p. 566–580.

Harvey, J. G., 1966, Large sand waves in the Irish Sea: Marine Geology, v. 4, p. 49–55.

Haywood, H. C., 1973, Depositional processes in the Casper sandstone of the southernmost Laramie Basin as indicated by settling velocities of light and heavy minerals [M.S. thesis]: Laramie, Wyoming Univ., 34 p.

Houbolt, J.J.H.C., 1968, Recent sediment in the southern bight of the North Sea: Geologie en Mijnbouw, v. 47, p. 246–273.

Hoyt, J. H., 1963, Permo-Pennsylvanian correlations and isopach studies in the northern Denver basin, in Rocky Mtn. Assoc. Geologists Guidebook to the geology of the northern Denver basin and adjacent uplifts, 14th Field Conf., Denver, Colo, 1963: p. 68–83.

Hubert, J. F., 1960, Petrology of the Fountain and Lyons Formations, Front Range, Colorado: Colorado School Mines Quart., v. 55, 242 p.

Inman, D. L., Ewing, G. C., and Corliss, J. B., 1966, Coastal sand dunes of Guerrero Negro, Baja California, Mexico: Geol. Soc. America Bull., v. 77, p. 787–802.

Jordan, G. F., 1962, Large submarine sand waves: Science, v. 136, p. 839–848.

Kirn, D. J., 1972, Sandstone petrology of the Casper Formation, southern Laramie Basin, Wyoming and Colorado [M.S. thesis]: Laramie, Wyoming Univ., 43 p.

Knight, S. H., 1929, The Fountain and the Casper formations of the Laramie Basin: A study of the genesis of sediments: Wyoming Univ. Pub. Sci., v. 1, p. 1–82.

——1960, Cross-lamination and local deformation in the Casper sandstone, southeast Wyoming, in Weimer, R. J., eds., Guide to the geology of Colorado: Denver, Rocky Mtn. Assoc. Geologists, p. 228–229.

Mallory, W. W., 1967, Pennsylvanian and associated rocks in Wyoming: U.S. Geol. Survey Prof. Paper 554–G, 31 p.

Marzolf, J. E., 1969, Regional stratigraphic variations in primary features of the Navajo Sandstone, Utah: Geol. Soc. America, Abs. with Programs for 1969, pt. 5 (Rocky Mountain Sec.), p. 50.

Maughan, E. K., and Wilson, R. F., 1960, Pennsylvanian and Permian strata in southern Wyoming and northern Colorado, in Weimer, R. J., and Haun, J. D., eds., Guide to the Geology of Colorado: Denver, Rocky Mtn. Assoc. Geologists, p. 34–42.

McKee, E. D., 1962, Origin of the Nubian and similar sandstones: Geol. Rundschau, v. 52, p. 551–587.

——1966, Structures of dunes at White Sands National Monument, New Mexico (and comparison with structures of dunes from other selected areas): Sedimentology, v. 7, p. 1–69.

McKee, E. D., and Bigarella, J. J., 1972, Deformational structures in Brazilian coastal dunes: Jour. Sed. Petrology, v. 42, p. 670–681.

McKee, E. D., Reynolds, M. A., and Baker, C. H., Jr., 1962a, Experiments on intraformational recumbent folds in crossbedded sand, in Short papers in geology, hydrology, and topography: U.S. Geol. Survey Prof. Paper 450–D, p. D155–D160.

——1962b, Laboratory studies on deformation in unconsolidated sediment, in Short papers in geology, hydrology, and topography: U.S. Geol. Survey Prof. Paper 450–D, p. D151–D155.

McKee, E. D., Douglass, J. R., and Rittenhouse, S., 1971, Deformation of lee-side laminae in eolian dunes: Geol. Soc. America Bull., v. 82, p. 359–378.

Miller, A. K., and Thomas, H. D., 1936, The Casper Formation of Wyoming and its cephalopod fauna: Jour. Paleontology, v. 10, p. 715–738.

Off, Theodore, 1963, Rhythmic linear sand bodies caused by tidal currents: Am. Assoc. Petroleum Geologists Bull., v. 47, p. 324–341.

Opdyke, N. D., and Runcorn, S. K., 1960, Wind directions in the western United States in the late Paleozoic: Geol. Soc. America Bull., v. 71, p. 959–971.

Parker, W. R., 1973, Folding in intertidal sediments on the west Lancashire coast, England: Sedimentology, v. 20, p. 615–623.

Pearson, E. F., 1972, Origin and diagenesis of the Owl Canyon and Lower Goose Egg Formations (Permian) of southeastern Wyoming and adjacent Colorado [Ph.D. thesis]: Laramie, Wyoming Univ., 170 p.

Pederson, S. L., 1953a, Stratigraphy of the Casper and Fountain formations, southeastern Wyoming and north-central Colorado, in Wyoming Geol. Assoc. Guidebook 8th Ann. Field Conf., Laramie Basin, Wyoming, and North Park, Colorado, 1953: p. 18–25.

——1953b, The stratigraphy of the Fountain and Casper Formations of southeastern Wyoming and north-central Colorado [M.S. thesis]: Laramie, Wyoming Univ., 87 p.

Pettijohn, F. J., 1957, Sedimentary rocks: New York, Harper & Brothers, 718 p.

Potter, P. E., and Pettijohn, F. J., 1963, Paleocurrents and basin analysis: New York, Academic Press, Inc., 296 p.

Pryor, W. A., 1971a, Petrology of the Permian Yellow Sands of northeastern England and their North Sea basin equivalents: Sed. Geology, v. 6, p. 221–254.

——1971b, Petrology of the Weissliegendes sandstone in the Harz and Werra-Fulda areas, Germany: Geol. Rundschau, v. 60, p. 524–551.

Rettger, R. E., 1935, Experiments on soft-rock deformation: Am. Assoc. Petroleum Geologists Bull., v. 19, p. 271–292.

Shotton, F. W., 1937, The lower Bunter sandstones of North Worchestershire and East Shropshire: Geol. Mag., v. 74, p. 534–553.

Stanley, K. O., Jordan, W. M., and Dott, R. H., 1971, New hypothesis of Early Jurassic paleogeography and sediment dispersal for western United States: Am. Assoc. Petroleum Geologists Bull., v. 55, p. 10–19.

Steidtmann, J. R., 1973, Structures in the moist, cold-climate sand dunes of southwestern Wyoming, in Wyoming Geol. Assoc. Guidebook 25th Ann. Field Conf., Greater Green River Basin, 1973: p. 209–213.

Stokes, S. L., 1968, Multiple parallel-truncation bedding planes — A feature of wind-deposited sandstone formations: Jour. Sed. Petrology, v. 38, p. 510–515.

Stride, A. H., 1963a, Current-swept sea floors near the southern half of Great Britain: Geol. Soc. London Quart. Jour., v. 119, p. 175–199.

——1963b, Northeast-trending ridges of the Celtic Sea: Ussher Soc. Proc., v. 1, p. 62–63.

Tenny, C. S., 1966, Pennsylvanian and Lower Permian deposition in Wyoming and adjacent areas: Am. Assoc. Petroleum Geologists Bull., v. 50, p. 227–250.

Terwindt, J.H.J., 1971, Sand waves in the southern bight of the North Sea: Marine Geology, v. 10, p. 51–67.

Thomas, H. D., Thompson, M. L., and Harrison, J. W., 1953, Stratigraphy of the Casper Formation: Wyoming Geol. Survey Bull. 46, pt. 1, p. 1–14.

Thompson, M. L., and Thomas, H. D., 1953, Systematic paleontology of fusulinids from the Casper Formation: Wyoming Geol. Survey Bull. 46, pt. 2, p. 15–56.

Thompson, W. O., 1949, Lyons Sandstone of Colorado Front Range: Am. Assoc. Petroleum Geologists Bull., v. 33, p. 52–72.

Visher, G. S., 1971, Depositional processes and the Navajo Sandstone: Geol. Soc. America Bull., v. 82, p. 1421–1424.

Walker, T. R., and Harms, J. C., 1972, Eolian origin of flagstone beds, Lyons Sandstone (Permian), type area, Boulder County, Colorado: Mtn. Geologist, v. 9, p. 279–288.

Whitaker, R. M., 1972, Grain size distributions and depositional processes of the Casper Formation (Pennsylvanian-Permian) southern Laramie Basin, Wyoming [M.S. thesis]: Laramie, Wyoming Univ., 59 p.

White, J. R., and Williams, E. G., 1967, The nature of a fluvial process as defined by settling velocities of heavy and light minerals: Jour. Sed. Petrology, v. 37, p. 530–539.

Wilson, W. H., 1950, The Casper Formation in the Red Buttes area [M.S. thesis]: Laramie, Wyoming Univ., 38 p.

MANUSCRIPT RECEIVED BY THE SOCIETY JANUARY 25, 1974
REVISED MANUSCRIPT RECEIVED MAY 6, 1974

14

Reprinted by permission of The Rocky Mountain Association of Geologists from *Mtn. Geologist* 9(2–3):279–288 (1972)

EOLIAN ORIGIN OF FLAGSTONE BEDS, LYONS SANDSTONE (PERMIAN),

TYPE AREA, BOULDER COUNTY, COLORADO[L]

T. R. WALKER: University of Colorado,

Boulder, Colorado

J. C. HARMS: Marathon Oil Company,

Littleton, Colorado

ABSTRACT: Typical Lyons Sandstone--the conspicuously cross-bedded flagstones of the type area--was deposited by eolian processes and represents a complex of ancient dunes resting on alluvial plain deposits. The most informative exposures are located in the Sterling quarry, near the village of Lyons, but similar features can be observed in other outcrops along the Front Range.

Large-scale steep sets of cross-strata are interpreted as accretion faces of dunes. Subsets of cross-laminae indicate that sand was drifted laterally across these faces, and avalanching on angle-of-repose slopes was not common or not commonly preserved. Major erosional surfaces bounding the large sets represent interdune blowouts, judging by their scoop-like, intersecting forms. Equally spaced coarse sand grains on some of these surfaces also suggest a deflation origin. Low parallel ripples are interpreted as wind-formed, and circular impressions as raindrop imprints. Avalanche structures are found on rare surfaces. Detailed preservation of animal tracks indicates that the sand was dry when these spoor were formed.

The Algodones Dunes in southeastern California may offer a reasonably accurate modern analog to the Lyons Sandstone.

INTRODUCTION

The depositional environment of the Lyons Sandstone has been interpreted in two ways, based upon sedimentary structures and other characteristics. Some writers (Fenneman, 1905, p. 23; Heaton, 1933, p. 149; Thompson, 1949) interpreted the formation as a nearshore marine deposit; whereas others (Vail, 1917, p. 90-93; Tieje, 1923) claimed that it represents ancient dunes.

The most comprehensive study of structures in the Lyons Sandstone is that of Thompson (1949), who concluded that it is an ancient beach deposit. His pioneering work, which included a detailed comparison to modern beaches in California, has been influential in the conclusions of later published work (Hubert, 1960, p. 234-235; Blood, 1970), and a beach origin has been widely accepted.

We believe that the typical Lyons Sandstone--that is, the conspicuously cross-bedded flagstones that are extensively quarried for building stone in the Boulder-Lyons region--was deposited by eolian processes and represents a complex of ancient dunes. This reinterpretation is based upon several characteristics of stratification, other sedimentary structures, and texture. A few of these characteristics of the Lyons are described here for the first time; others have been previously described but are here interpreted in new ways, based upon recently improved understanding of shoreline and eolian processes.

The most informative exposures of the flagstone beds of the Lyons are located in the Sterling quarry, at the northwest edge of the village of Lyons. Most of the photographs and diagrams in this paper are taken from these exposures. Similar features can be found at many points elsewhere along the foothills of the Front Range, but nowhere are they as well exposed as in the Sterling quarry.

Although we believe that the flagstone beds were deposited by eolian processes, these units are interbedded with other depositional facies. In the Lyons area, underlying sandstones occur in relatively thin, continuous, and tabular beds which differ significantly from the flagstone facies. These beds contain abundant horizontal stratification or trough cross-stratification, poorly sorted conglomerates, and layers of mudstone. Most of these beds and the conglomerate-filled channels noted by Thompson (1949) were probably deposited by fluvial processes, but we have not attempted to make a detailed study or interpretation of such units. However, close association of fluvial and eolian deposits is common in nature. The Lyons Sandstone is overlain on a sharp contact by the Lykins Formation, composed of red mudstones with some beds of algae-bearing dolomite and gypsum.

HIGH-ANGLE CROSS-STRATIFICATION

The Lyons Sandstone is characterized by large-scale cross-stratification, with high dips commonly

[1]Manuscript received and accepted, February 23, 1972.

Fig. 1.   View looking west in Sterling quarry. Broken line marks position of major unconformity between super-imposed dunes. Dunes are lettered to facilitate cross-referencing with other figures.

exceeding 25 degrees (Figs. 1, 2). Individual sets are thick, ranging from 10 to more than 40 ft, and commonly can be traced laterally for hundreds of yards. They are bounded by erosional surfaces that dip less steeply than the cross-strata. Laminations viewed on vertical faces parallel to the direction of maximum dip characteristically are straight (Figs. 1, 2, 3); viewed on major bedding surfaces, laminae are straight or slightly curved (Fig. 6). This kind of cross-stratification would be described as wedge-shaped planar sets by most geologists. That deposits having cross-stratification of such large scale and steep inclination are products of eolian, not aqueous, processes is supported by the fact that similar sets are commonplace in modern dunes (McKee, 1966), whereas they are very rare in modern water-laid deposits.

We wish to emphasize, however, that we have found no examples of cross-strata that dip at or close to the maximum angle of repose for sand of about 34 degrees. The largest dip we have measured in the quarry area is 28 degrees. We suggest that the absence of dips approaching 34 degrees is not detrimental to an eolian interpretation, although this is a commonly cited criterion for ancient dunes, and such high dips are common in some modern dune sands (Bagnold, 1941, p. 201; McKee, 1966, Figs. 6, 7, Bigarella and others, 1969, Figs. 9, 10). We attribute the absence of high dips in the Lyons Sandstone to two factors:

.1.   Some types of dunes, such as seif dunes, have relatively few angle-of-repose faces and these usually occur along the highest dune crests. The lower slopes of such dune masses, which are the most likely to be preserved, accrete mainly by sand driven laterally along the slope by variable winds, rather than by mass avalanching (Bagnold, 1941, Fig. 83; McKee and Tibbitts, 1964, Fig. 6). The rarity of avalanche markings, the abundance of ripples on lee slopes, and the geometry of sub-sets of cross strata in the Lyons Sandstone, discussed in more detail in following sections, suggest that much accretion was by nonavalanche mechanisms and that many of these dunes did not initially contain numerous angle-of-repose dips.

2.   We suggest that dips were somewhat flattened by post-depositional compaction and intrastratal alteration. Modern dune sands have porosities ranging from 40 to more than 50 percent (Pryor, 1971), and framework grains should have tangential contacts averaging about one contact per grain when viewed in thin section (Taylor, 1950; Gaither, 1953; Pettijohn, 1957, p. 674). Petrographic studies of the Lyons Sandstone between Lyons and Eldorado Springs by Veeh (1959, p. 42) showed that intergranular area between framework grains averages less than 20 percent, contacts are sutured or concavo-convex, and contacts per grain average 3 or more. In addition, other petrographic studies (Walker, unpublished data) show that unstable silicate grains such as feldspars in places have been altered to clay which, owing to its softness, has been squeezed into interstitial areas. These observations suggest that the Lyons has undergone substantial compaction. Because these post-depositional effects would reduce the dip of cross-laminae, it seems reasonable to conclude that the original dips were greater than those now observed. Indeed, calculations show that if an initial porosity of 40 percent were reduced to a final porosity of 20 percent, angle-of-response dips (34 degrees) would be decreased to about 27 degrees.

Cross-strata can be traced without interruption or change of dip over vertical distances of tens of feet; the thickest set in the Sterling quarry is 42 ft. This continuity suggests that the sets were deposited either entirely in air or completely submerged; they cannot reflect deposits that were partly exposed and partly submerged. Wind-blown dunes, because of the scale and simplicity of accretion surfaces, commonly contain such large sets. Beach and near-shore sand deposits on the other hand, are now known to contain numerous small sets of cross-strata with variable dips (Reineck, 1963; Hayes and others, 1969; Clifton and others, 1971). Large sets of cross-strata are relatively rare in deposits of unquestionable subaqueous origin, and such sets typically are

Fig. 2. .Steeply dipping foreset laminae. The cross-strata have restored dip of approximately 28° and can be traced across a vertical distance of 42 ft. View is looking south. The regional dip is eastward (left in photo) at 11°. Dunes are lettered to facilitate cross-referencing with other figures.

solitary and overlain and underlain by beds of contrasting textures and sedimentary structures. The Lyons cross-strata characteristically are tens of feet thick and, if water-laid, would imply maximum water depth exceeding tens of feet. Such depths, however, are incompatible with the other structures that commonly occur on the cross-lamination surfaces. Evidence of environment based on these structures is presented in later paragraphs.

TRANSPORT DIRECTION

The Lyons Sandstone typically is composed of sequences of superimposed sets of cross-strata that have different directions of dip and are separated by scoop-like erosional surfaces. In the Sterling quarry, the dip direction of both the cross-strata and the erosional surfaces is highly variable (Figs. 3, 4). Although we have not collected systematic directional data, our impression is that conspicuous large sets in the Lyons-Boulder area dip in a broad arc from southeast to west. Blood (1970, Fig. 10) found that the average dip direction in the Morrison area was to the southeast; Howard (1966, Fig. 3), in a study of directional features in the underlying Fountain Formation, showed inferred current directions ranging from northeast to southeast.

Transport of sand in a general southeasterly direction in the Sterling quarry area is consistent with regional reconstructions of probable source areas and basins of deposition. The highly variable local dip directions show that accretion occurred along faces of diverse orientation. These characteristics are compatible with a dune interpretation.

SCOOP-LIKE EROSIONAL SURFACES

Sets of cross-strata are separated by scoop-like erosional surfaces which have been extensively exposed in the quarries where overlying beds have been removed. These surfaces are large and broadly curved, and they

intersect along gently curved nicklines (Figs. 1, 2, 3, 6). These surfaces have been interpreted as beach cusps (Thompson, 1949), but their large scale, diverse orientation, and lack of arcuate concentrations of coarser material suggest that they are not swash-zone features. We interpret these scoop-like surfaces as interdune blowouts that have formed in a manner illustrated on Figure 4. Initially, a set of dunes migrated across a more or less flat erosional surface (Unconformity I) which truncates strata of older dunes (Figs. 4, 6). Individual dunes were separated by interdune areas in which the beveled surface of the older dunes was exposed in a manner similar to that commonly found in modern dune fields (Fig. 5). As the wind direction changed, the exposed surface in the interdune area continued to be eroded, forming a blowout; whereas the same surface was protected and preserved under the dunes until they, too, were partly removed by erosion. In this manner, a younger unconformity (Unconformity II) was formed which transects the older unconformity along the margin of the blowout (Fig. 4). The intersection of these two unconformities is reflected in a nickline which outlines the ancient scoop-like blowout. Examples of such nicklines are commonplace in the Sterling quarry (Figs. 2, 3, 6). The photograph in Figure 3 shows outcrop relationships that correspond to the left half of the diagram in Figure 4, from which dune C of Figure 3 was omitted for simplicity's sake.

SUBSETS OF CROSS-LAMINAE

The extensively stripped erosional surfaces in the Sterling quarry expose subsets of cross-laminae within large sets of high-angle cross-strata. The traces of laminae in these subsets revealed by the gently dipping erosional surfaces are commonly divergent and truncated. The photograph and sketch on Figures 6 and 7 illustrate the complex geometry of these subsets.

The south-dipping cross-strata of dune A (Fig. 1) did not build by uniform accretion along a planar surface. Rather, sand drifted with a lateral component along

Fig. 3.   Three sets of partially eroded dunes.   Compare with diagram in Fig. 4.   Dunes are lettered to facilitate cross-referencing with other figures.

these large-scale sloping surfaces, building slightly divergent subsets separated by surfaces of truncation (Figs. 6, 7). This geometry inescapably indicates that much of the sand was added as increments transported along the strike of the major depositional slopes and was not deposited by simple avalanching. The orientation of ripples, which invariably trend parallel to the dip of major cross-strata, also indicates transport along, rather than down, the slope. The interior of dune A, exposed by broad erosional surfaces (Figs. 1, 2, 3), shows that most subsets migrated westward across the south-dipping dune faces, but some also moved eastward.

We believe that these smaller subsets of cross laminae support an eolian interpretation. Eddying winds carry sand along the slopes of many modern dunes, particularly when the dunes are of seif or longitudinal form. Similar sets of cross-laminae are commonly observed on beveled dune surfaces (McKee, 1966, Plate VII) but have not been observed in modern beaches or sand bars (Reineck, 1963; Klein, 1970).

RIPPLES

Cross-stratification surfaces in the Lyons Sandstone commonly are ripple-marked (Fig. 8), and the ripples are invariably oriented parallel to the dip of the cross-laminations. The ripples are commonly so low as to be obvious only in low-angle illumination. The crests are straight to slightly curved, parallel, and mostly spaced at 4-5 in. The profiles are asymmetric; but because of the low relief, this asymmetry is difficult to observe except on the very best examples.

These ripples are strikingly similar to those that commonly occur on the lee slopes of modern dunes (Fig. 9), where they are caused by wind eddies that sweep laterally along the face. They are dissimilar to ripples formed by currents or waves in water, because they are so regular in pattern and have such high ripple indices (Harms, 1969). If the ripples like those shown on Figure 8 had been formed in shallow water by wave action, some sets of ripples should be oriented so that the crests are

not parallel to the dip of the slopes on which they formed. Moreover, if the ripples had been formed in water by waves acting on a sloping face, the spacing and height of the ripples would vary in response to depth changes down this slope. No such variations have been observed on rippled beds anywhere in the Lyons quarries.

RAINDROP IMPRESSIONS

In many places the upper surfaces of cross-strata of Lyons Sandstone are marked by abundant and more or less uniformly distributed, ring-shaped impressions which are strikingly similar to raindrop impressions on modern dune sands (cf. Fig. 10 with Fig. 11). These structures have been interpreted as bubble impressions by Thompson (1949), but we believe they cannot have this origin for the following reasons:

1. Bubble impressions in modern sands are not encircled by a raised rim (Twenhofel, 1950, p. 623), such as that which characterizes these structures in the Lyons Sandstone.

2. Modern bubble impressions commonly pass downward into vertical tubes through which the gas (usually air) that creates the bubble escapes to the surface. We have found no evidence of such tubes below the structures in the Lyons Sandstone; on the contrary, the laminae below are invariably uninterrupted.

3. Bubble impressions typically occur on the foreshore of beaches where energy is high, and erosion and redeposition are often repeated. The chances of preserving such delicate structures under these conditions seem almost impossible.

4. Bubble impressions would occur on a beach surface between tide limits. In the Lyons examples, impressions occur on surfaces that can be traced over vertical distances as large as 40 ft and would require enormous tidal ranges if interpreted as bubble-formed. Other features consonant with large tides, such as extensive tidal-flat deposits, are not recognized in the Lyons interval.

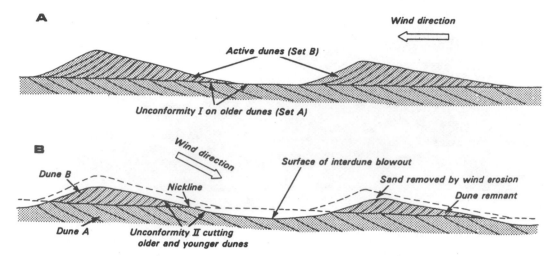

Fig. 4.   Diagram showing development of scoop-like erosional surfaces and nicklines. See text for explanation.

The preservation of these delicate raindrop impressions must have required a special environment and special post-depositional conditions. Although conjectural, we believe that they may imply the following sequence of events: The slopes on which the impressions occur were formed as the dunes migrated in response to periods of higher velocity winds. Such winds are sporadic and sand-size material is not moved in intervening periods of calm or gentle breeze. The raindrop impressions, in our opinion, were formed by rain falling on the dune surface during these calm intervals. The quiet wind conditions also allowed dust to settle from the atmosphere and form a thin film of clay. This clay film helped to protect these delicate structures and to produce a cleavable bedding surface between this and the succeeding sand lamina. Additionally, the dune surface probably was repeatedly moistened by water condensed from the atmosphere in the form of dew or fog, as suggested by McKee (1947). The moisture supplied in this manner may have initiated very weak cementation of the sand on the surface of the dunes, perhaps by hydrolysis of unstable silicates such as feldspars or by transfer of small amounts of soluble compounds. We do not envision a strong surface crust on the sand, but only sufficient cohesion to resist erosion when the next layer of sand was distributed across the surface.

ANIMAL TRACKS

The only fossil evidences of life that have been reported from the Lyons Sandstone are tracks of animals that are commonly preserved in remarkable detail (Fig. 12). Two types of tracks have been discovered in the Lyons quarries: *Laoporus coloradoensis* variously interpreted as an amphibian, a reptile, or a mammal-like reptile; and *Paleohelcura lyonensis* a scorpionid. These tracks

have been previously described and illustrated (Henderson, 1924; Gilmore, *in* Lee, 1927, p. 12, Pl. 33; Topelman and Rodeck, 1936; see also Thompson, 1949, p. 63-64 for discussion of history of classification of the tracks).

Were these tracks formed subaerially or subaqueously? Experimental studies with several animals indicate that the tracks could only be formed subaerially by animals walking on dry sand. The origin of tracks in the Coconino Sandstone of Arizona, which are almost identical with the tracks in the Lyons Sandstone (Thompson, 1949, p. 63), are discussed by McKee, (1947). He experimented with living reptiles and arthropods, attempting to duplicate various features of the tracks in the laboratory, and concluded that well-defined tracks could be formed in dry sand but not in wet or damp sand. He summarized (p. 27) that "... clearly defined tracks in fine sand such as those represented in the Coconino sandstone, are best explained as having been formed when the sand was dry." Equally important are laboratory experiments by Peabody (*in* McKee, 1947, p. 27) who studied tracks of living salamanders and found that "...it is practically impossible for a salamander to record a diagnostic trackway under water. The greater buoyancy of the salamander in water precludes a weight distribution similar to that which is present during locomotion on land. Under water the salamander finds it easier and more convenient to swim from place to place. When leaving the water, it swims to shore and then walks out on land." Peabody also reports: "Many observations on recent mudflats and the collection of numerous subaerial trackways therefrom have failed to discover a single instance in which the local amphibians, reptiles, or mammals left any underwater trackways. Consecutive kick marks of a swimming frog or wading trackways of birds such as the bittern may be found, but

Fig. 5. Beveled foreset laminae of partly eroded dune exposed in modern blowout. Active dunes are visible in distance. White Sands National Monument, New Mexico. (cf. with Fig. 6)

Fig. 6. Beveled laminae on surface of partially eroded older dune (Dune A). Two nicklines are shown. View is looking west. Compare with Fig. 5. (See Fig. 1 for location).

no occurrence of a diagnostic trackway recorded by a vertebrate completely immersed in water has been noted. Such 'wading' trackways as I have observed show the trackway pattern well enough but details of the individual footprints are poorly shown."

The tracks in the Lyons Sandstone are preserved in such detail that each claw mark on each footprint commonly is clearly discernible (Fig. 12). Tracks showing such excellent preservation of details could have been formed only in sand that was subaerially exposed and dry, based upon the studies quoted above (cf. Fig. 12 with McKee, 1947, Pl. 1, Figs. 1 and 2). Following their formation, the conditions that allowed the tracks to be preserved are thought to be like those that preserve the raindrop impressions discussed previously.

AVALANCHE STRUCTURES

Structures similar to those formed by avalanching of loose sand on modern dunes (Fig. 13) can be found in a few places in the Lyons quarries (Fig. 14). The structures in the Lyons Sandstone have two important characteristics that are nearly identical to those seen in the modern examples:

1. The margins are straight, extend down the maximum slope, and are step-like, and
2. Garland patterns cover the interior of the avalanche. Equally sharp outlines are uncommon on subaqueous slip faces. We conclude that these structures reflect subaerial avalanching and are additional evidence that the Lyons is an ancient dune deposit. However, because avalanche structures are rare, angle-of-repose slopes were apparently uncommon or rarely preserved, as noted in preceding discussions.

GRAIN SIZE AND LAG LAYERS

The Lyons Sandstone has a median diameter in the fine sand range and is well sorted. Because this sand-

stone is well indurated by quartz cement in the quarry area and many original feldspar grains have been altered by intrastratal solution, we have not attempted detailed size analyses. However, sands with similar textures are easily transported and sorted by eolian processes. We believe that the textures are compatible with a dune interpretation. Indeed, the absence of pebbles and granules and beds of silt- or clay-size material is best attributed to the powerfully selective soring processes of the wind. Most extensive modern dune sands are fine grained (Bagnold, 1941).

Coarse grains of sand are distributed uniformly on some of the scoop-like erosional surfaces and offer an informative exception to the textural homogeneity. These layers rest on truncation surfaces, are composed of medium to coarse feldspar or quartz grains which are rather uniformly spaced a few mm apart, and are only one grain thick. A representative photograph of such a layer is shown on Figure 15.

These coarse sand layers resemble lag deposits found in interdune blowouts in modern examples (Fig. 16; see also McKee and Tibbitts, 1964, Pl. I B). We believe that the layers in the Lyons were formed in just such a manner. The uniform spacing of coarser grains appears to be uniquely attributable to eolian processes, as illustrated and explained by Bagnold (1941, Figs. 10, 53). Saltation of wind-driven finer sand readily disperses coarser grains over a surface of deflation, as these few coarser grains are exposed and concentrated from an underlying deposit. Similar uniform dispersion of coarse sand grains by erosion under water has not been observed. Along beaches,

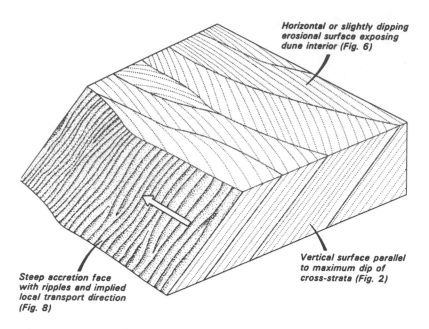

Fig. 7.   Block diagram of subsets of cross-laminae.

Fig. 8.   Ripple marks on accretion face in Dune A.
Note orientation parallel to dip, and compare with
Fig. 9. (See Fig. 1 for location.)

Fig. 9.   Ripples on the steep face of a modern dune.
Algodones Dunes near Yuma, Arizona.   The view faces
obliquely from the toe of the slope toward the dune crest
at the skyline.   The ripple crests are parallel to dip of
the face.   Note the lizard and rodent tracks.   Pencil is
6 in.

Fig. 10.   Raindrop impression on upper surface of foreset beds in Lyons Sandstone.   Compare with Fig. 11.
(see Fig. 2 for location).

Fig. 11.   Raindrop impressions on modern dune sand,
Sonoran desert, Baja California, Mexico.

coarser material is concentrated where breakers plunge;
in the swash zone, coarser grains occur in arcuate or
cluster arrangements.   The lag deposits on erosional
surfaces in the Lyons Sandstone indicate an eolian rather
than a shoreline origin.

SUMMARY OF INTERPRETATION

Several types of sedimentary structures are well
exposed in the Sterling quarry; similar features can be
seen in many other outcrops along the foothills, but not
as well or so completely. All of these sedimentary
structures, taken together, indicate that the cross-
stratified flagstone beds of the Lyons were deposited by
eolian processes in a dune complex. Some of the individual structures are incompatible with a beach or
nearshore marine origin. Exact modern analogs to the
Lyons may not exist, and we are certainly not familiar
with all possible examples. However, the Algodones
Dunes in southeastern California (Norris and Norris,
1961) exhibit some similarities in setting. Mountains
of substantial relief feed alluvial fans across which
these dunes have moved (Fig. 17). These modern components may be analogous to the Ancestral Front Range,
Fountain conglomerates, and Lyons sandstones.

ACKNOWLEDGMENTS

This study was partly supported by National
Science Foundation Grants GP-4060 and GA-1692.
William C. Bradley, John B. Hayes, David B.
MacKenzie, and Darwin R. Spearing reviewed the
manuscript. We gratefully acknowledge all of
this aid.

Fig. 12.   Tracks of *Laoporus coloradoensis* in Lyons
Sandstone, Lyons, Colorado. University of Colorado
Museum collection. Scale is 20 cm.

Fig. 13.　Avalanche on slip face of modern dune, Sonoran desert, Baja California, Mexico. Base of avalanche is at base of photo. The sun is striking the dune from the right. Jackknife gives scale.

Fig. 15.　Lag deposit of coarse sand on beveled surface of Dune A. Some grains have been removed by weathering, and their positions are marked by small pits. (See Fig. 2 for location of photographed surface.)

Fig. 14.　Avalanche structure on foreset bedding plane in Lyons Sandstone. This example is located in Dune C in northwestern part of the Sterling Quarry. Base of avalanche is at base of photo. The sun is striking the outcrop from the left. Jackknife gives scale.

Fig. 16.　Modern lag gravel on a serir bordering dunes, central Libya. The grains are coarser than the Lyons example (Fig. 15), but the even spacing caused by deflation and saltation of sand across this surface indicates that the eolian process was similar in both cases.

Fig. 17. View looking northeast on margin of Algodones Dunes near Yuma, Arizona. View depicts interpreted modern analogue of Lyons Sandstone (dunes), Fountain conglomerates (alluvial fans) and Ancestral Front Range (mountains in distance). (The nature of the animal trails has changed.)

REFERENCES

Bagnold, R. A., 1941, The Physics of blown sand and desert dunes: London, Methuen and Company, 265 p.

Bigarella, J. J., R. D. Becker, and G. M. Duarte, 1969, Coastal dune structures from Parana (Brazil): Marine Geology, v. 7, p. 5-55.

Blood, W. A., 1970, Upper portion of the Fountain Formation and the Lyons Formation at Morrison, Colorado: Mtn. Geologist, v. 7, p. 33-48.

Clifton, H. E., R. E. Hunter, and R. L. Phillips, 1971, Depositional structures and processes in the non-barred high-energy nearshore: Jour. Sed. Petrology, v. 41, p. 651-670.

Fenneman, N. M., 1905, Geology of the Boulder district, Colorado: U. S. Geol. Survey Bull. 265, 101 p.

Gaither, A., 1953, A study of porosity and grain relationships in experimental sands: Jour. Sed. Petrology, v. 23, p. 180-195.

Harms, J. C., 1969, Hydraulic significance of some sand ripples: Geol. Soc. America Bull., v. 80, p. 363-396.

Hayes, M. O., F. S. Anan, and R. N. Bozeman, 1969, Sediment disperal trends in the littoral zone; a problem in paleogeographic reconstruction: in Coastal Environments NE Massachusetts and New Hampshire, SEPM Field Trip, May 9-11, 1969, Coastal Research Group, Univ. Massachusetts, p. 290-315.

Heaton, Ross L., 1933, Ancestral Rockies and Mesozoic and Late Paleozoic stratigraphy of the Rocky Mountain region: Amer. Assoc. Petroleum Geologists Bull., v. 17, p. 109-168.

Henderson, J., 1924, Footprints in Pennsylvanian sandstones of Colorado: Jour. Geology, v. 32, P. 226-229.

Howard, J. C., 1966, Patterns of sediment dispersal in the Fountain Formation of Colorado: Mtn. Geologist, v. 3, p. 147-153.

Hubert, J. F., 1960, Petrology of the Fountain and Lyons Formations, Front Range, Colorado: Colorado School Mines Quart., v. 55, p. 1-242.

Klein, G. D., 1970, Depositional and dispersal dynamics of intertidal sand bars: Jour. Sed. Petrology, v. 40, p. 1095-1127.

Lee, W. T., 1927, Correlation of geologic formations between east-central Colorado, central Wyoming and southern Montana: U. S. Geol. Survey Prof. Paper 149, 80 p.

McKee, E. D., 1947, Experiments in the development of tracks in fine cross-bedded sand: Jour. Sed. Petrology, v. 17, p. 23-28.

_____, 1966, Structures of sand dunes at White Sands National Monument, New Mexico (and a comparison with structures of dune from other selected areas): Sedimentology, v. 7, p. 1-69.

_____, and G. C. Tibbitts, Jr., 1964, Primary structures of a seif dune and associated deposits in Libya: Jour. Sed. Petrology, v. 34, p. 5-17.

Norris, R. M., and K. S. Norris, 1961, Algodones Dunes of southeastern California: Geol. Soc. America Bull., v. 72, p. 605-620.

Pettijohn, F. J., 1957, Sedimentary Rocks, Second Ed.: New York, Harper and Bros., 718 p.

Pryor, W. A., 1971, Reservoir inhomogeneities of some Recent sand bodies: Soc. Petroleum Eng. 46th Ann. Mtg., New Orleans, paper SPE 3607.

Reineck, H. E., 1963, Sedimentgefuge im bereich der sudlichen Nordsee: Abh. senckenb. naturf. Ges. 505, p. 1-138.

Taylor, J. M., 1950, Pore-space reduction in sandstones: Am. Assoc. Petroleum Geologists Bull., v. 34, p. 701-716.

Thompson, W. O., 1949, Lyons sandstone of Colorado Front Range: Am. Assoc. Petroleum Geologists Bull., v. 33, p. 52-72.

Tieje, A. J., 1923, The redbeds of the Front Range in Colorado: Jour. Geology, v. 31, p. 192-207.

Toepelman, W. C. and H. G. Rodeck, 1936, Footprints in Late Paleozoic red beds near Boulder, Colorado: Jour. Paleontology, v. 10, p. 660-662.

Twenhofel, W. H., 1950, Principles of sedimentation, Second Ed.: New York, McGraw-Hill, 673 p.

Vail, C. E., 1917, Lithologic evidence of climatic pulsations: Science, New Ser., v. 46, no. 1178, p. 90-93.

Veeh, H. H., 1959, A petrographic study of the cementation in the Lyons sandstone: M.S. thesis, Univ. Colorado, 49 p.

15

Reprinted from pp. 351, 353–354, 356–363, 367–369 of *Jour. Geol.* **82:**351–369 (1974)

CORNSTONE (FOSSIL CALICHE)—ITS ORIGIN, STRATIGRAPHIC, AND SEDIMENTOLOGICAL IMPORTANCE IN THE NEW RED SANDSTONE, WESTERN SCOTLAND[1]

R. J. STEEL

Geologisk Institutt, University of Bergen, Norway

ABSTRACT

An examination of more than 200 cornstone (concretionary carbonate) profiles of New Red Sandstone age, in western Scotland, confirms that these carbonate accumulations are analogues to the caliche of semi-arid areas. New evidence of laminated, pisolitic and brecciated textures from the cornstone "caprock" strongly suggests that downward percolating water and non-tectonic fracturing were important during subaerial diagenesis.

The cornstones occur preferentially in floodplain deposits, and particularly within the upper parts of alluvial, fining-upwards cyclothems. The occurrence of cornstones in this part of the cyclothems and their persistence through such cyclothemic sequences suggest, contrary to a commonly accepted hypothesis, that climatic controls are important in the formation of such alluvial sequences.

The cornstones are stratigraphically more significant than the associated rock types in the monotonous New Red continental sequences. In particular, thick, mature cornstones indicate condensed sequences and are also a useful palaeoclimatic index, such caliche normally forming in semi-arid regions.

[1] Manuscript received June 8, 1973; revised November 10, 1973.

[*Editor's Note:* Material, including figure 1, has been omitted at this point.]

Fig. 2.—Incipient development of cornstone (type 1) as scattered carbonate nodules in fine-grained sandstone host-rock at Gruinard Bay. Ruler is 30 cm long.

FIELD APPEARANCE

Cornstone is an accumulation of authigenic carbonate. In the New Red Sandstone the host rocks are sandstone or siltstone and, occasionally, conglomerate. There is usually an upward increase in carbonate percentage through the zone of cornstone development (the cornstone profile).

The cornstone has been arbitrarily categorized into five types, the types being recognized by the gross form of the carbonate and the approximate proportion of the host rock still distinguishable:

Type 1.—Cornstone of this type usually appears as small (usually 1–6 cm in diameter), irregularly-shaped nodules which are occasionally crossed by small "veins" of the clastic host rock. The nodules compose less than 10% of the rock (fig. 2).

Type 2.—Here the carbonate nodules are larger (up to 10 cm. diameter) and often vertically elongate (up to 15 cm long), but occupy less than 50% of the rock in the upper part of the profile. There is usually a downward gradation into cornstone of Type 1.

Type 3.—The carbonate appears as nodules, vertical pipes or horizontal sheets. Carbonate occupies more than 50% of the rock but clastic sediment can still be clearly seen within the carbonate framework. There is a downward gradation into cornstone of Type 2 (fig. 3).

Type 4.—In this type of cornstone the carbonate exists as beds within which only rare patches of clastic sediment are seen. There is usually a downward gradation to Type 3 (fig. 4).

Type 4a.—Here there are distinct horizons of laminar, brecciated or pisolitic

FIG. 3.—A cornstone profile on the Island of Rhum. Cornstone of Type 3 with densely packed carbonate nodules grades down into Type 2, where nodules occupy less than half the volume of the rock (middle of hammer shaft). The hammer head rests on the coarse-grained member of an alluvial fining-upwards cyclothem. The coarse member of the overlying cyclothem is also visible.

FIG. 4.—A thick cornstone profile (bottom two-thirds of cliff face) on Inch Kenneth, Island of Mull. Bedded cornstone (type 4), occupying the top meter of the profile, grades down into nodules and vertical pipes of cornstone (type 3). The cliff face is 6 m high.

carbonate, usually as a capping to Type 4 (figs. 7, 8, and 12). Sometimes the carbonate is partly silicified and occasionally there are alternations of thin beds of carbonate (5 cm thick) and chert (2 cm thick).

This New Red Sandstone cornstone differs from most of the previously described cornstone in the nature of the bedded types. Devonian and Carboniferous cornstone, even when it is bedded, is apparently massive and structureless (Allen 1965, p. 155; Francis et al. 1970, p. 105). An exception is the Devonian cornstone described by Burgess (1961) which contains some of the structures and textures described here. The bedded cornstone here is often texturally complex

and displays carbonate lamination, carbonate or siliceous micro-breccia and cavities or beds of carbonate or chert pisolite, with relationships analogous to those reported from the laminated caprock of recent pedocals.

One of the most obvious features of the various types of cornstone is that they are related and have a preferred sequential arrangement. In other words, Type 4 cornstone is usually underlain by Types 3, 2, and 1 respectively, Type 3 by Types 2 and 1, Type 2 by Type 1, and Type 1 by normal host sediment. This suggests that each sort of cornstone belongs to a "series" in which the end members are Types 1 and 4a. Moreover, the stratigraphic order of the types, and evidence discussed below, suggests that any profile might have developed a Type 4a capping if the cornstone-forming conditions had

[Editor's Note: Tables 1 and 2 are not reproduced here.]

persisted. The fact that a variety of cornstone types exist at any one locality and that each cornstone profile is erosively overlain by clastic sediment probably indicates that the development of a Type 4a capping was at least partly controlled by "time," so that the longer the time interval between episodes of clastic sedimentation the greater the chance of massive cornstone developing. Hence there is a sense in which a cornstone profile of Type 1 may be referred to as immature and a Type 4a profile as mature.

Mature profiles showing the complete sequence of types from 1 through to 4a are rare but notable examples occur on the Islands of Mull (fig. 1, locations 19 and 20), Skye (location 11), and in Morven (location 18).

Cornstone profiles are normally less than 2 m thick. At some localities, however, any particular part of the profile may be extended to an abnormal thickness. Unusual thicknesses (10 m) of Type 2 occur at Gruinard Bay (location 2, fig. 1), while a very thick development (2 m) of Type 4a occurs at Inninmore (location 18). These abnormalities are discussed below.

An analysis of all cornstone profiles in terms of their maturity and the various features identified in the caprock are shown in tables 1 and 2.

Other types of cornstone.—At some localities the cornstone has not developed an asymmetric profile as described above but simply occurs as a massive carbonate blanket overlying or penetrating via cracks and joints, the host rock. This occurs where the host rock is relatively impermeable, usually at the base of the New Red Sandstone near the Pre-Cambrian unconformity. This type of cornstone can be explained in terms of the normal cornstone-forming process, the difference in form resulting from relative imperviousness of the host rock. The distribution of this type of cornstone at the base of the Mesozoic sequences in western Scotland is shown in table 3. These localities are

TABLE 3

An Analysis of the 'Unconformity-type' Cornstone in Terms of the Depth of Penetration Below the Unconformity and the Thickness of Cornstone Directly Overlying the Unconformity

Location	Max. Depth of Cornstone Below Unconformity (m)	Max. Thickness of Cornstone Overlying Unconformity (m)
1	C	0
2	C	0
3	0.3	0
4	C	0
5	C	0
5	C	0
6	0.3	0
7	—	—
8	C	0
9	C	0
10	C	3.0
11	C	0
12	2.0	2.7
13	1.0	0.6
14	0.7	0
15	—	—
16	3.0	2.5
17	—	—
18	C	0
19	C	0
20	1.0	0
21	2.5	2.0
22	C	0
23	C	0
24	—	—
25	—	—
26	3.0	0
27	3.0	0

Note.—C = 'clean' unconformity, i.e., no cornstone penetration below the surface of unconformity.
* See fig. 1 for localities.

compared with others which have a clean unconformity-surface.

MICROSCOPIC EVIDENCE

Most of the carbonate in the cornstone samples studied is authigenic; either replacing the original grains, leaving highly corroded grains, or infilling the pore space. A characteristic of most cornstone profiles is an upward change from pore-infilling to grain replacement. Along with calcite there is a variable amount of chert and dolomite.

Calcite is present in various generations and has a variable time relationship to the

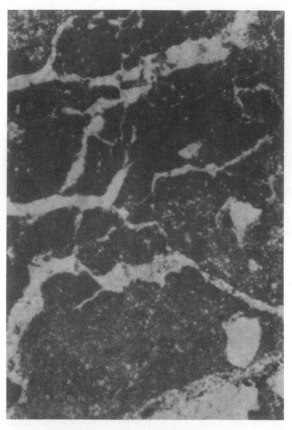

Fig. 5.—A mature stage of caliche development where only rare clastic grains from the original host-rock fabric remain. These clastic grains are floating in a carbonate cement which is crossed by sparry-calcite-filled dilation fractures. Photomicrograph: × 30.

other minerals. A reddish-brown micrite is often variably replaced by large plates of clear sparry calcite, and is often crossed by irregular, sparite-filled dilation fractures (fig. 5). Chert, when present, usually contains relics of calcite. Dolomite (fig. 6) is believed to have formed both before and after the sparry calcite.

Many mature cornstone profiles have a top horizon which, although containing the same minerals as lower levels show unique textures. The top horizons are characteristically laminated, brecciated or pisolitic and usually display some evidence of drusy cavity infilling.

Lamination.—Laminated micrite develops as a coating on the underside of nodular cornstone masses of Type 3, but the main laminated horizons, which are sub-horizontal (figs. 5 and 12), occur in Type 4a. The laminations are caused by differences in calcite coarseness, usually an alternation of micrite and micrite with local sparry calcite mosaic. The laminations are often emphasised by variation in amount of ferric oxide in adjacent bands, or by an interlamination of sandy sediment (fig. 7).

The carbonate may be partly or completely replaced by dolomite or silica. The former often occurs as large, zoned dolomite rhombs; the latter as microcrystalline granular quartz or as mozaic quartz (terminology of Wilson 1966).

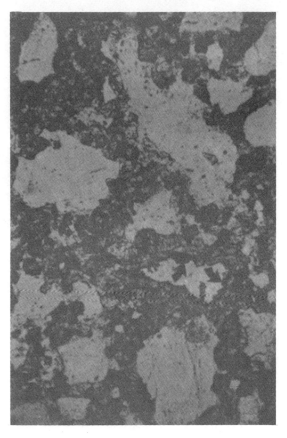

Fig. 6.—A mature stage of caliche development where the quartz grains of the host sandstone have been highly corroded by the carbonate. Most of the carbonate here is dolomite which has apparently replaced sparry calcite (arrowed) of an earlier generation. Photomicrograph: × 30.

Occasional zoned idiomorphic quartz crystals also replace the carbonate. Irregular cavities in these laminated bands, probably voids in the earlier carbonate fabric, are highlighted by chalcedonic overlays and spherulitic chalcedony cement. In most areas of silicification, apart from instances of chalcedony cementation, relic micrite patches often persist.

Pisolite.—Closely associated with laminated carbonate are bands or lenses of pisolite. The pisolites have either a single or composite nucleus of quartz, calcite or chert, and usually possess one or two enveloping micrite laminae (figs. 8 and 9). The laminae can be vertically asymmetric

(fig. 8) and may sometimes house perched inclusions (fig. 9). The pisolites are sometimes closely fitted into small cavities, giving the impression of *in situ* formation; in other instances individual pisolites are fractured or chipped, suggestive of transportation (fig. 9). At most localities some pisolite bands show a reverse grading, with the smallest pisolites lining the irregular cavity floors (fig. 10).

Pisolite bands are also silicified; some of the best preserved pisolitic fabrics being in this state. Idiomorphic quartz crystals occasionally replace complete carbonate pisolites, often fill pore space between pisolites and line cavities within silicified

Fig. 7.—Laminated carbonate from the top of a mature cornstone profile. Layers of sandy sediment alternate with, and occasionally erode into, the carbonate laminae. Actual size.

Fig. 8.—Pisolite filled cavities from the top part of a cornstone profile. The nucleii of the pisolites are usually enveloped by only one or two micrite laminae which is often thickest around the pisolite base (arrowed). Actual size. Compare with the caliche pisolites in plate 3 of Swineford et al. (1958, p. 106).

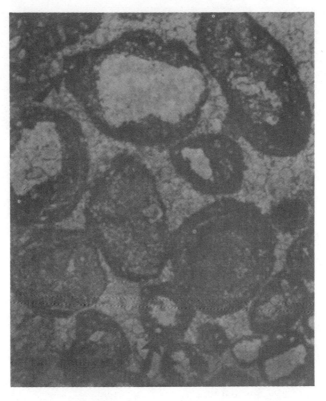

FIG. 9.—Examples of pisolites with single or composite nucleii of quartz (top centre), chert (left centre), or calcite (centre) enveloped by only a few micrite laminae. Perched inclusions (arrowed) are often included between laminae. Some pisolites (e.g., top centre) have been broken, probably during local transportation to their present site. × 30.

pisolites (fig. 11). Broken pisolites of chert and carbonate are sometimes present in the same cavity.

Brecciation.—Brecciated horizons commonly alternate with laminar carbonate or pisolite. The breccia is predominantly locally derived, with fragments merely rotated out of the plane of the parent carbonate layer (fig. 12). The breccia fragments can be of carbonate or chert, and the pore space may be filled with spherulitic chalcedony, mosaic quartz, sandy sediment or pisolites.

These micro-textures clearly show that the diagenetic history of mature cornstones has been complex, including recrystalization, dolomitization, pisolite formation and silicification.

HOST ROCK ENVIRONMENT

The preferred occurrence of cornstone in any particular lithofacies should provide evidence for details of the development of that lithofacies. Conversely, a general understanding of the depositional environment of the cornstone host rock (gained without reference to the cornstone) should provide information regarding the constraints of cornstone development.

Cornstone is much more common in floodplain sequences than in the alluvial fan (piedmont) sequences in the New Red Sandstone of western Scotland (fig. 1), typically occurring in the fine-grained member of cyclic fluviatile units (figs. 2, 3, and 4) which are now commonly known as

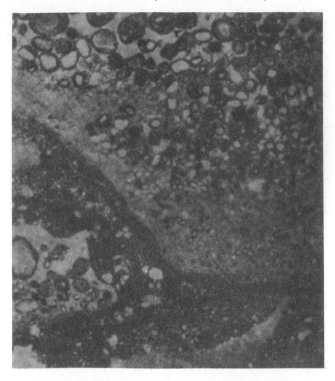

FIG. 10.—Reverse size-grading in a pisolite-filled cavity. × 10. Compare with the caliche pisolite size-grading in figure 7 of Bernoulli and Wagner (1971, p. 144).

fining-upwards cyclothems (Allen 1964, 1970; Moody Stuart 1966; Picard and High 1973). In addition, the base of any cyclothem normally shows an erosional contact with the cornstone of the underlying cyclothem. This indicates that the complex diagenetic processes involved in cornstone development occurred between individual periods of fluvial discharge, a fact of significance when considering the cause of cyclicity in alluvial sequences.

THE ANALOGY WITH CALICHE

There is an obvious similarity between cornstone and carbonate which accumulates in the soil profiles of semi-arid regions. The carbonate profiles in such soils are known as caliche or calcrete (America), kunkur (India and East Africa), croute calcaire (France), nari (Israel), kafkalla and havara (Cyprus), and Omdurman lime (Sudan).

Diagenetic carbonate, interpreted as pedogenic and texturally similar to cornstone is being increasingly described from the stratigraphic record. Fossil caliche has been recorded from Mesozoic sequences in Spain (Nagtegaal 1969), and Italy (Bernoulli and Wagner 1971), from Palaeozoic rocks in U.S.A. (Dunham 1969b), from Tertiary rocks in Germany (Nagale 1962) and U.S.A. (Swineford et al. 1958), as well as from the British sequences already mentioned.

As regards cornstone already described from the British continental sequences, studies have concerned the similarity of the profiles with the Cca horizons in semi-arid soils, and the evidence (e.g., brecciation and some diagenetic fabrics) that the profiles developed *in situ*. Attention is here drawn to a more detailed analogy between recent caliche and the ancient cornstone profiles, and particularly

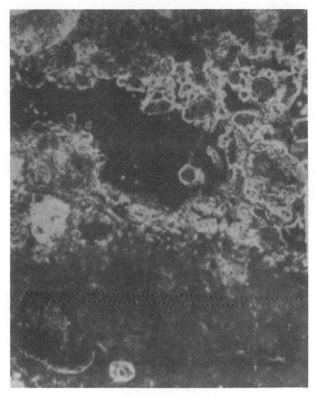

Fig. 11 —A cavity within silicified pisolite which is lined with idiomorphic quartz crystals and filled with chert. The ghost outline of pisolites (left bottom) is caused by areas of relic micrite. × 30.

to new evidence from the laminated zones of the cornstone profiles indicating that diagenesis took place sub-aerially by percolation of water through the vadose zone.

In describing the accumulation of authigenic carbonate in the profiles of desert soils, Gile et al. (1964, p. 74) proposed the terms *K-horizon* and *K-fabric*. The term *K-horizon* refers to the master soil horizon (a profile of sub-horizons) and the term *K-fabric* refers to the diagenetic fabric of the *K-horizon* in which, " . . . fine-grained authigenic carbonate occurs as an essentially continuing medium." The *K-horizon*, whose material takes such forms as massive, laminar, nodular, cylindrical, blocky and platy, is sub-defined on the basis of approximate volumes of K-fabric present (Gile et al. 1964).

A *K1-horizon* may occur near the top of the soil profile; it is relatively soft and nodular and contains at least 50% K-fabric. In the cornstone profiles such a horizon is also occasionally found above the main carbonate horizon. The *K2-horizon* in modern caliches is the most prominent part of the K-horizon, containing at least 90% K-fabric. Usually the K2-horizon is massive or blocky below and laminar above, comparing in detail with Types 3 and 4 (figs. 3 and 4) of the cornstone profiles. Type 4a (figs. 7 and 12) corresponds to the laminar top of K2. In the *K3-horizon*, carbonate content and volume of K-fabric decrease with increasing depth until a *C-horizon*, or bedrock, is reached. K3 corresponds approximately with Types 1 and 2 of cornstones.

There also occurs a detailed similarity between the textures of the caprock from

FIG. 12.—Laminated carbonate (base) has been brecciated and covered by sandy sediment. Younger carbonate laminae drape over the irregular surface. Actual size. Compare with the caliche laminar draping in figure 4 of Gile et al. (1966, p. 354).

the cornstone and caliche profiles. For example:

1. The usual lamination of the cornstone cap is matched by similar lamination in the caprock zone of most caliche (Bretz and Horberg 1949, p. 494), as well as by other laminated, carbonate crusts of sub-aerial origin (Multer and Hoffmeister 1968).

2. The close association of laminated carbonate with micro-breccia, the former often draping across the latter, is well known from caliche caprock (compare fig. 12 with fig. 4 of Gile et al. 1966, p. 354). Rotated breccia fragments have also been recorded in caliche breccia (Swineford et al. 1958, p. 106).

3. The pisolitic texture of the cornstone is analogous to that frequently recorded from caliche (Bernoulli and Wagner 1971, p. 142) as well as from other sub-aerial deposits such as bauxite (Gordon and Tracy 1952) and laterite (Mohr and van Baren 1954). In addition, details of the pisolite morphology from the cornstones are precisely matched in caliche. Pisolites with a composite nucleus (compare figure 9 with figure 9 in Dunham 1969*b*, p. 186), asymmetric "dripstone" pisolites (compare figure 8 with plate 3 in Swineford et al. 1958, p. 106), pisolites in reverse size-graded units (compare figure 10 with figure 7 in Bernoulli and Wagner 1971, p. 144) and a fitted, polygonal pisolite fabric all figure in the analogy.

4. The association of carbonate laminae and amorphous silica (fig. 11) has also been recorded from caliche (Reeves 1970, p. 354).

These textural analogies suggest that the cornstone developed by sub-aerial diagenesis in a zone above the water table.

[*Editor's Note:* Material, including figure 13, has been omitted at this point.]

caliche formation. Cornstone can form in areas with less than 10 cm annual precipitation (C. V. Dolliver, personal communication 1973) and Blatt et al. (1972, p. 258) state that the upper limit of precipitation is generally less than 60 cm annually. However, Reeves (1970, p. 353) suggests that caliche can form in an even wetter climate provided the temperature is relatively high or even over areas with relatively low evaporation provided infiltration is rapid and there are expanses of limestone to supply large amounts of calcium (Reeves, personal communication 1973). Despite these important instances it is probable that thick, mature cornstone profiles imply a semi-arid to arid climate.

The gradual disappearance of mature cornstone profiles towards the top of the New Red Sandstone successions suggests that the climate became progressively wetter with the approach of the Jurassic. Confirmation of such a climatic change is recorded in the sediments themselves, because their structural and textural characteristics necessarily altered with changing hydrologic conditions (Steel, in press).

CONCLUSIONS

A field and microscopic examination of some 200 cornstone profiles of New Red Sandstone age (post-Carboniferous/pre-Jurassic) in western Scotland suggests the following:

1. The sedimentary features of cornstone and the sequence of cornstone types in any one profile are analogous to those in caliches, which are presently forming in most semi-arid and arid regions.

2. Laminated, pisolitic, and brecciated textures from the top of many cornstone profiles suggest the importance of subaerial diagenesis in the cornstone-forming process, and probably imply that an alternation of downward percolating water and non-tectonic fracturing was of major importance. Similar vadose-zone processes occur during soil formation in semi-arid to desert areas today.

3. Cornstones preferentially occur in

CORNSTONE—A PALAEOCLIMATIC INDICATOR

It has been noted that Tertiary/Quaternary caliches are largely restricted to the semi-arid areas of the earth's surface (Goudie 1972, p. 449). To generalize, extreme aridity prevents soil leaching and the mobilization or accumulation of calcium carbonate, while on the other hand, too much moisture tends to leach out most of the solubles required for

New Red Sandstone floodplain deposits, and only occasionally in alluvial-fan areas. Assuming that precipitation, carbonate supply, and surface temperature are equal over areas of floodplain and piedmont, the association of cornstone with the former can be explained in terms of the lack of relief (lower slopes) and the greater amounts of surface moisture likely to be present on floodplains.

4. The presence of laterally extensive profiles at the top of the fine-grained members of alluvial fining-upwards cyclothems indicates that there was a prolonged period of exposure and soil development after the construction of each alluvial cyclothem. This, in turn, implies that precipitation and fluvial discharge were among the major factors in the initiation of any new floodplain-construction event. Hence the commonly accepted idea of the cyclic mechanism being wholly autocyclic is questionable in such floodplain sequences.

5. Cornstone is especially frequent and mature in thin stratigraphic sequences, i.e., it is a "condensed sequence" indicator.

6. Massive caliche requires, in most cases, a semi-arid climate for its development. Extreme aridity is inimical to pedocal soil development while excess rainfall causes leaching of the important soil solubles. Hence extensive, thick cornstone horizons are valuable palaeoclimatic indicators, suggesting an annual rainfall of less than 60 cm for those parts of the New Red Sandstone in which the thick profiles occur.

ACKNOWLEDGEMENTS.—I thank Professor A. J. Whiteman (University of Bergen), Dr. B. J. Bluck (University of Glasgow), and Dr. F. M. Broadhurst (University of Manchester) for helpful discussion and for criticism of the manuscript. The data was collected while I was in receipt of an NERC research studentship.

REFERENCES CITED

ALLEN, J. R. L., 1960, Cornstone: Geol. Mag. (Great Britain), v. 97, p. 43–48.
———— 1964, Studies in fluviatile sedimentation: six cyclothems from the Lower Old Red Sandstone, Anglo–Welsh Basin: Sedimentology, v. 3, p. 163–198.
———— 1965, The sedimentation and palaeogeography of the Old Red Sandstone of Anglesey, North Wales: Proc. Yorks. Geol. Soc., v. 35, p. 139–185.
———— 1970, Studies in fluviatile sedimentation: a comparison of fining-upwards cyclothems, with special reference to coarse member composition and interpretation: Jour. Sed. Petrology, v. 40, p. 298–323.
ARISTARAIN, L. F., 1970, Chemical analysis of caliche profiles from the High Plains, New Mexico: Jour. Geology, v. 78, p. 201–212.
BERNOULLI, D., and WAGNER, C. W., 1971, Subaerial diagenesis and fossil caliche deposits in the Calcare Massiccio Formation (Lower Jurassic, Central Apennines, Italy: Neues. Jahrb. Geologie u Palaeontologie Abh., v. 138, p. 135–149.
BLATT, H.; MIDDLETON, G.; and MURRAY, R., 1972, Origin of sedimentary rocks: Prentice-Hall, New Jersey, 634 p.
BRETZ, J. H., and HORBERG, L., 1949, Caliche in southeastern New Mexico: Jour. Geology, v. 57, p. 491–511.

BROWN, C. H., 1956, The origin of caliche on the northeastern Llano Estacado, Texas: Jour. Geology, v. 64, p. 1–15.
BRUCK, P. M.; DEDMAN, R. E.; and WILSON, R. C. L., 1967, The New Red Sandstone of Raasay and Scalpay, Inner Hebrides: Scottish Jour. Geology, v. 3, p. 168–180.
BUCKLAND, W., 1921, Description of the Quartz Rock of the Lickey Hills: Geol. Soc. London Trans., v. 5, p. 506–544.
BURGESS, I. C., 1961, Fossil soils of the Upper Old Red Sandstone of south Ayrshire: Geol. Soc. Glasgow Trans., v. 24, p. 138–153.
BUTLER, B. E., 1959, Periodic phenomena in landscapes as a basis for soil studies: CSIRO Australian Soil Pub. 14, 20 p.
DUNHAM, R. J., 1969a, Early vadose silt in Townsend Mound (Reef). New Mexico, in FRIEDMAN, G. M., ed., Depositional Environments in Carbonate Rocks: SEPM Special Publication 14, p. 139–181.
———— 1969b, Vadose pisolite in the Capitan Reef (Permian), New Mexico and Texas, in FRIEDMAN, G. M., ed., Depositional Environments in Carbonate Rocks: SEPM Special Publication 14, p. 181–191.
FRANCIS, E. H.; FORSYTH, I. H.; READ, W. A.; and ARMSTRONG, A., 1970, The geology of the Stirling district, Scotland: Mem. Geol. Survey, Scotland, p. 117–136.

GILE, L. H.; PETERSON, F. F.; and GROSSMAN, R. B., 1964, The K-Horizon: a master soil horizon of carbonate accumulation: Soil Sci., v. 99, p. 74–82.

——; ——; — —— 1966, Morphological and genetic sequences of carbonate accumulation in desert soils: Soil Sci, v. 101, p. 347–360.

GORDON, M., JR., and TRACEY, J. I., JR., 1952, Origin of the Arkansas bauxite deposits, *in* Problems of clay and laterite genesis—a symposium: Am. Inst. Mining Metall. Engineers, p. 12–34.

GOUDIE, A., 1972, The chemistry of world calcrete deposits: Jour. Geology, v. 80, p. 449–463.

LOWE, M. J. B., 1965, Some aspects of the stratigraphy and sedimentation of the Triassic rocks of the Hebrides and North-West Highlands of Scotland: Unpub. Ph.D. thesis, St. Andrews Univ, Scotland.

MOHR, E. C. J., and VAN BAREN, F. A., 1954, Tropical Soils: The Hague N.Z. Vitgeverij W. Van Hoeve, 498 p.

MOODY-STEWART, M., 1966, High and low-sinuosity stream deposits, with examples from the Devonian of Spitzbergen: Jour. Sed. Petrology, v. 36, p. 1102–1124.

MULTER, H. G., and HOFFMEISTER, J. E., 1968, Subaerial laminated crusts of the Florida Keys: Geol. Soc. Amer. Bull., v. 79, p. 183–192.

NAGALE, E., 1962, Zur Petrographie und Entstehung des Albstens: Neues. Jahrb. Geologie u. Palaeontologie Abh., v. 115, p. 44–120.

NAGTEGAAL, P. J. C., 1969, Microtextures in recent and fossil caliche: Leidse Geol. Med., v. 42, p. 131–142.

PICARD, M. D., and HIGH, L. R., 1973, Sedimentary structures in ephemeral streams: Elsevier, London, 223 p.

PICK, M. C., 1964, The stratigraphy and sedimentary features of the Old Red Sandstone, Portishead coastal section, north-east Somerset: Geologists Assoc. Proc., v. 75, p. 199–221.

PRICE, W. A., 1958, Sedimentology and Quaternary geomorphology of south Texas: Gulf Coast Assoc. Geol. Soc. Trans., p. 47–75.

REEVES, C. C., JR., 1970, Origin, classification and geologic history of caliche on the southern High Plains, Texas and eastern New Mexico: Jour. Geology, v. 78, p. 352–362.

RIGHTMIRE, C., 1967, A radiocarbon study of the age and origin of caliche deposits: Unpub. thesis, Univ. Texas, Austin.

RUHE, R. V., 1967, Geomorphic surfaces and surficial deposits in southern New Mexico: New Mexico Bur. Mines and Mineral Resources, Memoir, 18, 66 p.

SELLEY, R. C., 1970, Ancient sedimentary environments: Chapman and Hall, London, 237 p.

STEEL, R. J., 1971a, Sedimentation of the New Red Sandstone in the Hebridean province, Scotland: Unpub. Ph.D. thesis, Glasgow Univ., Scotland, 396 p.

—— 1971b, New Red Sandstone movement on the Minch Fault: Nature, Phy. Sci., v. 234, p. 158–159.

—— (in press), New Red Sandstone floodplain and piedmont sedimentation in the Hebridean province, Scotland: Jour. Sed. Petrology.

SWINEFORD, A.; LEONARD, A. B.; and FRYE, J. C., 1958, Petrology of the Pliocene Pisolitic Limestone in the Great Plains: State Geol. Survey Kansas Bull., v. 130, p. 98–116.

WILSON, R. C. L., 1966, Silica diagenesis in Upper Jurassic limestones of southern England: Jour. Sed. Petrology, v. 34, p. 1036–1049.

WHITEMAN, A. J., 1971, The geology of the Sudan Republic: Oxford Univ. Press, London, 290 p.

16

Copyright © 1975 by the Geological Society of America

Reprinted from *Geol. Soc. America Bull.* **86**(3):371–376 (1975)

Chemistry and Mineralogy of Precambrian Paleosols in Northern Michigan

J. KALLIOKOSKI *Department of Geology and Geological Engineering, Michigan Technological University, Houghton, Michigan 49931*

ABSTRACT

Paleosol profiles can be recognized on a basement terrain of granodiorite, peridotite, and diabase that underlie the upper Keweenawan (?) Jacobsville sandstone. The main characteristics of the granodiorite and peridotite profiles are a substantial loss in SiO_2, Al_2O_3, and alkaline earths; a smaller loss in Fe_2O_3 and NiO; the concentration of dolomite and quartz in paleocaliche horizons; and the strong pigmentation of some of the primary minerals at the lower fringe of weathering. The granodiorite paleosol consists of a mixture of secondary and primary minerals: sericite, chlorite, vermiculite, biotite, muscovite, K-feldspar, dolomite, quartz, hematite, and rutile. The paleosol on peridotite contains chlorite, vermiculite, dolomite, quartz, hematite, and talc. The paleosol profiles resemble present-day ones developed under semi-arid to arid conditions. *Key words: paleosols, chemical analysis, paleoclimatology, paleogeography, Precambrian, weathering.*

INTRODUCTION

Hamblin (1958) noted that in northern Michigan the upper Precambrian[1] Jacobsville sandstone overlies an irregular erosion surface with up to 400 ft (130 m) of local relief and that this erosion surface contains local concentrations of weathered debris. On Presque Isle Point at Marquette, Michigan, the physical appearance and general mineralogical nature of one of these paleosol occurrences has been described by Krimmel (1941) and Gair and Thaden (1968). Krimmel interpreted it to have formed by weathering and Gair and Thaden, by weathering and hydrothermal alteration of peridotite in pre-Jacobsville time (Fig. 1).

This paper reports the results of a new study of the Presque Isle paleosol occurrence. The significance of this study is that a distinctive pre-Jacobsville weathering profile has been found on each of three rock types: granodiorite, serpentinized perido-

[1] Most recent workers assign this information to the upper Keweenawan Bayfield Group rather than the Cambrian (for example, White, 1966, p. 28).

tite, and diabase. The work documents that by about 1 b.y. B.P., the soil-forming environment was capable of producing a soil that resembles a modern type.

For the record, few Precambrian paleosols have been studied in the detail reported here (Pienaar, 1963), although many workers have noted their occurrence (Baker, 1939; J. A. Donaldson, personal commun.; James and others, 1961; Thwaites, 1931; Wright, 1965, p. 1002).

This study is based on the mineralogical and analytical work of Lewan (1972), supplemented by a complete restudy of all of the available material and data, as well as the collection and chemical analysis of additional samples. I take full responsibility for this reinterpretation of Lewan's data and recognize the fact that all of the alteration profiles have a hydrothermal origin is at some variance with Lewan's view. This difference in our interpretations in no way detracts from the excellence of the chemical data assembled by Lewan, which undoubtedly will be studied by many future workers.

BASEMENT ROCKS

The most important rock for this study is a medium-grained granodiorite, with white feldspar in a reddish matrix, that outcrops on the west side of the point and is interpreted to be in fault contact with peridotite. In this region all granitic rocks are considered to be of Archean age, in the lack of evidence to the contrary (Bodwell, 1972). In thin section of the freshest material, the K-feldspar margins and interstices show a distinctive micrographic intergrowth with quartz. The feldspar, originally microperthite, is altered almost completely to a clouded mass of fine-grained mica (sericite). The interstitial material is opaque and consists of a mixture of primary and secondary minerals: biotite, muscovite, chlorite, and vermiculite, heavily stained by fine-grained hematite. I consider the sericite and the interstitial alteration and pigmentation to be a weathering phenomenon.

A greenish-black serpentinized peridotite occupies approximately three-fourths of the study area and is well exposed locally in cliffs 30 m high. Modal analysis of three

samples gives an average composition of 57.5 percent serpentine, 9.2 percent augite, 8.9 percent magnetite and chromite, 8.6 percent olivine, 6.6 percent chlorite, 5.6 percent calcic plagioclase, with less than 1 percent biotite, hornblende, and hematite. Talc is a minor but variable constituent.

The peridotite shows a rather characteristic structural feature. In the eastern outcrops, it grades from a massive rock to one in which spheres of highly serpentinized peridotite, 30 cm or more in diameter, are rimmed by networks of dolomite and (or) quartz veinlets. In the direction of less altered peridotite, such spheres become larger (up to 3 m) and more angular, with the rimming veinlets becoming correspond-

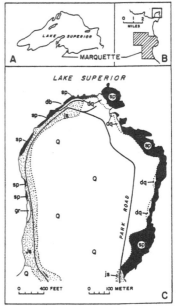

Figure 1. Geology of Presque Isle, Michigan: sp, serpentinized peridotite; gr, granodiorite; db, diabase; dq, dolomite-quartz rocks; js, Jacobsville sandstone; Q, glacial overburden and beach gravel.

ingly less prominent. In the opposite direction, the spheres become smaller and more platelike, imparting to the rock a crude foliation.

The overall pattern of planar, to blocky, to massive peridotite (shown in modified form in Fig. 2) resembles closely the marginal phases of peridotite bodies from Vermont ascribed by Jahns (1967) to the process of peridotite emplacement. Such features have controlled the distribution of what are considered to be the earlier hydrothermal dolomite-quartz veins at Presque Isle.

The largest of the hydrothermal veins are several centimeters thick and can be traced for a meter or more. These consist of several generations of white- to tan-weathering, fine-grained to distinctly crystalline dolomite, some in banded layers, interlayered with and replaced by several generations of light-colored, fibrous, colloform, or fine-grained quartz. Thin-section and field examinations show that some of the veins have inherited the fibrous structure from pre-existing cross-fiber serpentine.

As will be noted later, the distinction between the above veinlets considered to be hydrothermal and those considered to be supergene is by no means resolved. A second problem arises from the fact that the platy facies of peridotite has a subhorizontal attitude and hence subparallels the paleosol horizons that also have various kinds of platy structures.

The third rock type is diabase that cuts the peridotite and outcrops in the northwestern part of the point in a low cliff as a 1-m wide vertical dike, striking N. 78° W. In all likelihood the dike is of Keweenawan age and from its strike, possibly lower Keweenawan.

PROFILE DEVELOPED ON GRANODIORITE

The lowermost samples of the profile (Fig. 3) are of fairly hard granodiorite with clouded perthite in micrographic intergrowth with quartz. As noted earlier, the sericite-chlorite-vermiculite alteration in these samples is considered to be a reflection of the chemical effects of weathering. These minerals may be diagenetic, but the rock has probably not undergone much metasomatism during diagenesis.

The rock is overlain by a slightly less coherent TiO_2-rich saprolitic material, with recognizable white feldspar pseudomorphs and tiny carbonate veinlets in a reddish matrix. (Most of the paleosol is "pedolith" following the usage of Yaalon [1971, p. 27].) Higher in the section, the feldspars are represented by 1 × 0.2 mm clay flecks, and the material has a weak subhorizontal folia-

tion. The top 50 cm of the section is a red-brown, loose, soil-like material, with lumps of saprolite in a sandy matrix.

The profile contains two dolomite-cemented zones, about 30 cm apart, near the base of the paleosol horizon that probably represent paleocaliche horizons. Below the lower one is a layer a few centimeters thick of subhorizontal, crenulated colloform veinlets of reddish fine-grained quartz and chalcedony. From its general association, its horizontal attitude, fine grain size, and simple mineralogy, it is interpreted to represent the opaline veinlets that are found in arid-land soils.

Chemical analyses of the freshest material resemble those of average granodiorite. The main differences are that the present material shows a low content of Na_2O and CaO and a higher MgO content. The former probably denotes leaching and the latter the addition of supergene MgO, now fixed in carbonate.

The specific gravity of the material in the profile is relatively constant, ranging only from 2.4 to 2.75 g/cm³. Thus, the TiO_2 content provides a good direct index of residual concentration and hence of the quantity of elements leached from the rocks. Three upward-increasing cycles for TiO_2 content in the profile are noted: one starting from the freshest rocks and two extending upward from the two dolomitic paleocaliches (Fig. 3). These cycles may be a product of variation in soil permeability. The fact that the TiO_2 content in the paleocaliche layers is about the same as that in the granodiorite suggests that the layers were formed before the granodiorite soil had undergone much collapse. The profile showing TiO_2 enrichment was derived from about a 9-m thickness of granodiorite. Some of the compaction must have been lithostatic.

If one recalculates the chemical analyses so that each sample has 0.45 g TiO_2 per original unit volume, one obtains the abundance relations shown in Figure 4. The remarkably constant proportions among the major elements confirm that in the upper part of the profile, these elements are fixed in illite, hematite, potash feldspar, and quartz, as determined from x-ray diffraction work, and they suggest that the abundances of the mineral species are fairly constant.

The quantities of major elements lost during the soil-forming process are indicated by the differences between the values in the soil horizon (Fig. 4) and in the least altered granodiorite. Quite clearly, the soil has lost large amounts of SiO_2 and Al_2O_3 and lesser quantities of Fe_2O_3 and K_2O (and by inference, Na_2O and CaO). The twofold increase in Fe_2O_3 near the surface, in Figure 3 (15.8 wt percent versus 6.3 percent) is

true only in a relative rather than an absolute sense, and instead there appears to be a slight absolute enrichment near the base of the saprolite zone (Fig. 4).

The data suggest that because both SiO_2 and Al_2O_3 have been lost, the ground waters that caused this leaching must have been alkaline, with a pH near 9 (Mason, 1966, Fig. 6.6). Such alkaline waters have been described from recent playas. The fact that the paleosol is red, and that only about a third of the iron became lost, suggests that the soil-forming conditions must have been fairly oxidizing (James and others, 1961, p. 76–77) and hence that the soil did not support appreciable organic life or contain much organic material.

PROFILE DEVELOPED ON PERIDOTITE

The paleosol profile on peridotite consists at the top of a cemented, ferruginous residual paleosol overlying a leached zone, which in turn overlies a supergene dolomite-quartz zone developed in places on a peridotite saprolite (Fig. 2). As noted earlier, complexities arise from the fact that the paleosol is derived from a nonhomogeneous source rock; the peridotite contains earlier hydrothermal dolomite-quartz veinlets, and the irregularly developed blocky structures subparallel the unconformity. As a result, the profiles through the peridotite show considerable unsystematic variation. An added complexity arises from the fact that nowhere is a complete section exposed extending up from fresh peridotite. Thus the profiles had to be reconstructed from several measured and studied sections.

Figure 2. Schematic east-west section through Presque Isle, showing general relations: 1, peridotite; 2, blocky peridotite; 3, granodiorite; 4, diabase; 5, paleosol; a, incipient weathering; b, dolomite-quartz paleocaliche; c, dolomite-leached horizon; d, upper soil horizon; 6, Jacobsville sandstone. Horizontal distance about 700 m; vertical about 70 m. Peridotite structure adopted from Jahns (1967).

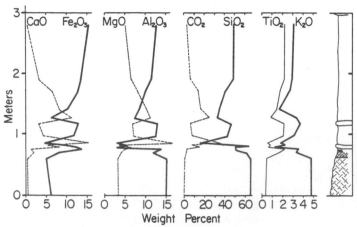

Figure 3. Chemical variation in vertical profile through paleosol on granodiorite.

The weathering of the peridotite was controlled by pre-existing joints, fractures, and vein borders. Thus, there are spherical weathered surfaces that resemble angular blocks that have been rounded by weathering, but in all likelihood, they represent merely the spherical peridotite blocks with a weathered rind. However extensive, such weathering results in the breakdown of the peridotite into grit and a change in color from dark gray to brown.

As with the granodiorite, the first weathering effect shown by the peridotite in thin section is the pigmentation of the minerals; olivine and pyroxene become clouded by an opaque substance. However, in addition to the pigmentation, such darkened rocks also show intricate veining by dolomite and quartz on the microscopic scale. The initial stages are represented by peridotite with darkened olivines rimmed by carbonate, but where the carbonate is more abundant, it may occur as myriad tiny veinlets cutting the rock into a kind of breccia. Where carbonate is leached, the rock forms a granular sand. Some of the veinlets show

cross-fiber structure or remnants of asbestos. However, because of its association with darkened olivine and pyroxene, this carbonate is considered to be supergene, perhaps representing the B2 paleosol horizon. The carbonate veinlets contain variable quantities of quartz as a later replacement or infilling, some a most delicately colloform-banded chalcedony, white, brownish, or red color. In places the slabby foliation in peridotite is further enhanced by the parallel arrangement of dolomite-quartz veinlets. On the east shore about 600 m south of the point of land is a synclinal trough of thinly laminated peridotitic material now intricately impregnated by dolomite-quartz. The trough is formed over a fissure in the massive structureless unveined peridotitic floor that shows here about 3 m of relief. The situation resembles the setting for a pocket of soft iron ore in the Mesabi Range. I think this represents a pocket of caliche-impregnated residual paleosol on an irregularly weathered peridotite surface. At this locality the contact between weathered and unweathered peridotite is abrupt, in contrast to the general situation on Presque Isle.

Dolomite and calcite are common minerals in arid-land soil horizons (Hausenbuiller, 1963, p. 368), but in recent peridotite profiles, the common carbonate seems to be magnesite (Bodenloss, 1950; Rice, 1957, p. 396; Sinclair, 1939, p. 10). Perhaps the dolomite is analogous in origin to such magnesites, but the ratio of Ca^{++}/Mg^{++} was greater than 10^{-3} (Garrels and Christ, 1965, Fig. 10.18) in the ground water.

Chalcedonic or jaspery veinlets are common supergene features in younger weathered peridotites (Cumberlidge and Chace, 1968, p. 1658; Rice, 1957, p. 396). In nickel laterite deposits, they more generally form an opaline boxwork (Cumberlidge and Chace, 1968, p. 1662; Rogers, 1973, p. 15), not recognized at Presque Isle. The precipitation of abundant silica in veins may have been controlled by a prevailing pH that was considerably above 6, in the range in which Krauskopf found silica precipitation by electrolytes to be rapid (Krauskopf, 1956).

The horizon veined by dolomite-quartz grades upward rather abruptly into a dark-gray, brownish, or reddish zone that is about 1 m thick. This material is similar structurally and texturally to the dolomite-quartz rock, except that carbonate is lacking from the veins. This leached zone is thought to represent the B1 paleosol horizon.

The top of the paleosol profile is represented by a distinctly horizontally laminated horizon, some 30 to 90 cm in thickness. This is now a fairly hard, fine-grained rock, dark reddish-brown in color, with a

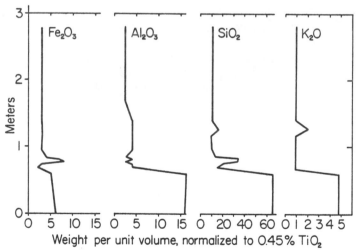

Figure 4. Chemical variation in vertical profile through paleosol on granodiorite, normalized to 0.45 g TiO_2 per unit volume.

local greenish cast, made up of a matrix of mixed-layer chlorite-vermiculite and hematite in which are set numerous subhorizontal veinlets of purplish, white, or jaspery quartz. Locally the veinlets exceed the matrix in volumetric abundance. From similarities in appearance, it is obvious that all of this material is an unbedded residual concentration of the underlying leached horizon, with a consequent increase in the amount of detrital quartz vein material, TiO_2 and Cr_2O_3 (Fig. 5).

In some places a few of the quartz veinlets are vertical rather than horizontal. These tend to be much more contorted than the horizontal ones. Before any chemical analyses were made, I interpreted the geometry of the contortions to denote about threefold compaction of the upper horizon.

Chemical analyses show several interesting features. On Figure 5 is plotted a profile through weathered peridotite that has varying amounts of dolomite-quartz. These plots show absolute elemental abundances to the degree that one can presume that the TiO_2 content, varying between 0.4 and 0.6 wt percent, is primary. It is surprising that despite the seemingly abundant quartz veins, more than half of the SiO_2 is gone. From field observations it is obvious that in some places quartz content is higher; nevertheless, the analyses suggest that the quartz has been locally derived and is probably supergene.

The profile shows the presence of two dolomite-rich zones, represented by increases in the abundance of CaO, MgO, and CO_2. These are interpreted to be veinform paleocaliche features. Between these is a quartz-rich zone. This similarity to the granodiorite profile suggests that this interpretation may be a reasonable one.

Analyses of material from the leached (B_1) and the uppermost horizons (A) are plotted in the upper part of Figure 5. The latter represents material that occurs as the matrix between the abundant residual quartz veins, but it obviously must include vein material because it contains some free quartz. The chemical analyses show TiO_2 and Cr_2O_3 abundances that are 2.6 to 4.0 times higher than in the fresh peridotite, suggesting about 3.5-fold residual enrichment. It is interesting to recall that the vertical, contorted quartz veinlets suggested a similar threefold compaction. In the upper part of Figure 5 are plotted both the uncorrected analyses and those normalized by the 3.5 factor. Except for a possible slight enrichment of iron near the surface, the normalized chemical profiles maintain their characteristics to the unconformity.

The main clay minerals in the weathered peridotite represented by the analyses in Figure 5 are chlorite and vermiculite, the latter probably an alteration product of the

former. These minerals show no systematic distribution pattern in the profiles, except that chlorite was found to have been most extensively altered to vermiculite in the upper part of the laminated trough described earlier from the east shore.

A set of 16 analyzed samples representative of the range of dolomite-quartz veining (Figure 6) shows a strong correlation between the abundances of alumina and nickel, with a correlation coefficient of 0.972. This suggests strongly that nickel is tied up in an aluminous phase, probably chlorite. However, the paleosol profile does not resemble one over a lateritic nickel deposit. Figure 6 shows that there is also a strong correlation between alumina and chrome abundances, suggesting that chrome also occurs in an aluminous phase, such as chlorite or a spinel. This is somewhat problematic because the enrichment pattern for Cr_2O_3 near the top of the profile resembles that for TiO_2.

In summary, the peridotite profile can be characterized as showing (1) no nickel concentration; (2) a slight loss in iron and alumina and a greater loss in silica, as in the granodiorite profile; (3) an upward-increasing loss in magnesia and in talc; and (4) a near-surface concentration of TiO_2 and Cr_2O_3.

The profile contains two paleocaliche horizons with an intervening quartz-rich zone. In the loss of magnesia at the surface and carbonate enrichment at depth, this profile resembles those developed in areas with slight to moderate rainfall (for example, Samoylova, p. 176, in Yaalon, 1971).

The lack of any appreciable TiO_2 or Cr_2O_3 enrichment lower in the profile suggests that much of the peridotite retained its physical strength, perhaps through recementation, during the weathering process. Hence it did not undergo any appreciable collapse, except near the surface.

The loss of SiO_2, Al_2O_3, and MgO suggest that the water had a pH near 9, similar to that in the granodiorite profile. However, the presence of the dolomite-free leached zone suggests that at some time the water must have had a lower pH. The overall pattern resembles that in the granodiorite but is more complex on the detailed scale.

PROFILE ON DIABASE

In the northern end of the area, a 1-m wide vertical diabase dike occurs in the dolomite-quartz and leached horizons in peridotite. Although the dike does not exhibit any megascopic discoloration or vein-

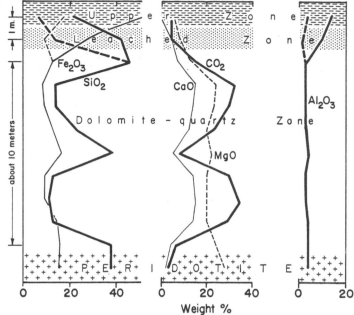

Figure 5. Chemical variation in a composite vertical profile through peridotite, dolomite-quartz paleocaliche, leached, and upper horizons.

ing suggestive of weathering, in thin section dusty hematite completely replaces the mafic minerals, and quartz and chlorite-vermiculite replace the plagioclase laths. The diabasic texture remains, but the rock obviously has been intensely altered, presumably by weathering. In its resistance to the hammer and in its clouded minerals, the dike resembles those I saw in residual soil profiles both on the savannas and in the rain forests of Venezuelan Guayana, (Kalliokoski, 1974, unpub. data). Perhaps because of their fine grain size and lack of porosity, diabase dikes have been fairly resistant to chemical weathering.

In modern residual soil profiles, diabase dikes grade upward into a mass of residual boulders and soil, quite commonly capped by spherical residual boulders. These do not occur above the Presque Isle dike. Unquestionably the diabase dike was being weathered almost at the same rate as the peridotite host, but the diagnostic weathering products have become lost. Now the dike is truncated by the basal conglomerate that drapes over the dike, indicating that the weathered dike has undergone less compaction than the peridotite paleosol on either side of it.

COVER ROCKS

The paleosol on both the peridotite and the granite is covered by the late Precambrian Jacobsville sandstone. The base of the sandstone immediately overlying the paleosol is a distinct basal congloermate, from a few centimeters to a few tens of centimeters thick, consisting of close-packed, irregular quartz vein material. The pieces are 1 to 5 cm thick and as much as 10 cm or more in longest dimension. Some show sharp corners, whereas others have been abraded. They occur in a dark-red,

fine-grained matrix grading from material that resembles the uppermost peridotite paleosol to a slightly more sandy substance. In some places this conglomerate occurs in shallow scours eroded into the paleosol. In others the lower contact is more gradational but easily recognizable by the abrupt increase in the abundance of quartz vein fragments.

The jaspery conglomerate is overlain by beds of red feldspathic sandstone, characteristic of the Jacobsville, and in places it interfingers with the sandstone, thereby forming a second bed of conglomerate. The conglomerate also overlies the granodiorite and the diabase dike. From festoon cross-bedding orientation, it is apparent that the general direction of flow was from peridotite toward granite. In other words, higher land was formed on peridotite. This is interesting, because this greater resistance to weathering of peridotite is characteristic of unglaciated landscapes today.

It is quite clear that the quartz vein material in the basal conglomerate was derived from the peridotite soil profile. Where unbedded, the conglomerate resembles, in general appearance and abundance of quartz blocks, the colluvial mantle of residual cherty iron formation and quartz that blankets the slopes of iron formation ridges in the Venezuelan savannas.

The fact that at Presque Isle Point the quartz vein material also occurs at distinct beds interbedded with sandstone suggests that the colluvium was further reworked as this small hill was gradually buried by the fluviatile sands of the Jacobsville sandstone.

CONCLUSIONS

The late Precambrian paleosol profiles described here resemble those that are now being produced in a semiarid climate under

conditions of low vegetative cover. In such arid areas the relatively few plants add little CO_2 to the soil, thereby permitting the development of alkaline conditions. Thus it is not certain to what degree the Precambrian paleosols may be the products of a purely arid climate or to what degree the products of soil formation in an even more humid climate — but during a geologic period when all forms of land organisms were far less abundant.

In this connection, Yaalon (1963) outlined four distinct geological stages through which soils have evolved. His description of the second stage applies rather well to the Presque Isle Point profiles:

Stage 1. The primary residual regolith — protosoils — developed under relatively anaerobic weathering conditions.

Stage 2. An incipient regolith covered with incompletely leached primitive soils prevailed until the appearance of land plants. The bare land of commonly low relief resulted in a slow circulation of elements. K became fixed in illite, which was the dominant clay mineral. The more mobile elements, especially Mg and Ca, reached the sea and precipitated as carbonates.

Stage 3. A rudimentary pedosphere was established (since Devonian time).

Stage 4. A fully evolved pedosphere developed.

The rate of weathering is an unknown. In the probable absence of soil organisms, they may have been considerably below present-day rates under comparable climatic and geographic conditions; the data of Cawley and others (1969) from Iceland suggests that it may have been one-fifth of present-day rates.

ACKNOWLEDGMENTS

Field work by M. Lewan was supported by funds from Bear Creek Mining Company and my work by funds from Michigan Department of Natural Resources. All chemical analyses were made under the general supervision of W. I. Rose, Jr., and A. P. Ruotsala, using facilities in the department, in the Institute of Mineral Research, and in the Department of Metallurgy.

REFERENCES CITED

Baker, M. D., 1939, The floor of the Paleozoic in Canada: Royal Soc. Canada Trans., 3d sec., v. 33, sec. 4, p. 10–18.
Bodenloss, Alfred J., 1950, Geology of the Red Mountain magnesite district, Santa Clara and Stanislaus Counties, California: California Jour. Mines and Geology, v. 46, no. 2, p. 223–278.
Bodwell, Willard A., 1972, Geologic compilation and nonferrous metal potential, Precambrian section, northern Michigan [M.S.

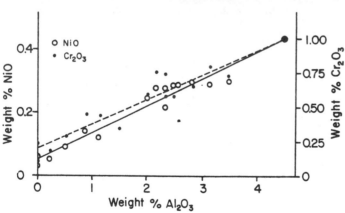

Figure 6.   NiO:Al$_2$O$_3$ and Cr$_2$O$_3$:Al$_2$O$_3$ relations in paleocaliche on peridotite.

thesis]: Houghton, Michigan Tech. Univ., 71 p.

Cawley, J. L., Burruss, R. C., and Holland, H. D., 1969, Chemical weathering in central Iceland: An analog of pre-Silurian weathering: Science, v. 165, p. 391–392.

Cumberlidge, John T., and Chace, Frederick M., 1968, Geology of the Nickel Mountain mine, Riddle, Oregon, in Ridge, John D., ed., Ore deposits in the United States 1933–1967: Am. Inst. Mining and Metallurgy, p. 1650–1672.

Gair, J. E., and Thaden, R. E., 1968, Geology of the Marquette and Sands quadrangles, Michigan: U.S. Geol. Survey Prof. Paper 397, p. 56–57.

Garrels, R. M., and Christ, R. L., 1965, Solutions, minerals, and equilibria: New York, Harper & Row, 450 p.

Hamblin, W. K., 1958, Cambrian sandstones of northern Michigan: Michigan Dept. Conservation, Geol. Survey Div., Pub. 51, 149 p.

Hausenbuiller, R. L., 1963, Principles of soil science: Bombay, Orient Longsmans, 440 p.

Jahns, R. H., 1967, Serpentinites of the Roxbury district, Vermont, in Wyllie, P. J., ed., Ultramafic and related rocks: New York, John Wiley & Sons, Inc., 464 p.

James, H. L., Clark, L. D., Lamey, C. A., and Pettijohn, F. J., 1961, Geology of central Dickinson County, Michigan: U.S. Geol. Survey Prof. Paper 310, p. 76–77.

Krimmel, P., 1941, The serpentinite of Presque Isle [M.S. thesis]: Evanston, Ill., Northwestern Univ., 57 p.

Krauskopf, K. B., 1956, Dissolution and precipitation of silica at low temperatures: Geochim. et Cosmochim. Acta, v. 10, p. 1–26.

Lewan, Michael D., 1972, Metasomatism and weathering of the Presque Isle serpentinized peridotite, Marquette, Michigan [M.S. thesis]: Houghton, Michigan Tech. Univ., 55 p.

Mason, Brian, 1966, Principles of geochemistry (3d ed.): New York, John Wiley & Sons, Inc., 329 p.

Pienaar, P.J., 1963, Stratigraphy, petrology, and genesis of the Elliot Group, Blind River, Ontario, including the uraniferous conglomerate: Geol. Survey. Canada Bull. 83, 140 p.

Rice, Salem J., 1957, Nickel, p. 391–399, in Wright, L. A., ed., Mineral commodities of California: Calif. Div. Mines Bull. 176, 736 p.

Rogers, Robert K., 1973, Element distribution in

the Rio Tuba nickeliferous laterite deposit, Philippines [M.S. thesis]: Houghton, Michigan Tech. Univ., 81 p.

Sinclair, W. E., 1939, Magnesite in South Africa: Mining Mag., v. 60, no. 1, p. 9–12.

Thwaites, F. T., 1931, Buried Precambrian of Wisconsin: Geol. Soc. America Bull., v. 42, p. 719–750.

White, Walter S., 1966, Geologic evidence for crustal structure in the western Lake Superior basin, in Steinhart, J. S., and Smith, T. J., eds., The Earth beneath the continents: Am. Geophys. Union Geophys. Mon. 10, p. 28–41.

Wright, C. M., 1965, Syngenetic pyrite associated with Precambrian iron ore deposits: Econ. Geology, v. 60, p. 998–1019.

Yaalon, Dan H., 1963, Weathering and soil development through geologic time: Israel Research Council Bull., sec. G, v. 11, no. 3, p. 149–150.

——ed., 1971, Paleopedology; origin, nature, and dating of paleosols: Internat. Soc. Soil Sci. and Israel Univ. Press, 350 p.

MANUSCRIPT RECEIVED BY THE SOCIETY JANUARY 28, 1974
REVISED MANUSCRIPT RECEIVED JULY 8, 1974

Part II

DELTAIC AND PARALIC COMPLEXES

Editor's Comments
on Papers 17 Through 21

DELTAIC DEPOSITS

Just as modern deltas have received more attention than any other facies from alluvial fan to the paralic realm, so description of ancient deltaic complexes have dominated studies devoted to ancient non-marine to paralic rocks. This is due not only to the fact that deltaic complexes are common in the stratigraphic record but also because they are large reservoirs of petroleum. Significantly, compared with large modern deltas that prograde toward major ocean basins, most of the ancient ones record relatively limited detrital supply and construction in shallow embayments on unstable cratons (Ferm 1970). Miall (1976a) has assembled a very useful review of deltaic models based on depositional processes on modern deltas, with an annotated bibliography of ancient ones in Canada.

Among the papers on prehistoric deltas that compare them with the anatomy of modern ones, some have dealt mainly with fluvial deposits of the delta plain and their cyclic sedimentation. These include the

Devonian Catskill deposits of New York (Allen and Friend 1968), Carboniferous sediments in the central Appalachians (Meckel 1969) and in western Germany (Casshyap 1975a), and late Cretaceous deposits in eastern Utah (Van de Graaff 1972). Studies that consider both marine and nonmarine deltaic facies describe Carboniferous deposits in Great Britain (Wright 1967; Williams 1968; Elliott 1975), late Carboniferous sequences in the central Appalachians (Donaldson 1974; Williams 1974), late Cretaceous deposits in Colorado (Masters 1967), a Paleocene and Eocene succession in Arctic Canada (Miall 1976b), and Eocene sediments in northern Spain (Van Eden 1980). Paralic and marine facies in deltaic complexes have been emphasized in studies of Devonian deposits in New York (Johnson and Friedman 1969; Walker and Harms 1971; Mazzullo 1973; and Sutton and Ramsayer 1975), of Carboniferous sediments in southeastern United States (Ferm et al. 1972; Kreisa and Bambach 1973), in England (DeRaaf et al. 1965; Collinson 1969) and in Spain (Van de Graaff 1971), and of Cretaceous deposits in northeastern Mexico (McBride et al. 1975) and Alberta (Jansa 1972). Most of these papers discuss the role of alternations of marine and nonmarine facies called *cyclothems.*

Belt (1975) has presented a balanced review of the different ways deltaic coal-measures cyclothems have been defined. The American school placed the base below sandstone channeled into marine shale and considered the overlying fining-upward sequence (sandstone-siltstone-shale-coal-marine limestone and shale) to be the result of transgression followed by eustatic lowering of sea level that caused scouring and initiated another cyclothem. The European school placed the base between the coal (or rooty bed) and overlying marine deposits and commonly called on sedimentational control of the cyclic development. In contrast, the Delta school that Belt favors places the boundary between the transgressive marine limestone and the overlying progradational mudstone or coarsening-upward sequence and emphasizes the role of varying sediment input and compaction in producing cyclic sedimentation. In his analysis of early Carboniferous deposits in Scotland, Belt (1975) sorted the vertical cyclothem patterns of transgressive, progradational, and aggradational phases into six coded paleogeographic categories. By this method he was able to organize a great deal of information and reconstruct a detailed evolution of the delta system. Casshyap (1975b) applied the Markov method to deltaic cyclothems in the late Carboniferous coal measures in western Germany, using data from 32 boreholes. He found that the preferential transition path is coal-siltstone-sandstone-siltstone-rooty bed-coal in delta-plain deposits that lack marine limestone.

Each of the five selections describing ancient deltaic complexes illustrates a somewhat different aspect of analysis. Glaeser (1974) developed a three-dimensional picture of the distribution of prodelta, coastal-margin, and alluvial-plain facies of the constructional (prograding) late Devonian Catskill delta in northeastern Pennsylvania (Paper 17). His study provides an unusually good comparison of outcrop and subsurface characteristics of the sedimentary rocks in a deltaic complex.

Collinson (1968) focused on two facies of sandbodies in Carboniferous coal measures in midland England (Paper 18). Large tabular cross-bedded sets of coarse sandstone, initially interpreted as the foreset facies of a delta-top environment, probably were large bars in distributary channels of a constructional delta (J. D. Collinson, personal comm., 1976). In contrast, medium-scale cross-bedding developed in migrating fluvial channels on the upper delta plain. Elliott (1976) has shown that the coarsening-upward cycles in Carboniferous coal measures in southwestern England (Paper 19) reflect renewed progradation of elongate river-dominated constructive deltas into a repeatedly deepened basin, supplemented by differential compaction. Each cycle consists of a progradational sequence of prodelta to mouth bar and distributary channel (bar-finger) deposits; an aggradational sequence of beach, interdistributary-bay, and crevasse deposits; and local-abandonment (destructive) facies of bioturbated mudstone and coal.

Ferm (1970) constructed a three-dimensional model of late Carboniferous deltaic coal measures in West Virginia and Pennsylvania (Paper 20). It is a large lens composed of landward mudstone and progradational sandbodies enclosed in a veneer of abandonment peat, rooty clay, and marine limestone. Lateral shift of delta lobes produced *en echelon* arrangement of the successive detrital lenses.

Hubert et al. (1972) have described a Late Cretaceous wave-dominated (high-destructive) delta in Wyoming (Paper 21). Sandstone that accumulated on the outer delta platform developed abundant flattened pillow structures by downslope sliding. Sandstone on the inner delta platform was transported by strong waves and subordinate tidal currents and deposited in beaches, swash bars, and subtidal areas swept by longshore drift. Overlying delta-plain sediments include carbonaceous mudstone, lenticular channel sandstone, and minor lignite. Compaction of thick prodelta mud lowered the delta surface and initiated a transgressive destructional marine sandstone and shale.

REFERENCES

Allen, J. R. L., and Friend, P. F. 1968. Deposition of Catskill facies, Appalachian region: With notes on some other Old Red Sandstone basins. *Geol. Soc. America Spec. Paper 106*, pp. 21–74.

Belt, E. S. 1975. Scottish Carboniferous cyclothem patterns and their paleoen-vironmental significance. In *Deltas, Models for Exploration,* ed. M. L. Broussard. Houston, Texas: Houston Geol. Soc., pp. 429–449.

Casshyap, S. M. 1975a. Lithofacies analysis and paleogeography of Bochumer Formation (Westphal A 2), Ruhrgebiet. *Geol. Rundschau* **64:**610–640.

———— 1975b. Cyclic characteristics of coal-bearing sediments in the Bochumer Formation (Westphal A 2), Ruhrgebiet, Germany. *Sedimentology* **22:**237–255.

Collinson, J. D. 1969. The sedimentology of the Grindslow shales and the Kinderscout Grit: A deltaic complex in the Namurian of northern England. *Jour. Sed. Petrology* **39:**194–221.

Cotter, E. 1975. Deltaic deposits in the Upper Cretaceous Ferron Sandstone, Utah. In *Deltas, Models for Exploration,* ed. M. L. Broussard. Houston, Texas: Houston Geol. Soc., pp. 471–484.

DeRaaf, J. F. M.; Reading, H. G.; and Walker, R. G. 1965. Cyclic sedimentation in the Lower Westphalian of North Devon, England. *Sedimentology* **4:**1–52.

Donaldson, A. C. 1974. Pennsylvanian sedimentation of Central Appalachians. *Geol. Soc. America Spec. Paper 148,* pp. 47–78.

Elliott, T. 1975. The sedimentary history of a delta lobe from a Yoredale (Carboniferous) cyclothem. *Yorkshire Geol. Soc. Proc.* **40:**505–536.

Ferm, J. C.; Milici, R. C.; and Eason, J. E. 1972. Carboniferous depositional environments in the Cumberland Plateau of southern Tennessee and northern Alabama. *Tennessee Div. Geology Rept. Inv.* **33:**1–32.

Jansa, L. 1972. Depositional history of the coal-bearing Upper Jurassic-Lower Cretaceous Kootenay Formation, southern Rocky Mountains, Canada. *Geol. Soc. America Bull.* **83:**3199–3222.

Johnson, K. G., and Friedman, G. M. 1969. The Tully clastic correlatives (Upper Devonian) of New York State: A model for recognition of alluvial, dune (?), tidal, nearshore (bar and lagoon), and offshore sedimentary environments. *Jour. Sed. Petrology.* **39:**451–485.

Kreisa, K. D., and Bambach, R. K. 1973. Environments of deposition of the Price Formation (Lower Mississippian) in its type area southwestern Virginia. *Am. Jour. Sci.* **273A:**326–342.

Masters, C. D. 1967. Use of sedimentary structures in determination of depositional environments, Mesaverde Formation, Williams Fork Mountains, Colorado. *Am. Assoc. Petroleum Geologists Bull.* **51:**2033–2043.

Mazzullo, S. J. 1973. Deltaic depositional environments in the Hamilton Group (Middle Devonian), southeastern New York State. *Jour. Sed. Petrology* **43:**1061–1071.

McBride, E. F.; Weide, A. E.; and Wolleben, J. A. 1975. Deltaic and associated deposits in Difunta Group (Late Cretaceous to Paleocene), Parras and La Popa basins, northeastern Mexico. In *Deltas, Models for Exploration,* ed. M. L. Broussard. Houston, Texas: Houston Geol. Soc., pp. 485–522.

Meckel, L. D. 1967. Origin of Pottsville Conglomerate (Pennsylvanian) in the Central Appalachians. *Geol. Soc. America Bull.* **78:**223–258.

Miall, A. D. 1976a. Facies models 4. Deltas; *Geosci. Canada* **3:**215–227.

———— 1976b. Sedimentary structures and paleocurrents in a Tertiary deltaic succession, northern Banks Basin, Arctic Canada. *Canadian Jour. Earth Sci.* **13:**1422–1432.

Sutton, R. G., and Ramsayer, G. R. 1975. Association of lithologies and sedimentary structures in marine deltaic environments. *Jour. Sed. Petrology* **45:**799–807.

Van de Graaff, F. R. 1972. Fluvial-deltaic facies of the Castlegate Sandstone (Cretaceous), east-central Utah. *Jour. Sed. Petrology* **42:**558–571.

Van de Graaff, W. J. E. 1971. Three Upper Carboniferous, limestone-rich, high destructive, delta systems with submarine fan deposits, Cantabrian Mountains, Spain. *Leidse Geol. Meded.* **46:**157–235.

Van Eden, J. G. 1970. A reconnaissance of deltaic environment in the Middle Eocene of the south-central Pyrenees, Spain. *Geol. Mijnb.* **49:**145–160.

Walker, R. G., and Harms, J. C. 1971. The "Catskill Delta": A prograding muddy shoreline in central Pennsylvania. *Jour. Geol.* **79:**381–399.

Williams, E. G. 1974. Controls of early Pennsylvanian sedimentation in western Pennsylvania. *Geol. Soc. America Spec. Paper 148,* pp. 135–152.

Williams, P. F. 1968. The sedimentation of Westphalian (Ammanian) measures in the Little Haven-Amroth coalfield, Pembrokeshire. *Jour. Sed. Petrology* **38:**332–362.

Wright, M. D. 1967. Comparison of Namurian sediments of the central Pennines, England, and recent deltaic deposits. *Sed. Geology* **1:**83–115.

17

Reprinted from pp. 1, 8–15, 17–23, 24, 61–64 of *Pennsylvania Geol. Survey General Geol. Rept.* **63**, 4th ser.:1–89 (1974)

UPPER DEVONIAN STRATIGRAPHY AND SEDIMENTARY ENVIRONMENTS IN NORTHEASTERN PENNSYLVANIA[1]

by

J. Douglas Glaeser

ABSTRACT

This report uses surface and subsurface data to develop a three-dimensional picture of the distribution of the complex sedimentary environments comprising the Upper Devonian rocks in northeastern Pennsylvania. Fourteen deep wells and the Lehigh River valley outcrop section are utilized because they represent the complete preserved record of Upper Devonian sedimentation and each can be related to a time-stratigraphic datum, the Tioga metabentonite. Stratigraphic units defined within the Upper Devonian are thought to represent depositional environments preserved during coastline growth resulting from a major progradational event recorded as the classic Catskill delta.

The sedimentary environments are envisaged as responses to complex geomorphically controlled processes and encompass deposits formed in the prodelta, coastal margin and alluvial plain. These depositional processes may be observed in modern deltaic areas where distinct sediment types are recognized accumulating in the various subregions of prograding coastal plains. Sediment types in these subregions have physical properties remarkably similar to those of the Devonian rocks of northeastern Pennsylvania.

Prodelta, coastal-margin and alluvial deposits comprising the three major depositional environments of the Catskill delta system are distinguished by grain-size distributions, sorting, primary sedimentary structures, lithologic assemblages, vertical sequence and stratigraphic context. These criteria are described for each environment and are used to construct a three-dimensional time-stratigraphic correlation diagram and isopach maps which show the regional relationships among the delta's geomorphic-environmental components deposited during Upper Devonian time in the northeastern part of the Appalachian basin.

---

[1] This report was completed in June, 1971. All drafting and editorial preparations for publication were complete in June, 1972, when a flood destroyed the Pennsylvania Geological Survey's facilities. Another year elapsed before this report could be reassembled to its pre-flood status. Although new information and ideas have been published since the 1971 completion, this report does not consider work published after that date.

[Editor's Note: Material has been omitted at this point.]

SEDIMENTARY ROCK TYPES IN THE UPPER DEVONIAN OF NORTHEASTERN PENNSYLVANIA

SEDIMENTOLOGIC CHARACTER OF UPPER DEVONIAN ROCKS IN OUTCROP

The formations and members used in this report are those previously mapped and described at the fully exposed section of Upper Devonian rocks found along the Lehigh River in northeastern Pennsylvania (Epstein and others, 1974). The units have also been mapped in several other quadrangles in Carbon and Monroe Counties, Pennsylvania,[4] but nowhere are they all exposed in one place as at Lehigh River. The Lehigh River section is therefore doubly important, because the sedimentary features and stratigraphic relationships there were the basis for determining the types of deposits and their sedimentary origins, and because these interpretations were later applied to the rocks in the subsurface.

There are two Upper Devonian formations at Lehigh River, the Trimmers Rock Formation and the Catskill Formation. The Catskill Formation has nine members. The general thicknesses of the formations and members at Lehigh River and their sedimentary origins are shown in Figure 2.

The Trimmers Rock Formation there consists of 1168 feet of gray siltstone. Graded bedding and other primary sedimentary structures indicate it is a turbidite deposit. Bed thickness and grain size increase upward in the formation. Marine fossils (mainly brachiopods and pelecypods) are present but not abundant. Flaser structures occur in the uppermost beds. Because of the evidence of increasingly shallower water conditions upward in the Trimmers Rock Formation, it has been interpreted as a prodelta deposit.

The upper beds of the Trimmers Rock Formation pass into the basal member of the Catskill Formation, the Towamensing Member. The

---

[4] Christmans, Pohopoco Mountain, Hickory Run and Blakeslee (Sevon, in preparation) ; Kunkletown (Epstein and Sevon, in preparation); Brodheadsville (Berg, in preparation) — all are 7½-minute quadrangles in Carbon and Monroe Counties, Pennsylvania.

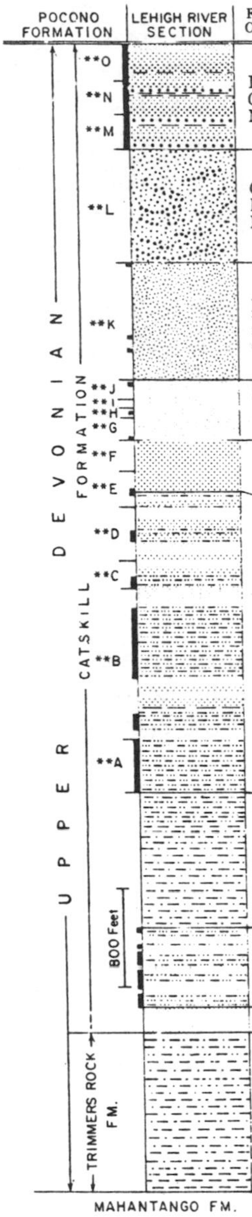

POCONO FORMATION	LEHIGH RIVER SECTION	FORMATION OR MEMBER	THICK-NESS	DESCRIPTION	ORIGIN
**O, **N, **M		DUN-CANNON MEMBER	968'	Red conglomerates, sandstones and siltstones arranged in fining-upward cycles. Cross-beds common. Siltstones and shales, often rippled and mud-cracked, are fissile. Scour and intertongued beds common in lower third; above, fining-upward sequences lack conglomerates and dominance of red color.	Meandering rivers
**L		CLARK'S FERRY MEMBER	906'	Nonred, medium- to coarse-grained sandstones and conglomerates. Massive, planar cross-beds. Bimodal-bipolar vectors. Bed tops often strewn with fine-grained rock fragments. Giant ripples and parting lineations common.	Braided rivers
**K		BERRY RUN MEMBER	985'	Nonred, medium- to fine-grained sandstones dominate. Few red siltstones up to 40 feet thick in upper part. Minor conglomerates. Cross-beds common. Leached limonite-rich zones with shale and round carbonate fragments.	Braided rivers
**J, **H, **G		SAWMILL RUN MEMBER	424'	Nonred, dominantly fine- and very fine grained sandstones, commonly cross-bedded. Some sandstones calcareous. Basal, middle and upper red siltstones. Minor shale. Leached limonite-rich zones with shale and round carbonate fragments.	Braided rivers
**F, **E		PACKER-TON MEMBER	416'	Dominantly medium- to fine-grained, well-sorted sandstones; commonly cross-bedded. Minor fine conglomerates, siltstones and shales. Some calcareous sandstones. Leached limonite-rich zones with shale and round carbonate fragments.	Braided rivers
**D, **C					
**B		LONG RUN MEMBER	2363'	Predominantly red, parallel-sided beds of massive, fine- to very fine grained sandstones and red siltstones and fissile shales. Sandstones more prominent upward. Several upward-fining successions in upper part.	Delta plain
**A					
		BEAVER-DAM RUN MEMBER	963'	Nonred. Principally siltstone with some very fine grained sandstones. Parallel-sided, massive and fissile beds alternate. Ball-and-pillow structures prominent; some slumped. Tentaculites and crinoid columnals at bases of many massive beds.	Prodelta
		WALCKS-VILLE MEMBER	649'	First red bed sequence above Catskill base. Parallel-sided, massive and fissile, non-red and red siltstones and sandstones. Abrupt lateral color and grain-size changes common. Some upward-fining successions.	Delta plain
		TOWAMEN-SING MEMBER	190'	Gray, fine- to very fine grained sandstones and coarse siltstones. Many undulatory and scoured surfaces. Shale-clast conglomerates. Poorly sorted sandstones contain carbonaceous debris. Churned and burrow-mottled siltstones. Some burrowed zones graded, shale-capped, and rippled or mud-cracked with infilling of overlying sediment. Near middle, some graded beds.	Delta front
		TRIM-MERS ROCK FORMA-TION	1168'	Gray siltstones. Alternating massive graded (coarse- to fine-grained siltstone) and parallel-laminated beds. Cross-laminations minor. Fine siltstones at tops of graded beds commonly burrow mottled. Upward bed thickness increases and ball-and-pillow structures (2' x 8') occur. Brachiopods associated with shale chips at base of some graded beds and in lenticular zones "scouring" underlying bed. Carbonaceous material often in fine siltstone at top of graded beds.	Prodelta

MAHANTANGO FM.

Figure 2. Description of Upper Devonian rocks at Lehigh River outcrop.
Black bars to left of patterns denote red beds.
**Informal members (Glaeser, 1963)

219

Towamensing is 190 feet thick, and consists of massive, parallel-laminated sandstone beds. The sandstones are medium to very fine grained, and often contain carbonaceous material. Some beds contain shale chips, and some beds have pelecypods and their burrows. The Towamensing sands are believed to have been deposited as an apron on the finer prodelta detritus near the river mouths that were supplying sediment to the basin (Epstein and others, 1974).

The Walcksville Member lies above the Towamensing. The Walcksville is 649 feet thick, and contains the lowest red beds in the Catskill Formation. It is dominated by siltstones and upward-fining units and is believed to be a delta-plain deposit, a broad alluvial deposit built seaward upon the marine sediments of the underlying delta platform. In the delta-plain environment, sand is periodically swept across the plain, where burrowed muds and carbonate fragments are common. The preserved sand has the mud fragments and carbonate fragments at its base. The sandstone has parallel-laminated beds which grade upward into evenly bedded red siltstones, and further upward into intensely burrowed finer siltstones and shales (Epstein and others, 1974).

Above the Walcksville, there is a nonred sandstone unit which passes upward into crinoid-bearing siltstones. Both units are part of the Beaverdam Run Member. The gray siltstones are the dominant lithology in this member. They are arranged in alternating massive and fissile beds. Crinoids are found in thin layers in the massive beds, usually at the base of the beds, indicating graded bedding. Slump structures and ball-and-pillow structures are common throughout the member. Although the sedimentary structures are not as complete as those of the Trimmers Rock Formation, the Beaverdam Run is believed to have a similar prodelta origin. The sandstone at the base of the Beaverdam Run has the character of a delta-front deposit. It is logical, according to Walther's (1894) law of facies, that this delta-front sandstone should be situated between the Walcksville delta-plain deposit below and the prodelta siltstones forming the main part of the Beaverdam Run deposit above, since the delta-front deposit would be shoreward of the prodelta deposit and seaward of the delta-plain deposit. The Beaverdam Run Member therefore represents a transgression, the only one observed at the Lehigh River section, which otherwise shows an overall regressive pattern.

A thick succession of red beds and upward-fining lithologies lies above the Beaverdam Run prodelta deposits. This member, the Long Run, is 2363 feet thick and is lithologically similar to the Walcksville Member. It has been interpreted as a delta-plain deposit (Epstein and others, 1974). The basal beds of this member are not exposed, so it is not possible to determine whether there is a transition or a depositional hiatus between the delta-plain deposits and the underlying prodelta deposits.

In the lower few hundred feet of the Long Run Member at its type locality, red siltstones and shales are the only lithologies present, and the upward-fining beds with their associated basal shale and round carbonate fragments are not seen (Epstein and others, 1974). Higher in the type section, however, sandstone increases and upward-fining successions become the dominant sedimentary feature.

The succeeding four members of the Catskill Formation consist mainly of sandstone and are believed to be braided-river deposits. They would be considered as one unit in mapping, but there are red beds in the succession which permit them to be divided into distinct members. The members are, from oldest to youngest, the Packerton, Sawmill Run, Berry Run and Clark's Ferry. The aggregate thickness of the four is 2731 feet. The sandstones are believed to be braided-river deposits because they have a repetitive bedding assemblage with three components that represent preserved migrating sand waves. Shale, round carbonate fragments and abundant limonite are found in the sand-wave trough. Above them, steep foreset beds dip into the trough, and planar parallel-laminated beds at the top dip gently toward the opposite pole from the steeper foresets (Epstein and others, 1974). The sand-wave crest is preserved, and shows the transition from steep foresets to gentle, planar, parallel-laminated beds thought to represent upper regime flow deposits. The preserved sand waves themselves are products of lower regime flow.

Sedimentary features such as those described above can be formed in any alluvial system where repeated floods occur. However, the near-absence of fine-grained detritus in the 2731 feet of section indicates that winnowing has taken place. Winnowing is typical of braided rivers, wherein much of the stream cross section is influenced by current activity, at least during flood stage. Mid-channel bars and other "islands" within the braided river are transitory and fines have little opportunity to accumulate. In meandering streams, however, overbank muds are deposited and commonly preserved.

The Clark's Ferry Member is the uppermost of the braided-river sandstones. It is 906 feet thick and consists almost entirely of sandstone and sandstone conglomerates. Primary current lineations and current crescents occur, and linguoid ripples, rib-and-furrow structures and lag gravels are found on bedding surfaces. Parallel laminations are abundant in both the planar beds and cross-beds. The cross-bedding in the Clark's Ferry Member is the most important clue to its depositional origin. Vector data show a bipolar-bimodal distribution of cross-bed orientations (Epstein and others, 1974) resulting from channel-bar migration.

Above the Clark's Ferry Member is the uppermost member of the Catskill Formation at Lehigh River, the Duncannon Member, which is

968 feet thick. Its coarse conglomerates, upward-fining cycles, desiccation features, dominant red color, and northwestward-oriented cross-bed vectors all imply a fluvial origin. Cycles of coarse to fine sediments repeated throughout the outcrop section indicate repeated preservation of channel, overbank and flood-plain deposits. These deposits are apparently of meandering-river origin and not of braided-river origin. The differences between the two are based largely on content of fines and do not necessarily represent a change in geomorphic setting. Apparently, braided streams and meandering streams can occur within the same stream system (Leopold and Wolman, 1957).

The uppermost fluvial sequence of the Catskill grades into the transgressive basal deposits of the Mississippian Pocono Formation (Sevon, 1968, 1969). Thus the spectrum of Upper Devonian environments described above — from the Trimmers Rock turbidites formed in the prodelta region through the fluvial deposits at the top of the Catskill Formation — relates a story of significant progradation of nearshore and nonmarine sediments over marine deposits. These deposits, now preserved as 9032 feet of rock at one location along the Lehigh River, record the basin-filling processes which occurred, from open marine through alluvial sedimentation. Because this stratigraphic record represents several depositional environments, one preserved upon the other at one point in space and in a vertical time sequence, there is clear evidence that the Lehigh River area was subsiding less rapidly than the rate of sediment input. However, it did subside at a sufficient pace to permit one environment to overstep and bury the one adjacent to it shoreward without obliterating the latter by processes of reworking. Even the single transgression observed within the formation appears to have been a transition of nearshore over shoreward sediments. Thus the entire spectrum of preserved deposits suggests that a delicate balance existed between input and subsidence within the area of shoreline migration, so that geomorphically contiguous environments are arranged in a predictable stratigraphic order. No major hiatuses resulting from nonaccumulation, nonpreservation or reworking appear to have occurred there during the Upper Devonian.

The environmental units comprising the Lehigh River section can be considered within the context of geomorphic subdivisions of the Catskill delta. Table 1 lists the environments and four geomorphic units distinguishable in the Upper Devonian rocks at the Lehigh River section. Their three-dimensional arrangement is shown in Figure 3. Braided- and meandering-river sediments are grouped together in one alluvial environment. Their vertical superposition at Lehigh River suggests that, in a three-dimensional context, they probably occupied parts of the same broad physiographic region, one overstepping the other during the last progradational interval of Catskill sedimentation.

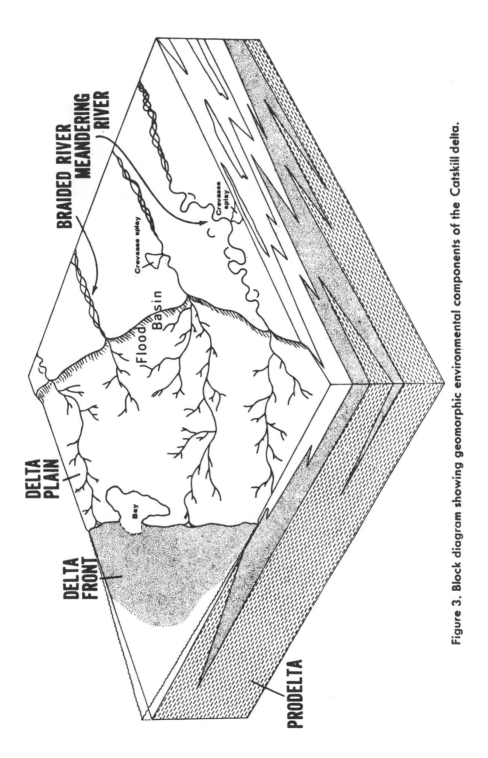

Figure 3. Block diagram showing geomorphic environmental components of the Catskill delta.

Table 1. *Outcrop Properties Indicating Depositional Environments*

PRODELTA

Graded beds

Turbidite structures

Upward increase in grain size and bed thickness

Flaser bedding near top

Marine shells on bed surfaces, scattered within beds and as "channel-shaped" concentrations; brachiopods, pelecypods and crinoids

Burrow mottling

Ball-and-pillow structure

Example: Trimmers Rock Formation

DELTA FRONT

Predominantly sandstone

Massive bedding, commonly internally laminated

Plant remains, some concentrated in layers

Rounded granule-size quartz fragments in few beds

Shale-chip conglomerates common

Some scoured bedding surfaces

Pelecypods and their preserved burrows

Lag gravels and "beach" cusps

Prominently developed primary current lineations

Bipolar-bimodal distribution of cross-bed vectors

Example: Towamensing Member of Catskill Formation

DELTA PLAIN

Upward-fining successions with planar basal contacts, planar bedding and absence of cross-beds in sandstone portions

Round carbonate clasts and few plant fragments and shale chips in basal beds

Color changes from gray to red near sandstone-siltstone transition; olive green commonly occurs in upper few inches of siltstone

Burrow mottling common in siltstone

Abrupt lateral changes occur in grain size and color

Example: Walcksville and Long Run Members of Catskill Formation

ALLUVIAL PLAIN

BRAIDED RIVERS

Predominantly sandstones

Wide range in grain size

Three components of bedding:

1. Basal lens of shale containing round carbonate fragments, shale chips and abundant limonite;

MEANDERING RIVERS

Roughly equal amounts of coarse and fine terrigenous detritus

Fining-upward cyclothems

Evidence of scour at cycle bases with red shale and siltstone chips from underlying unit incorporated as intraformational conglomerate into coarse portion of cycle

Table 1. (Continued)

BRADED RIVERS	MEANDERING RIVERS

BRAIDED RIVERS

2. Steep foreset cross-beds, some extending down into 1;

3. Broad planar parallel-laminated beds, some bevelling steep foresets of 2.

Few crests of sand waves preserved

Massive planar beds occur between above types and, in places, antidune backsets preserved

Example: Packerton, Sawmill Run, Berry Run and Clark's Ferry Members of Catskill Formation

MEANDERING RIVERS

Tabular and trough cross-beds at or near cycle bases

Desiccation structures and burrow mottling at tops of cycles

Example: Duncannon Member of Catskill Formation

SEDIMENTOLOGIC CHARACTER OF THE UPPER DEVONIAN ROCKS IN SUBSURFACE

In the complex spectrum of sedimentary environments at the Lehigh River, it has been possible to recognize the progradation of seaward followed by more landward environments. It is, therefore, predictable that in sections penetrating the preserved record of Upper Devonian sedimentation in areas near the Lehigh River section, there are laterally contiguous deposits that were formed at the same time as the major environmental units at the Lehigh River.

Because the Lehigh River section is situated on the southernmost margin of the Upper Devonian outcrop belt in northeastern Pennsylvania, all data for this study lie to the northeast, north and northwest of that site. All known paleogeographic data for shallow-water deposits of middle and late Paleozoic age in the Central Appalachian basin suggest that the paleoslope was toward the northwest (Meckel, 1970). Therefore, all Upper Devonian sections to the north and northwest of the Lehigh River section can be considered to have been situated basinward of those deposits at the Lehigh River. The correlations among rocks in northeastern Pennsylvania should consequently represent various basinward-oriented slices through the three-dimensional body of sediments comprising the Catskill delta. These correlations permit a conceptualization of the depositional events and processes occurring in various parts of the Upper Devonian delta during its history of progradation.

Fourteen exploratory wells in the six northeasternmost counties of Pennsylvania (Figure 1) represent the only nearly complete record of Upper Devonian sedimentation in the region adjacent to the Lehigh River section. Table 2 lists the names and locations of the fourteen wells,

[*Editor's Note:* Table 2 is not reproduced here.]

and the Upper Devonian thicknesses penetrated by them. The wells were wildcat hydrocarbon explorations. One, the L. G. Richards #1, produced significant amounts of gas, and four others had one or more gas shows in the Upper Devonian rocks penetrated. With the exception of three, they all reached the Middle-Upper Devonian boundary, which in the subsurface is placed at the Tully Limestone. In outcrop, the Tully occurs just beneath the boundary of the Middle Devonian Mahantango Formation (dark-gray shales and siltstones) and the Upper Devonian Trimmers Rock Formation (gray siltstones characterized by turbidites).

Cable tool samples are available for the fourteen wells and descriptions have been made by the Geolog Corporation. Descriptions from the wells drilled by the Transcontinental Pipe Line Corporation were also made available to the Pennsylvania Geological Survey. Some of the fourteen wells were reexamined and described by Wagner (1963a). The writer examined in detail samples from several wells in order to determine whether his observations of grain size, sorting and color differed from those recorded by Geolog or Transcontinental personnel. Differences were not significant; therefore, Geolog descriptions were used in studying most of the wells.

The lithologic properties of well cuttings from each well site allow the subsurface Upper Devonian rocks to be divided into three lithologic assemblages, distinguished from each other by persistence of one lithology or repetitive sequences of lithologies through many feet of penetrated section. The lithologic characteristics described for the subsurface samples are considered to be the products of depositional environments and geomorphic settings related to those already discussed for the Upper Devonian rocks exposed in the Lehigh River section, where primary sedimentary structures lend added support to interpretations of sedimentologic origins. Because distinctive lithologic assemblages can be recognized in the subsurface sample logs, they are used for the first time in this report as stratigraphic units.

A list of the lithologic assemblages observed in logs of the fourteen wells and the environments in which they formed is given in Tables 8A-8N (Appendix). The three major assemblages recognized are prodelta, coastal-margin, and alluvial-plain deposits. Alluvial-plain deposits may be further subdivided into delta-plain, braided-river, and meandering-river deposits.

Criteria for Recognition of Prodelta, Delta-Front and Bay Deposits in Subsurface Data

One of the most important requirements in establishing the progradational sequence of deposits in the Catskill delta is that the subsurface sections should penetrate the Middle-Upper Devonian boundary and at least be sufficiently complete to permit recognition of the first major sandstone unit above the thick gray siltstones and shales which show the same upward stratigraphic changes that were observed in the Trimmers Rock Formation at the Lehigh River (Epstein and others, 1974). Beds coarsen upward, shale content diminishes, and sand grains increase in percentage. The coarser siltstone lithologies become more persistent and frequent upward, suggesting increasing massiveness of bedding. In addition, detailed observations of every sampled interval (usually 10 to 20 feet) in several wells in this portion of the subsurface section revealed grain-size relationships that can be attributed to graded bedding. There is a repetitive occurrence in each sampled interval of minor amounts of coarse and fine silt-size fractions associated with the dominant size fraction, medium-grained silt. This graded-bed sequence is assigned to the prodelta geomorphic "niche" in the deltaic model conceptualized in this report. The term "distal prodelta" has been applied to those portions of the prodelta succession which are made up entirely of gray marine siltstone and shale. Where minor quantities of sandstone (usually very fine grained) are associated with the siltstone and shale in a ratio of less than one, the term proximal prodelta has been assigned. The terms do not necessarily connote the distal and proximal turbidites described by Walker (1967), since primary sedimentary structures used in his assignment of such terms are not observable in the subsurface samples. Thirteen of the fourteen wells examined have both distal- and proximal-prodelta deposits immediately above the Middle-Upper Devonian boundary (Tables 8A-8N, Appendix).

Above the dominantly siltstone prodelta lithologies, samples are marked by an abrupt increase in sand size and sand percent. This abrupt occurrence of sandstone above a thick interval (more than 100 feet) of siltstone with graded-bed characteristics has been interpreted as the in-troduction of delta-front detritus. This environment is so designated where the ratio of sandstone to siltstone plus shale exceeds one. The fact that the sandstone is in immediate contact with rocks considered to be prodelta in origin is obviously critical to its interpretation as a delta-front deposit. However, the sandstone unit also has several physical features similar to those observed in outcrop which support the delta-front interpretation. Sand grain sizes range from medium to very fine. Scattered fragments of quartz granules or pebbles are found in some horizons (i.e., interval 2590'-3156' in Table 8C; interval 3340'-4000' in Table 8D; interval 3262'-3441' in Table 8F). The delta-front sandstones commonly have carbonate cement and carbonaceous fragments. Despite the grain size range, there is no evidence of upward-fining or cyclicity; in fact, many sampled intervals show a persistence of sandstone without interruption by shale or siltstone. This type of sequence in the sub-surface samples suggests massive sandstone beds, a feature already ob-served to be typical of the delta-front Towamensing Member of the Catskill Formation in the Lehigh River section (Epstein and others, 1974). In the Lally #1 well, glauconite was noted in the uppermost delta-front sandstone unit (Table 8G, interval 360'-1290').

Immediately above several delta-front sandstone units are relatively thin horizons of alternating siltstone and shale lithologies (Tables 8B, 8C, 8E and 8M). Sandstone content is minor and commonly absent. The stratigraphic position of these fine-grained deposits, the lithologic con-trast to the immediately underlying delta-front deposits and the pres-ence of fossils (noted at interval 980'-1115' in the Sheehan #1 well, Table 8L) suggest that they can be considered to be bay deposits. In modern deltas, bay deposits develop when there is a shift in active supply of detritus from one part of the delta platform to some other part. The area that no longer receives detritus begins to subside, transgression occurs and bay sediments accumulate.

Criteria for Recognition of Alluvial Deposits in Subsurface Data

Delta-Plain Deposits

A very prominent lithologic assemblage made up of a repetitive se-quence of gray sandstones with upward-increasing amounts of red silt-stone is found in wells east of the Lackawanna syncline (Figure 4 and Tables 8B, 8C, 8F, 8I, 8J and 8K). These lithologies appear as a series of upward-fining successions on graphic logs. Carbonate fragments are associated with most of the gray sandstone units. Similar upward-fining, with carbonate fragments in the basal sandstone bed and with a color change from gray to red near the sandstone-siltstone transition, was

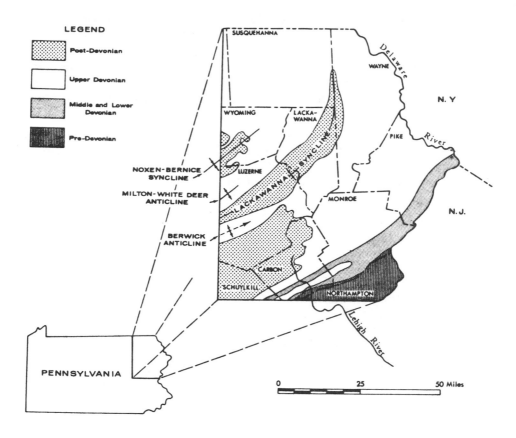

Figure 4. Geologic map showing location of Lackawanna syncline, Noxen-Bernice syncline, Berwick anticline and Milton-White Deer anticline.

observed in the Walcksville and Long Run Members of the Catskill Formation at the Lehigh River. There, evidence suggested that these members accumulated on the delta plain (Epstein and others, 1974). The same evidence, with the exception of the primary sedimentary structures (not recoverable in cable tool drilling) occurs in the subsurface. Thus, these units with upward-fining successions *and* carbonate fragments in the basal coarse bed are also considered to be delta-plain deposits.

When many vertically contiguous samples from any one well are set out to be examined, multiple upward-fining successions grading from coarse through fine detritus and nonred through red coloration can easily be recognized. Each succession begins with a relatively coarse-grained sandstone, above which the sand fraction becomes finer in succeeding sampled intervals. The sand acquires a red coloration often by

229

the addition of minor amounts of siltstone. Siltstone then becomes more dominant until the samples are comprised entirely of red siltstone. The siltstone becomes shaly upward. In overlying samples, the first signal of the next cycle is the presence of relatively coarse sand as an admixture with the red shale through a 10-20-foot interval. Above, the nonred sandstone again dominates and so the cyclic pattern begins again.

Meandering-River, Crevasse-Splay, and Flood-Basin Deposits

There is another type of upward-fining unit comprising a less prominent portion of the Upper Devonian column in several wells (Tables 8B, 8C, 8F, 8J and 8K) which does not have carbonate at the base of the cycles. In all other lithologic respects, the succession of properties is the same, including the color change from gray to red near the sandstone-siltstone boundary. The absence of carbonate is not simply a random circumstance of preservation or recovery from the well; it is apparently fundamental to this type of upward-fining cycle, as can be seen in Table 8B (interval 3773'-3920'), Table 8C (intervals 3156'-3517'; 602'-883'), Table 8F (intervals 1950'-2127'; 707'-1222'), Table 8J (intervals 2016'-2653'; 848'-1720'; 357'-715') and Table 8K (interval 1878'-2535'). All material present in these intervals comprises upward-fining cycles. In contrast, intervals designated as delta-plain deposits both in outcrop (Walcksville and Long Run Members described in Epstein and others, 1974) and those recognized in subsurface (Tables 8B, 8C, 8F, 8J, and 8K described above) are not made up entirely of upward-fining successions.

This pattern of upward-fining cycles in which no carbonate occurs and in which all material comprises the upward-fining units was observed at the Lehigh River section in the Duncannon Member of the Catskill Formation. The Duncannon has the elements of alluvial origin described by Allen (1965); each basal sandstone rests on a scoured surface of the underlying cycle, and the sandstone fines upward into siltstone, commonly acquiring red color as the admixture of fines increases. In addition, most basal units above the scoured surface in the Duncannon Member contain red shale fragments. These red shale fragments in the basal beds of some cycles probably were deposited after local scour and are preserved as "intraformational conglomerates" (Allen and Friend, 1968, p. 19-20).

The upward-fining cycles in the Duncannon Member are interpreted to be products of a meandering-river environment. Those units recognized in the subsurface which have upward-fining cycles and no carbonate at cycle bases are also considered to be products of meandering-river environments.

There are two variations on the alluvial cyclothem pattern of nonred sandstone succeeded upward by red siltstone. Both are thought to be depositional products which accumulated away from the main activity of the meandering rivers that deposited the upward-fining cyclothems. The first has all of the lithologic elements of the cyclothems but lacks the upward fining of grain size. Instead, sandstone alternates with siltstone and shale, commonly red (Table 8B, intervals 2950'-3132', 3202'-3289', and 0'-272'; and Table 8E, interval 0'-650'). Because of the repeated occurrence of the sandstone alternating with red siltstones and shale in these successions, it is thought that they represent crevasse-splay deposits. The red siltstones and shales apparently are suspension deposits within the flood basin which buried the sands that were deposited as a result of repeated crevassing of the meandering-river levees during floods.

The second variation associated with the alluvial cyclothems is repeated red siltstone and shale units forming easily distinguishable stratigraphic intervals in the subsurface (Table 8B, interval 3132'-3202'; Table 8C, interval 2406'-2590'; Table 8F, interval 0'-158'; Table 8J, interval 0'-173'). These fine terrigenous beds are thought to have been deposited from suspension after being carried over the meandering-river banks during floods. In these flood-basin deposits, the siltstone and shale alternate through a considerable thickness of sampled intervals without interruption by other lithic types and without the occurrence of carbonate fragments. In some thick sequences there are a few cycles in which siltstone grades upward into shale. However, this cyclicity may be more apparent than real since interval thicknesses sampled at the well sites control the degree of mixing of siltstone and shale amounts that appear in each sample.

There are few differences in the units composed entirely of siltstone and shale to help determine whether they are flood-plain or bay deposits. The geographic association between bays and meandering rivers on the surface of the delta need not differ from that of flood basins and the river system higher in the alluvial valley. In this report, stratigraphic context is the important factor in distinguishing the two. Bay deposits are associated with delta-front sandstones, whereas flood-basin deposits are associated with alluvial deposits. This distinction can best be observed in a single well, Schreiber #1, in which intervals 4630'-4903' and 5010'-5184' are bay deposits and interval 3132'-3202' is a flood-basin deposit (Table 8B).

Braided-River Deposits

One of the important distinguishing characteristics of braided-river deposits is the development of one sandstone body upon another, resulting in thick successions dominated by sandstones. The sandstone dominance which characterizes the surface exposures of the Packerton, Sawmill Run, Berry Run and Clark's Ferry Members of the Catskill Formation along the Lehigh River is also found in the subsurface intervals of many of the wells. Braided-river deposits in subsurface stratigraphic profile are comprised of complexly interbedded sandstones of variable size ranges. They contrast with the typical meandering-river deposits, which have both a coarser bedload fraction and finer suspension detritus.

Thick sandstone intervals occur in many of the wells examined (viz., Table 8B, intervals 975'-1437' and 1685'-2298'; Table 8K, intervals 0'-695' and 1517'-1878'). The grain size of the sandstone ranges from very coarse to very fine. Carbonate fragments, some coaly plant material, pyrite and siderite are present, as well as minor amounts of siltstone and shale, commonly red. Frosted quartz pebbles occur in interval 1517'-1878' of the Schreiber #1 well (Table 8B). The sandstones are interpreted as braided-river deposits and are distinguished from delta-front deposits by the common presence of carbonate fragments, frequency of red in the associated fine detritus, wide range in sand grain size and, most importantly, stratigraphic position. Sandstones interpreted as braided-river sediments do not occur stratigraphically adjacent to the siltstones assigned to the prodelta environment, whereas delta-front sandstones are invariably associated with prodelta deposits.

Table 3 lists the criteria used in this report to distinguish each of the three principal sedimentary environments recognized in subsurface samples of the Upper Devonian rocks in northeastern Pennsylvania. It is important to emphasize that the interpretation of origin of these units can be debated and is subject to refinement and modification as further work produces critical clues for recognizing geomorphic settings in ancient deltas. However, the rock units themselves are physically distinctive and can be easily recognized in examinations of lithologic logs of the wells. Therefore, although environmental interpretations are used to characterize the individual rock-stratigraphic units and are the bases for the correlation network to be introduced in the following section, one should not lose sight of the fact that these environmental interpretations are based upon objective recognition of physically distinctive rock units which are used to reconstruct the internal organization and spatial geometry of the Catskill delta.

[*Editor's Note:* Material has been omitted at this point.]

Table 3. *Criteria for Distinguishing Prodelta, Coastal Margin and Alluvial Deposits in Subsurface*

Geomorphic Setting	Depositional Environment*	Lithologic Criteria
Alluvial deposits	Braided rivers	Dominantly sandstones, commonly with carbonate fragments; no fining-upward cycles.
	Meandering rivers	Fining-upward cycles starting with sandstone and lacking carbonate fragments.
	Flood basin	Siltstone and shale, usually red, with or without fining-upward cycles.
	Crevasse splays	Sandstones with red shale and siltstone interbeds.
	Delta plain	Upward-fining successions with carbonate fragments at base.
Coastal-margin deposits	Bay	Siltstone and shale alternations; sandstone minor to absent.
	Delta front	Dominantly sandstones with siltstone and shale present; sandstone/(shale + siltstone) more than 1; no cycles present.
Prodelta marine deposits	Proximal prodelta	Dominantly siltstone and shale with sandstone present; sandstone/(shale + siltstone) less than 1. Graded beds.
	Distal prodelta	Siltstone and shale; no sandstone present.

*Each of these environments is shown on Figure 3.

[*Editor's Note:* Material has been omitted at this point.]

REFERENCES

Allen, J. R. L. (1964), *Studies in fluviatile sedimentation: six cyclothems from the Lower Old Red Sandstone, Anglo-Welsh Basin*, Sedimentology, v. 3, no. 3, p. 163-198.

———— (1965), *Late Quaternary Niger Delta and adjacent areas; sedimentary environments and lithofacies*, Am. Assoc. Petroleum Geologists Bull., v. 49, no. 9, p. 547-600.

Allen, J. R. L. and Friend, P. F. (1968), *Deposition of the Catskill facies, Appalachian region, with notes on some other Old Red Sandstone basins*, Geol. Soc. America Special Paper 106, p. 21-74.

Arndt, H. H., Wood, G. H., Jr. and Trexler, J. P. (1962), *Subdivision of the Catskill Formation in the west part of the anthracite region of Pennsylvania*, U.S. Geol. Survey Prof. Paper 450-C, p. C32-C36.

Ayrton, W. G. (1963), *Isopach and lithofacies map of the Upper Devonian of northeastern United States*, in Shepps, V. C., ed., *Symposium on Middle and Upper Devonian stratigraphy of Pennsylvania and adjacent states*, Pa. Geol. Survey, 4th ser., Gen. Geology Rept. 39, p. 3-6.

Berg, T. M. (in prep.), *Geology and mineral resources of the Brodheadsville 7½-minute quadrangle, Monroe County, Pennsylvania*, Pa. Geol. Survey, 4th ser., Atlas 205a.

Caster, K. E. (1934), *The stratigraphy and paleontology of northwestern Pennsylvania: Part I, Stratigraphy*, Bull. Am. Paleontologist, v. 21, no. 71, p. 1-185.

Coleman, J. M. and Gagliano, S. M. (1964), *Cyclic sedimentation in the Mississippi River deltaic plain*, Trans. Gulf Coast Assoc. Geological Society, vol. XIV, p. 67-80.

———— (1965), *Sedimentary structures: Mississippi River deltaic plain*, in G. V. Middleton, ed., *Primary sedimentary structures and their hydrodynamic interpretation*, Soc. Econ. Paleontologists and Mineralog. Spec. Publ. 12, p. 133-148.

Conlin, R. R. and Hoskins, D. M. (1962), *Geology and mineral resources of the Mifflin-town quadrangle*, Pa. Geol. Survey, 4th ser., Atlas 126.

Dennison, J. M. (1961), *Stratigraphy of Onesquethaw Stage of Devonian in West Virginia and bordering states*, W. Va. Geol. Survey, Bull. 22, 87 p.

——— (1969), *Tioga bentonite and other isochronous units in the Devonian Oriskany to Tully interval of the Appalachian Basin* [abs.], Geol. Soc. America, Northeastern Section, 4th Annual Meeting, Program, p. 13.

Dennison, J. M. and Textoris, D. A. (1970), *Devonian Tioga tuff in northeastern United States*, Bull. Volcanolagique, Tome XXXIV, p. 289-294.

Dyson, J. L. (1954), *Relation of stratigraphy and structure to uranium occurrences near Mauch Chunk, Pennsylvania*, Proc. Pennsylvania Acad. Sci., v. 28, p. 124-134.

——— (1963), *Geology and mineral resources of the northern half of the New Bloomfield quadrangle*, Pa. Geol. Survey, 4th ser., Atlas 137ab.

——— (1967) *Geology and mineral resources of the southern half of the New Bloomfield quadrangle, Pennsylvania*, Pa. Geol. Survey, 4th ser., Atlas 137cd.

Epstein, J. B. and Sevon, W. D. (in prep.), *Geology of the Kunkletown quadrangle, Monroe County, Pennsylvania*, U.S. Geol. Survey Geol. Quad. Map.

Epstein, J. B., Sevon, W. D. and Glaeser, J. D. (1974, in press), *Geology and mineral resources of the Lehighton and Palmerton 7½-minute quadrangles, Pennsylvania*, Pa. Geol. Survey, 4th ser., Atlas 195cd.

Fettke, C. R. (1953), *Oil and gas developments in the Appalachian Basin, past and present*, Pa. Geol. Survey, 4th ser., Min. Resource Rept. 37.

Fisher, W. L. and others (1969), *Delta systems in the exploration for oil and gas*, Bur. of Economic Ecology, Univ. of Texas at Austin, 78 p.

Fisk, H. N. (1961), *Bar-finger sands of Mississippi Delta*, in Peterson, J. A. and Osmond, J. C., eds., *Geometry of sandstone bodies*, Am. Assoc. Petroleum Geologists, p. 29-52.

Fletcher, F. W. and Woodrow, D. L. (1970), *Geology and mineral resources of the Pennsylvania portion of the Milford and Port Jervis 15-minute quadrangles*, Pa. Geol. Survey, 4th ser., Atlas 223, 64 p.

Frakes, L. A., Glaeser, J. D., Wagner, W. R. and Wietrzychowski, J. F. (1963), *Stratigraphy and structure of Upper and Middle Devonian rocks in northeastern Pennsylvania*, Guidebook, 28th Ann. Field Conf. Pa. Geologists, 44 p.

Friedman, G. M. and Johnson, K. G. (1966), *The Devonian Catskill deltaic complex of New York, type example of a "tectonic delta complex,"* in Shirley, M. L., ed., *Deltas and their geologic framework*, Houston Geol. Soc., p. 171-188.

Gagliano, S. M. and McIntire, W. G. (1969), *Delta components*, in *Recent and ancient deltaic sediments: a comparison*, A symposium conducted by J. M. Coleman and others, April 1-5, 1969, Louisiana State Univ., p. 1-7.

Glaeser, J. D. (1963), *Catskill reference section and its correlation to other measured surface sections in northeast Pennsylvania*, in Shepps, V. C., ed., *Symposium on Middle and Upper Devonian stratigraphy of Pennsylvania and adjacent states*, Pa. Geol. Survey, 4th ser., Gen. Geology Rept. 39, p. 51-62.

——— (1967), *Catskill stratigraphy at the Lehigh Gap reference section and its relation to type locations in northeastern Pennsylvania*, Geol. Soc. America Spec. Paper 115, p. 266.

——— (1969), *Geology of flagstones in the Endless Mountains region, northern Pennsylvania*, Pa. Geol. Survey, 4th ser., Info. Circular 66, 14 p.

——— (1970), *Correlation of Catskill sedimentary environments and its economic significance*, Geol. Soc. America Abs. with Programs, v. 2, no. 1, p. 21.

Gray, Carlyle and others (1960), *Geologic map of Pennsylvania*, Pa. Geol. Survey, 4th ser., Map 1, 1:250,000.

Grow, G. C., Jr. (1964), *Gas production near hard coal areas of Pennsylvania*, World Oil, v. 159, no. 4, p. 79-81.

Kehn, T. M., Glick, E. E. and Culbertson, W. C. (1966), *Geology of the Ransom quadrangle, Lackawanna, Luzerne, and Wyoming counties, Pennsylvania*, U.S. Geol. Survey Bull. 1213, 81 p.

Kelley, D. R. (1967), *Geology of the Red Valley sandstone in Forest and Venango Counties, Pennsylvania*, Pa. Geol. Survey, 4th ser., Min. Resource Rept. 57, 48 p.

Klemic, Harry, Warman, J. C. and Taylor, A. R. (1963), *Geology and uranium occurrences in the northern half of the Lehighton, Pennsylvania quadrangle and adjoining areas*, U.S. Geol. Survey Bull. 1138, 97 p.

Krajewski, Stephen (1971), *The relationship between the landforms, occurrence, composition, texture and physical properties of the Upper Devonian flagstones (bluestone) from northeastern Pennsylvania*, unpub. M.S. thesis, Pennsylvania State Univ., 187 p.

Leopold, L. B. and Wolman, M. G. (1957), *River channel patterns, braided, meandering and straight*, U.S. Geol. Survey Prof. Paper 282-B, p. 39-73.

Lesley, J. P. (1892), *A summary description of the geology of Pennsylvania*, Pa. Geol. Survey, 2nd ser., Final Rept., v. II.

McCauley, J. F. (1961), *Uranium in Pennsylvania*, Pa. Geol. Survey, 4th ser., Min. Resource Rept. 43.

Meckel, L. D. (1970), *Paleozoic alluvial deposition in the Central Appalachians*, in Fisher, G. W. and others, eds., *Studies of Appalachian geology, Central and Southern*, p. 49-67.

Montgomery, Arthur (1969), *The mineralogy of Pennsylvania, 1922-1965 — supplementing and updating Gordon's "The mineralogy of Pennsylvania (1922),"* Acad. Nat. Sci. Philadelphia Spec. Pub. 9, 104 p.

Oliver, W. A., Jr. (1971), *Isopach and lithofacies maps of the Devonian in the Appalachian basin*, Pa. Geol. Survey, 4th ser., Prog. Rept. 182.

Prosser, C. S (1892), *The Devonian System of eastern Pennsylvania*, Amer. Jour. Sci., 3 ser., v. 44, p. 210-221.

――― (1894), *The Devonian System of eastern Pennsylvania and New York*, U.S. Geol. Survey Bull. 120.

Rogers, H. D. (1858), *The geology of Pennsylvania*, Pa. Geol. Survey, 1st ser., Phila., v. 1.

Rose, A. W. (1970), *Atlas of Pennsylvania's mineral resources: Part 3, Metal mines and occurrences in Pennsylvania*, Pa. Geol. Survey, 4th ser., Min. Resource Rept. 50, 14 p.

Scott, A. J. and Fisher, W. L. (1969), *Delta systems and deltaic sedimentation*, in Fisher, W. L. and others, *Delta systems in the exploration for oil and gas*, Bur. of Econ. Geol., Univ. of Texas at Austin, p. 10-29.

Scruton, P. C. (1960), *Delta building and the deltaic sequence*, in Shepard, F. P. and others, eds., *Recent sediments, northwestern Gulf of Mexico*, Am. Assoc. Petroleum Geologists, Tulsa, Okla., p. 82-102.

Sevon, W. D. (in press), *Geology and mineral resources of the Christmans and Pohopoco Mountain 7½-minute quadrangles, Pennsylvania*, Pa. Geol. Survey, 4th ser., Atlas 195ab.

Shepps, V. C., ed. (1963), *Symposium on Middle and Upper Devonian stratigraphy of Pennsylvania and adjacent states*, Pa. Geol. Survey, 4th ser., Gen. Geology Rept. 39, 301 p.

Wagner, W. R. (1963a), *Correlation of Susquehanna Group in part of northeastern Pennsylvania*, in Shepps, V. C., ed., *Symposium on Middle and Upper Devonian stratigraphy in Pennsylvania and adjacent states*, Pa. Geol. Survey, 4th ser., Gen. Geology Rept. 39, p. 63-79.

――― (1963b), *Gas exploration in northeastern Pennsylvania 1930-1963*, Guidebook, 28th Ann. Field Conf. Pa. Geologists, p. 37-42.

Walker, R. G. (1967), *Turbidite sedimentary structures and their relationship to proximal and distal depositional environments*, Jour. Sed. Petrology, v. 37, no. 1, p. 25-43.

Walther, J. (1894), *Lithogenesis der Gegenwart, Beobachtungen uber die Bildung der Gesteine an der heutigen Erdoberflache, dritten Teil einer Einleitung in die geologische historische Wissenschaft*, Verlag von Gustav Fischer, Jena, section 8, p. 621.

236

Weeks, L. G. (1958), *Habitat of oil and some factors that control it*, in Weeks, L. G., ed., *Habitat of oil — a symposium*, Am. Assoc. Petroleum Geologists, p. 1-61.

Wheeler, H. E. (1963), *Catskill and the Acadian discontinuity*, in Shepps, V. C., ed., *Symposium on Middle and Upper Devonian stratigraphy in Pennsylvania and adjacent states*, Pa. Geol. Survey, 4th ser., Gen. Geology Rept. 39, p. 103-114.

White, I. C (1881), *The geology of Susquehanna County and Wayne County*, Pa. Geol. Survey, 2nd ser., Rept. G 5.

———— (1882), *The geology of Pike and Monroe Counties*, Pa. Geol. Survey, 2nd ser., Rept. G 6.

Wietrzychowski, J. F. (1963) , *Subsurface correlation of Upper Devonian lithostratigraphy in northeastern Pennsylvania and southeastern New York*, in Shepps, V. C., ed., *Symposium on Middle and Upper Devonian stratigraphy in Pennsylvania and adjacent states*, Pa. Geol. Survey, 4th ser., Gen. Geology Rept. 39, p. 43-50.

Willard, Bradford (1936), *Continental Upper Devonian of northeastern Pennsylvania*, Geol Soc. America Bull., v. 47, no. 4, p. 565-608.

———— (1939), *The Devonian of Pennsylvania*, Pa. Geol. Survey, 4th ser., Gen. Geol. Rept. 19, 481 p.

Willard, Bradford and Stevenson, R. E. (1950) , *Northeastern Pennsylvania and central New York petroleum probabilities*, Am. Assoc. Petroleum Geologists Bull., v. 34, p. 2269-2283.

Williams, H. S. (1900), *Discussion on the differing age of the Catskill Formation in Pennsylvania*, Geol. Soc. America Bull., v. 11, p. 594-595.

Winslow, A. (1896), *The Lehigh River section*, Pa. Geol. Survey, 2nd ser., Annual Rept., Pt. 4, p. 1331-1371.

Woodrow, D. L. and Fletcher, F. W. (1967), *Late Devonian paleogeography in southeastern New York and northeastern Pennsylvania*, Internat. Symposium on the Devonian System, v. 2, Calgary, Alberta Soc. Petroleum Geologists, p. 1327-1334.

18

Reprinted from pp. 233, 236, 241–249, 254 of *Sedimentology* **10**(2):233–254 (1968)

DELTAIC SEDIMENTATION UNITS IN THE UPPER CARBONIFEROUS OF NORTHERN ENGLAND

JOHN D. COLLINSON

Department of Geology and Mineralogy, Oxford (Great Britain)[1]
(Received March 16, 1968)

SUMMARY

In the Kinderscout Grit (Namurian, R_1c) of the southern half of the Central Pennine Basin, England, there occur tabular, isolated, cross-bedded sets of coarse sandstone, between 4 and 40 m thick. The sets are up to 1 km wide and the foresets, in plan, are convex in the direction of dip. The sets interfere laterally to form extensive sheets. Foresets dip at up to 27° and are sometimes themselves internally cross-bedded in trough-shaped sets, here termed "intrasets". The large sets are thought to be deltaic sedimentation units rather than sandwaves. In the delta top environment in which these sets were deposited, a high rate of bed-load sediment supply and a sudden deepening were required to initiate the sets.

Medium-scale cross-bedding, thought to have been laid down in migrating, fluviatile channels overlies the large-scale sets and represents the topset member of the sedimentation units. From the spatial arrangement of the large- and medium-scale cross-bedding, it is possible to distinguish those areas where deposition took place during the deepening, from those where it took place essentially after the deepening. The causes of the deepening, which must have been on at least a basinal scale, may have been eustatic or tectonic.

[*Editor's Note:* Material, including figures 1 and 2, has been omitted at this point.]

---

[1] Present address: Institute of Physical Geography, Uppsala, Sweden.

Ed. Note: Further study reveals that the large cross-bedded sets probably were bars in the major distributary channels of the delta (J. D. Collinson, personal comm., 1976).

FACIES DESCRIPTION

Two major facies constitute the sand bodies and will be described and interpreted in turn. Both are composed of extremely coarse and pebbly sandstone, with a modal grain size in the range 1–2 mm and with pebbles up to 3 cm long. It is poorly sorted and arkosic in composition.

Large-scale cross-bedding

Description. The extremely coarse pebbly sandstone is cross-bedded in isolated, tabular sets, from 4 to 40 m thick. The sets are of great lateral extent in the direction of foreset dip. On the northern edge of the Kinderscout Plateau (NGR[1] SK 097898 to 073898), a single set, between 30 and 35 m thick, can be traced as a continuous exposure for at least 2.4 km in the direction of foreset dip (Fig.3). Perpendicular to the direction of foreset dip, the sets of about 30 m thickness are apparently

Fig.3. Northern Edge of the Kinderscout Plateau, with Fairbrook on the left. The exposure of large-scale cross-bedding is 26 m high, the foresets having a uniform dip to the west. This set extends for 2.4 km in the plane of the exposure.

about 1 km across. The dip directions curve in broad arcs which are generally convex in the direction of dip (Fig.5).

The largest of the large-scale sets (12–40 m thick) have foresets which are concave upwards in vertical sections. They are tangentially based with dips up to 27° in the tops of the sets and the sets are often directly underlain by horizontally

---

[1] British National Grid Reference.

[*Editor's Note:* Figure 4 has been omitted.]

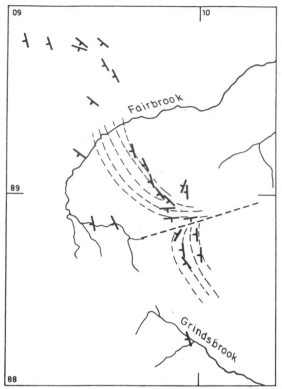

Fig.5. Foreset dip directions on the Kinderscout Plateau near Fairbrook, plotted as strike lines, the ticks being on the down dip sides. Each dip measurement is from one distinct foreset. Individual foresets may sometimes be traced laterally for a few tens of meters before being lost under peat cover. The convex curvature in the direction of overall dip is shown by the light broken lines in the vicinity of the sharp discontinuity, indicated by the heavier line. The discontinuity is thought to represent the margin between two lobes. The rather variable direction in the northwest may be due to a degree of concave curvature in the foreset plan or to another, less clearly defined discontinuity.

bedded coarse sandstone, with no intervening fine-grained horizon (Fig.6). Smaller sets (4–15 m thick) have more planar foresets which give a marked angular divergence at the base of the set. The tops of all sets are truncated by horizontal erosion surfaces (Fig.7).

The foresets themselves are most commonly almost parallel sided and up to 10 cm thick. Internally they may be massive, graded or inversely graded. In the largest sets (i.e., those with tangential foresets), the foresets may themselves be cross-bedded in medium-scale trough-shaped sets up to 30 cm thick (Fig.8). It is proposed that these medium-scale sets be termed "intrasets".

Measurement of directions of intraset trough axes is difficult because bedding surfaces are infrequently exposed. A histogram of the large-scale foreset dip directions is presented (Fig.9A). Relations between intraset directions and foreset dip direction show a tendency for the intraset troughs to indicate currents pre-

Fig.6. Woodhead Tunnel, Longendale (N.G.R. SK 114999). A large-scale set of cross-bedding 12 m thick, is overlain by medium-scale cross-bedding in the top of the quarry, the two being separated by a horizontal erosion surface. The large foresets are asymptotically based. Within the base of the large-scale set "intraset" medium-scale cross-bedding (not visible in the photograph) is directed up the foreset slope.

dominantly down the foreset slope (Fig. 9C). Intrasets may occur at any level within the large-scale sets, even in the top, where the dip of the large-scale foresets is of the order of 25°. The orientation of the intrasets appears to be independant of their level within the set. However, the only two examples of intraset cross-bedding, which is directly opposed in direction to that of the foreset dip, occur in the lower 2 m of 12-m sets (Fig.6).

Occasional erosion surfaces, which can sometimes be traced over the full thickness of some large sets, separate groups of foresets, but always truncate the underlying foresets. The overlying group of foresets dips parallel to the erosion surface (Fig.10).

Bands of fine, frequently silty, sandstone and siltstone also occur within the predominantly coarse and pebbly sandstone foresets. These bands may be up to at least 4 m thick and dip parallel to the coarse grained, large-scale foresets. The fine sandstone may be ripple cross-laminated and the siltier beds are frequently bioturbated.

Fig.7. Bamford Edge (N.G.R. SK 211846). A large-scale set of at least 8.5 m thickness, with planar foresets. These are truncated by a horizontal erosion surface, which is overlain by medium-scale trough cross-bedding.

Interpretation. Cross-bedding on a scale of 4–40 m is usually ascribed to the migration of large aeolian dunes (POTTER and PETTIJOHN, 1953; MCKEE, 1966). Neither the geological context, the grain size of the sediment nor the presence of burrows in the finer grained bands supports an aeolian origin for this facies and a water-lain origin is more likely. Water-lain cross-bedding of this scale could, theoretically, be produced by the migration of two types of structure, large-scale sandwaves or deltaic sedimentation units. In the case of sandwaves the whole of the bedform is submerged in water which must be considerably deeper than the sandwave height. In the case of a deltaic sedimentation unit, the foreset slope is fed by streams migrating and diverting on the upper surface and at any given instant, only a relatively small area of the delta top need be submerged and actively carrying sediment. The following differences occur between the deposits of the two structures.

(1) Scale. To deposit cross-bedded sets of up to at least 40 m thickness, a migrating sandwave would have to be of least that height itself and, under likely conditions of net sedimentation, considerably more. That sandwaves of that scale should occur under conditions other than very low rates of net sedimentation is

Fig.8. Dovestone Tor, Derwent Edge (N.G.R. SK 197897). "Intraset" cross-bedding within a 25-m set. The large-scale dip is away from the camera, to the south, being indicated by the bedding plane just below the 1-m tape.

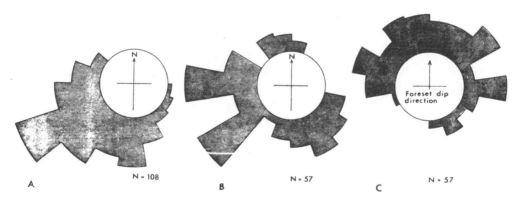

Fig.9. Histograms of the directions of the "intraset" cross-bedding and the large-scale foresets. A shows the dips of the large-scale foresets in sets with and without intrasets; B shows the true intraset directions and C the divergences between the intraset and large-scale foreset dip directions, when both are known.

Fig.10. Mouselden Quarry (N.G.R. SK 208868). The face is about 25 m high, exposing erosion surfaces within the large-scale sets. The lowest foresets dip to the right and towards the camera. The prominent erosion surface which truncates them, dips to the left, and the overlying foresets dip parallel to it. A further erosion surface (not really visible on the photograph) truncates the second group of foresets and the highest group of foresets, seen in the top left of the figure, dip parallel to this surface and towards the camera.

highly unlikely, unless they were migrating down a pre-existing slope. For deltaic sedimentation units, the water depth need only equal the sum of the delta height and the depth of the channel on its upper surface. Under conditions of continued subsidence, all the sedimentation unit will be preserved.

(2) *Shape.* Extrapolating ALLEN's (1963) fig.4 to sandwaves of 40 m height, the overall dip of the foreset slope should be of the order of 4° and for sandwaves 4 m high the value should be about 10°. The foresets of a deltaic sedimentation unit with planar foresets will dip at angles close to the angle of rest of the sediment, which, in the case of coarse sand is around 30°. Asymptotically based delta foresets will give a somewhat lower overall angle.

(3) *Organisation.* Solitary sets of the planar type are best explained by delta building (ALLEN, 1963, p.188) while sandwaves would be expected to give a coset of more than one set.

(4) *Context.* The delta-top of a prograding deltaic complex (READING, 1964; WALKER, 1966; COLLINSON, 1968) is not a likely site for large-scale sand-wave migration. In such an environment, strong currents, such as would be needed

to generate and move large-scale sandwaves, would be confined to channels and there is no evidence to suggest that the large-scale cross-bedding is an in-channel deposit. Indeed, its widespread extent (Fig.3) suggests a wide, flat area of deposition.

In the light of the differences in the processes of sandwave and deltaic advance, it seems that the latter process is the more satisfactory to explain the large-scale cross-bedding. The internal features of large-scale sets will be interpreted later in the paper.

Medium-scale cross-bedding

Descriptions. This facies is formed of similar sandstone to that forming the large-scale sets. The sets are from 2 m to 10 cm thick. They may be tabular or trough-shaped and occur in cosets up to 5 m thick. The cosets form extensive sheets with horizontal bounding surfaces, the lower one of which is always erosive. Within a coset of medium-scale cross-bedding, there is frequently an upward change in the character of the sets from tabular in the base to trough-shaped in the top, frequently accompanied by a decrease in set size. Trough sets have a maximum thickness of about 30 cm.

Interpretation. Cosets of medium-scale cross-bedding have been attributed, on geometrical grounds by ALLEN (1963) to the migration of large-scale ripple marks (dunes) under conditions of net sedimentation, tabular and trough-shaped sets being attributable to straight-crested and linguoid (or lunate) forms, respectively. HARMS and FAHNSTOCK (1965) suggested a similar genesis based on field observations, though with a less rigid geometrical association. A coset which rests on a horizontal erosion surface and shows an upwards decrease in set size, has been attributed to deposition from dune migration in a laterally migrating fluviatile channel (ALLEN, 1964; VISHER, 1965) and the present author favours a similar interpretation for this facies. Whether the pattern of shifting channels in this case was braided or meandering is uncertain.

RELATIONSHIP OF THE FACIES

Description

The two facies are vertically associated with each other in a fixed way (Fig.11). The medium-scale cross-bedded facies always occurs directly above the large-scale set with a horizontal erosion surface between them. The surface sometimes has a slight relief but more commonly is remarkably flat. There may be a thin pebble conglomerate immediately above the erosion surface. The two facies together, in the relationship, constitute what will be referred to below as the "facies association".

The dip directions of the foresets of medium-scale tabular sets and the axial directions of the medium-scale trough sets are always extremely close to the dip

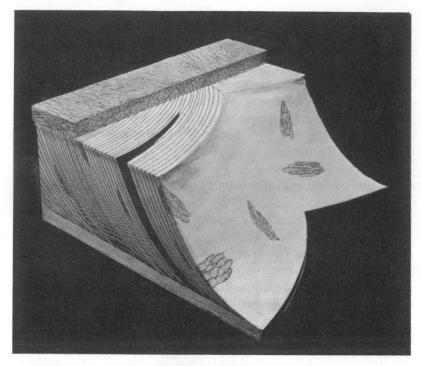

Fig.11. A model, with 4 × vertical exaggeration, showing the principle features of the facies association. The medium-scale cross-bedding is separated from the large-scale set by a horizontal erosion surface. Intrasets, a fine-grained band within the large set and the convex and interfering nature of the foresets in plan view are shown. While the intrasets on the model show down-slope directional components in the top of the slope and up-slope components in the bottom of the slope, the reverse cases may also be found in the field. While there is an overall prefered intraset direction down the slope, there is wide spread around this with no obvious relationship between direction and level within the set.

direction of the underlying large-scale foresets. The maximum divergence between the large-scale foreset dip direction and that of the bottom-most medium-scale set is 10°. Divergence increases upwards within the coset and at 2.5 m above the erosion surface, the maximum observed divergence is 30°.

Interpretation

The close vertical sequential and palaeocurrent relationship between the two facies suggests that the two are genetically closely related. Accepting that the large-scale set is the result of the advance of a deltaic sedimentation unit, the medium-scale cross-bedded facies then becomes the "topset" division of GILBERT's (1883) scheme, and represents the deposits of migrating streams on the delta top. Such channels fed the large-scale foreset slopes with bedload material. A slowly rising base level is needed for the accretion of topset beds (NEVIN and TRAINER, 1927; MCKEE, 1957; JOPLING, 1966b) without the initiation of a further deltaic sedimentation unit. What proportion of the topset bed was deposited by essentially vertical topset accretion and what by lateral channel migration, it is impossible to say in most cases. Erosion surfaces with pebble conglomerates suggest the latter ·mechanism.

The initiation of a deltaic sedimentation unit requires, in the first instance, the rapid feeding of bedload material into a body of standing water. To deposit such beds in an area which had previously been under very shallow water, rapid deepening of the water is required. Slow deepening would produce a more gradual accretion.

INTERPRETATION OF INTERNAL FEATURES OF THE LARGE-SCALE SETS

Intrasets

The genesis of intraset cross-bedding, within the large-scale foresets, presents an intriguing problem. Two broad types of current may have been responsible.

(*1*) Currents within the body of standing water. Considering the sedimentological context of the deltaic sedimentations units (Fig.2), it is quite possible that tidal currents and waves might have had considerable influence in reworking the foresets. From the evidence seen in the underlying Grindslow Shales, however, it seems that both these factors were of only slight importance during the prograding of the main, fine-grained delta slope (COLLINSON, 1967; 1968). There is no compelling reason to suppose that they should have increased in strength later. None of the palaeocurrent data (Fig.9) suggests, particularly, tidal or wave activity, nor do the sedimentary structures within the fine-grained bands.

(*2*) Currents associated with the inflowing, fluviatile currents. These depend on the velocity and depth of the fluviatile currents, the depth of the basin, the suspended sediment load and the temperature and salinity differences between the basinal and fluviatile waters. At one extreme, of low density contrast, there may be flow separation at the lip of the foreset slope and the development of a vortex system in front of the channel mouth (Fig.12). The vortices may impinge on the foreset slope at any level, depending on the flow conditions and the depth ratio (Fig.12B), or they may impinge on the sediment surface in front of the foreset slope (Fig.12C). In the situation illustrated in Fig.12B, there is the possibility of cross-bedding being formed which may point both up and down the slope. If the point of impact of the locus of zero velocity is high on the slope, most of the cross-bedding developed on the foreset slope will point down the slope (cf. Fig.9C). In the situation illustrated in Fig.12C, cross-bedding will be directed exclusively up the slope. The two examples of up-slope directed cross-bedding, found in the bases of 12-m sets may be the results of this flow pattern. The patterns shown in Fig. 12B and 12C are only two-dimensional representations in the vertical plane. In the horizontal plane, eddies with vertical axes will develop on either side of the main

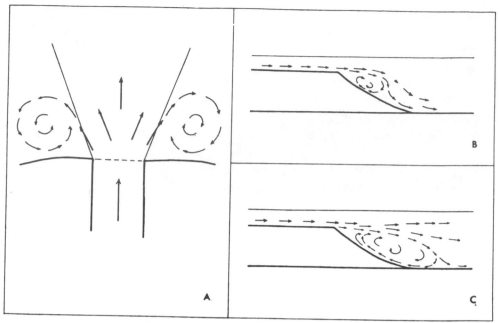

Fig.12. Schematic diagrams of the separation vortices which may be developed in front of a channel flowing into a body of standing water when the density contrast between the inflowing water and the basinal water is very low. A. shows in plan view, eddies with vertical axes on either side of the mouth. B. shows the vertical section with the locus of zero velocity impinging high on the slope. C. shows the vertical section with the locus of zero velocity in front of the slope. For a given density relationship, high influx velocity and low basinal water depth will tend to favour the development of the situation shown in C. When the density of the inflowing water is much greater than that of the basinal water, an extreme case of the situation figured in B. may develop, when there is no separation and the flow is in the form of a density underflow down the foreset slope.

flow (V. Axelsson, personal communication, 1967; Fig.12A). These may be very powerful and interference between these and the eddies with horizontal axes may lead to complex current patterns on the foreset slope. The eddies with vertical axes may also serve to sweep aside fine-grained sediment and thereby inhibit the development of a fine-grained, bottomset horizon. Currents within the basin may further deflect the flow pattern and add to the scatter of intraset cross-bedding directions.

When the density contrast between the basinal and fluvial water is higher, the fluvial being the denser, underflows may develop with no flow separation at the delta lip to give the hyperpycnal flow of BATES (1953). These currents may possibly be strong enough to move bedload material down the slope and this will tend to reduce the angle of foreset dip in the lower parts, giving tangentially based foresets (AXELSSON, 1967, p.23). The development of density currents down the foreset slope could, therefore, lead to the development of intrasets with the palaeocurrent pattern seen in Fig.9C.

[Editor's Note: Material has been omitted at this point.]

REFERENCES

ALLEN, J. R. L., 1963. Asymmetrical ripple marks and the origin of water laid cosets of cross-strata. *Liverpool Manchester Geol. J.*, 3:187–234.

ALLEN, J. R. L., 1964. Studies in fluviatile sedimentation; six cyclothems from the Old Red Sandstone, Anglo-Welsh Basin. *Sedimentology*, 3:163–198.

AXELSSON, V., 1967. The Laiture Delta; A study of deltaic morphology and processes. *Geograf. Ann. Stockholm, Ser. A*, 49:1–127.

BATES, C. C., 1953. A rational theory of delta formation. *Bull. Am. Assoc. Petrol. Geologists*, 37:2119–2162.

COLLINSON, J. D., 1966. Antidune bedding in the Namurian of Derbyshire, England. *Geol. Mijnbouw*, 45:262–264.

COLLINSON, J. D., 1967. *Sedimentation of the Southern Part of the Grindslow Shales and the Kinderscout Grit*. Thesis, University of Oxford, 192 pp. (unpublished).

COLLINSON, J. D., 1968. The sedimentology of the Grindslow Shales and the Kinderscout Grit. A deltaic complex in the Namurian of northern England. *J. Sediment. Petrol.* (in press).

COLLINSON, J. D. (in preparation). Deep channels, massive beds and turbidity current genesis in the Central Pennine Basin.

COLLINSON, J. D. and WALKER, R. G., 1967. Namurian sedimentation in the High Peak. In: R. NEVES and C. DOWNIE, (Editors), *Geological Excursions in the Sheffield Region*. Univ. of Sheffield, Sheffield, pp.80–88.

GILBERT, G. K., 1883. The topographic features of lake shores. *U.S. Geol. Surv. Ann. Rept.*, 5:69–123.

HARMS, J. C. and FAHNSTOCK, R. K., 1965. Stratification, bed forms and flow phenomena (with an example from the Rio Grande). In: G. V. MIDDLETON (Editor), *Primary Sedimentary Structures and their Hydrodynamic Interpretation — Soc. Econ. Paleontologists Mineralogists, Spec. Publ.*, 12:84–115.

JOHANSSON, C. E., 1965. Structural studies of sedimentary deposits. *Geol. Fören. Stockholm Förh.*, 87:3–61.

JOPLING, A. V., 1966a. Some principles and techniques used in reconstructing hydraulic parameters of paleoflow regime. *J. Sediment. Petrol.*, 36:5–49.

JOPLING, A. V., 1966b. Some applications of theory and experiment to the study of bedding genesis. *Sedimentology*, 7:71–102.

McKEE, E. D., 1957. Flume experiments in the production of stratification and cross stratification. *J. Sediment. Petrol.*, 27:129–134.

McKEE, E. D., 1966. Studies of dunes of White Sands National Monument, New Mexico (and a comparison with structures of dunes from other selected areas). *Sedimentology*, 7:3–69.

NEVIN, C. M. and TRAINER JR., D. W., 1927. Laboratory studies in delta building. *Bull. Geol. Soc. Am.*, 38:451–458.

POTTER, P. E. and PETTIJOHN, F. J., 1963. *Paleocurrents and Basin Analysis*. Springer, Berlin, 269 pp.

READING, H. G., 1964. A review of the factors affecting the sedimentation of the Millstone Grit (Namurian) in the Central Pennines. In: L. M. J. U. VAN STRAATEN (Editor), *Deltaic and Shallow Marine Deposits*. Elsevier, Amsterdam, pp.340–346.

VISHER, G. S., 1965. Use of the vertical profile in environmental reconstruction. *Bull. Am. Assoc. Petrol. Geologists*, 49:41–61.

WALKER, R. G., 1966. Shale Grit and Grindslow Shales: Transition from turbidite to shallow water sediments in the Upper Carboniferous of Northern England. *J. Sediment. Petrol.*, 36:90–114.

Reprinted from *Geol. Soc. London Jour.* **132**, Pt. 2:199–208 (1976)

Upper Carboniferous sedimentary cycles produced by river-dominated, elongate deltas

T. ELLIOTT

SUMMARY

Coarsening upwards cycles in the upper Carboniferous Bideford Group of north Devon reflect the repeated progradation of river-dominated, elongate deltas. The cycles consist of *progradational sequences* (large-scale, pro-delta to mouth bar coarsening upwards sequences, and major distributary channel sandstones), *aggradational sequences* (small-scale beach-spit coarsening upwards sequences and fining upwards crevasse channel sequences), and sporadically developed *abandonment facies* (thin, bioturbated siltstones and a coal horizon). Variations between cycles are governed principally by the location of bar fingers, and axial, lateral and (?) distal bar finger successions are distinguished.

MODERN river deltas exhibit considerable morphological variation, primarily in response to the interaction of fluvial and marine processes which defines the regime of the depositional area. For example, the morphology of the river-dominated Mississippi delta may be contrasted with that of the wave influenced Rhone and Nile deltas, the tide influenced Brahmaputra delta, and the tide and wave influenced Niger delta (Scruton 1960, van Andel & Curray 1960, Allen 1965, Coleman & Wright 1973, Wright & Coleman 1973). Extensive core drilling of the Mississippi, Rhone and Niger deltas has demonstrated that delta morphology is of prime importance in determining facies distribution and sand body characteristics of deltaic sediments (Fisk 1955, Oomkens 1967, 1974, Weber 1971). It therefore follows that no single delta model is capable of summarising facies and sequence distribution and that studies of ancient deltaic sediments should aim to elucidate the regime and morphology of the palaeo-delta.

Fisher (1969) outlined general criteria for the recognition of delta type in ancient successions, distinguishing high-constructive (lobate and elongate) deltas from high-destructive (wave- and tide-dominated) deltas. These criteria have been applied to selected ancient deltaic systems but have not so far been applied to upper Carboniferous coarsening upwards cycles frequently attributed to delta progradation (Reading 1967, Williams 1968, Read *et al.* 1971). The present study describes a 1200 m succession of upper Carboniferous deltaic cycles in north Devon with a view to determining the overall delta morphology represented by the cycles. The cycles are repeatedly exposed in a single, dip-aligned coastal section (Reading 1965) and the conditions of the study are therefore partly analogous to a borehole investigation; a significant point considering the importance of deltaic sediments in productive oil basins and coal-bearing successions.

1. Upper Carboniferous cycles at Westward Ho!, north Devon

The stratigraphy was described by Prentice (1960, 1962) and later modified to its existing form by de Raaf *et al.* (1965, fig. 1). Precise dating of these strata is hindered by the paucity of fauna which is confined to the Abbotsham Formation, of Westphalian age, possibly occurring within the *lenisulcata* zone. Six coarsening upwards cycles were recognized in the Abbotsham Formation (de Raaf *et al.* 1965) by means of a facies analysis in which observed relationships between facies were attributed to the repeated progradation of a complex deltaic shoreline. In a review of upper Carboniferous sedimentary sequences, Reading (1967) stressed that the Bideford Group comprised nine predictable, though complex, coarsening upwards sequences whereas the underlying Westward Ho! Formation appeared to be a random association of facies. Walker (1969) described the basal cycle of the Northam Formation (the Rock Nose cyclothem) and demonstrated a passage from basinal mudstones with turbidites into shallow, agitated water siltstones and shorelines facies, but did not discuss the overall depositional system. Walker (1970) challenged Reading's assertion that the Westward Ho! Formation was random, and defined three coarsening upwards sequences representing abrupt transitions from deep water turbidites into shallow water facies, induced by repeated basin-wide fluctuations in sea level. The upper Carboniferous in this area was therefore deposited in a moderately deep and subsiding basin which was repeatedly infilled by copiously supplied prograding shorelines.

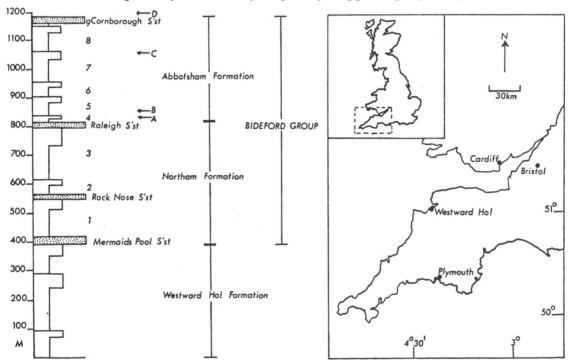

FIG. 1. Location map and stratigraphical section of the upper Carboniferous at Westward Ho! Numbers 1–9 denote the Bideford Group cycles; letters A–D refer to faunal occurrences: A—*Carbonicola* ? *cf. bellula* (Bolton)— ? relatively high *lenisulcata* zone; B—minute goniatites; C—non-marine bivalves—*lenisulcata* zone; D—non-marine bivalves—*ovalis (communis)* zone. The *Carbonicola* fauna at A was previously regarded as upper Namurian, but Eagar (pers. comm. 1973) now considers it to be (?) high *lenisulcata* zone.

2. Description of the Bideford Group cycles

De Raaf *et al.* (1965) distinguished constructive facies, comprising thick intervals of clastic sediments deposited during delta progradation, from destructive facies represented by thin, bioturbated siltstone beds accumulated after delta abandonment (the "abandonment facies" of Elliott 1974a). The present study confirms the distinction between constructive and destructive (abandonment) facies, and distinguishes progradational and aggradational sequences in the former (Fig. 2).

PROGRADATIONAL SEQUENCES

Delta progradation into the basin is represented by large-scale (50–150 m) coarsening-upwards sequences and major channel sandstones.

The coarsening-upwards sequences reflect shoreline progradation and commence with thick intervals of mudstones and finely laminated siltstones with sporadic turbidites, deposited in deep basin and pro-delta environments. In Cycles 2, 5 and 7 these facies dominate the sequences, with coarser facies confined to 2–5 m of irregularly laminated (wave-influenced?) siltstones and sandstones. In contrast the deep basin and pro-delta facies in Cycles 1 and 8 pass upwards into substantial developments of rapidly interbedded mudstones, siltstones and sandstones (also prominent in Cycle 3). The fine grained facies exhibit thin laminae and isolated ripple form sets, and represent wave-agitated background sediment in which a complex variety of thin sandstone beds is set. Generally the sandstones are ripple- or flat laminated with sharp, planar bases and upper surfaces occasionally reworked into symmetrical or quasi-symmetrical ripple forms. These beds represent discrete incursions of sediment laden currents into the wave-agitated basin. The currents represent river flood incursions and exceptional floods may have developed into density currents which deposited turbidite sands in the deep basin and pro-delta environments (the "river generated turbidites" of Collinson 1970). The interbedded facies records an interaction between river and wave processes and compares favourably with intermediate depth mouth bar facies of the modern Mississippi delta (Fisk 1961, Wright & Coleman 1974). The progradational sequence in Cycle 8 terminates in this facies, but in Cycle 1 the coarsening trend continues into a 15 m sandstone member (the Rock Nose Sandstone). This member is dominated by thick sheet sandstones (0.5–2.0 m) with climbing ripple lamination, flat lamination and dispersed mudflakes, indicating rapid deposition from currents with a substantial sediment load. As the member occurs above the intermediate depth mouth bar facies it probably represents a proximal mouth bar environment dominated by river flood incursions.

In Cycles 3 and 9 composite (multi-storey) channel sandstones form part of the progradational sequences. In Cycle 3 a 20 m sandstone member (the Raleigh Sandstone) cuts into intermediate mouth bar facies and is dominated by medium-scale, unidirectional foresets and flat laminated sandstones with primary current lineation. The upper part of the channel sequence comprises interbedded siltstones and sandstones with irregular ripple- and climbing ripple lamination. In Cycle 9 the channel sandstone (Cornborough Sandstone) is 26 m thick and

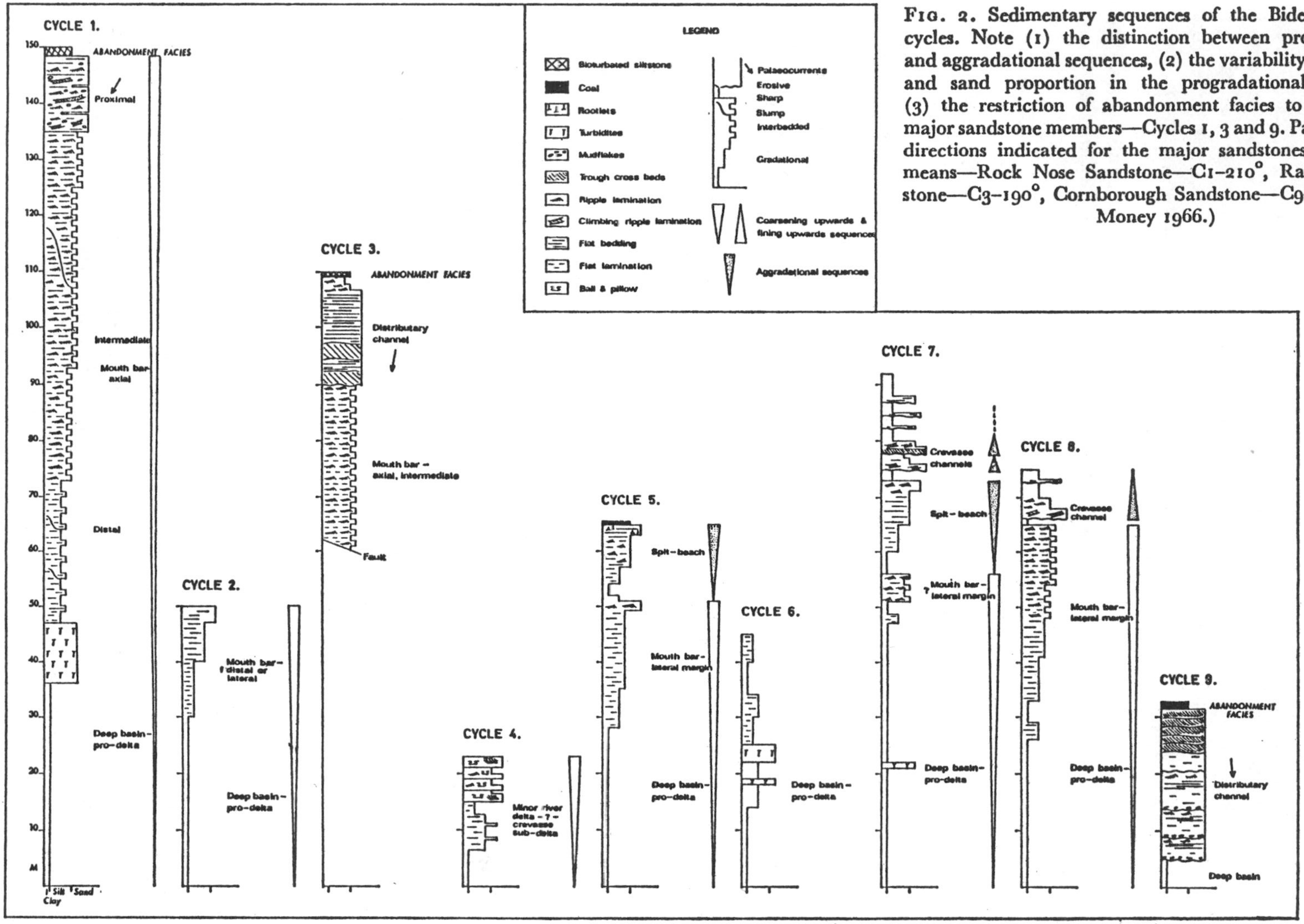

FIG. 2. Sedimentary sequences of the Bideford Group cycles. Note (1) the distinction between progradational and aggradational sequences, (2) the variability in thickness and sand proportion in the progradational sequences, (3) the restriction of abandonment facies to sites above major sandstone members—Cycles 1, 3 and 9. Palaeocurrent directions indicated for the major sandstones are vector means—Rock Nose Sandstone—C1–210°, Raleigh Sandstone—C3–190°, Cornborough Sandstone—C9–173° (from Money 1966.)

cuts into basinal mudstones. The channel fill includes repeated waning flow sequences (erosive base–vaguely bedded sandstone with dispersed mudflakes–flat lamination–ripple lamination), with cosets of unidirectional trough cross beds in the upper part of the succession indicating downstream migrating dunes (channel bars?). A fluvial origin for the channels is indicated by the unidirectional palaeocurrents and is supported by the complete absence of features generally associated with tidal channels (e.g. current reversals, mud drapes, marine fauna). The channels are regarded as major fluvial distributaries which eroded into offshore facies as progradation continued, and compare favourably with distributary channel sequences described from Recent deltaic sediments (Fisk 1961, Oomkens 1967, 1974).

AGGRADATIONAL SEQUENCES

The progradational sequences of Cycles 5, 7 and 8 are overlain by small-scale coarsening-upwards sequences and/or fining-upwards channel sequences (Fig. 2). In the former, mudstones and fine siltstones pass upwards into ripple laminated coarser siltstones, and in one case continue into flat bedded sandstones with irregular laminations, occasional wave ripple laminations and sporadic rootlets, capped by coal (Cycle 5). The coarse (channel) members of the fining-upwards sequences range in thickness from 1.5–2.5 m and are characterised by unidirectional trough cross beds and climbing ripple lamination. The fine member mudstones and siltstones yield abundant plant debris and include numerous thin, sharp-based sandstone beds with parallel- and current ripple lamination.

These smaller scale sequences are explicable in terms of an interdistributary bay environment (Elliott 1974b). The coarsening-upwards sequences resemble sequences produced by low wave-energy beach-spits, comparable to those in interdistributary bays of the Mississippi delta, whereas the size of the coarse member in the fining-upwards sequences suggests that they represent minor crevasse channels, overlain by fine grained bay sediments with crevasse splay sands.

ABANDONMENT FACIES

These facies are restricted to cycles which terminate in a major sandstone unit (Cycles 1, 3, 9). Thin horizons of bioturbated siltstone at the top of Cycles 1 and 3 represent prolonged periods of non-deposition and biogenic reworking after abandonment, whilst Cycle 9 is capped by a thick bed of soft, impure Culm indicating persistent peat accumulation (initially drifted) in the upstream area of the abandoned delta. Other cycles have a sharp planar contact with the basinal mudstones of the overlying cycle, but no distinct lithological marker horizon.

3. Delta morphology of the Bideford Group cycles

The characteristics of the progradational sequences and abandonment facies reflect delta regime and morphology. In the progradational sequences the river-generated turbidites of the deep basin–pro-delta, abundant river flood incursions in the intermediate and proximal mouth bar, and fluvial nature of the distributary channels suggest that the deltas were river-dominated (high-constructive type of Fisher 1969). There is no evidence for tidal activity, and the effects of wave processes were confined to intermediate shoreline depths and sites remote from the main sediment supply (Cycles 2, 5, 7). Furthermore the vertical distinction between thick constructive (progradational and aggradational) facies and thin or absent abandonment facies demonstrates that marine processes did not substantially rework sediment after delta abandonment. This supports the river-dominated argument as, in wave- and tide-dominated deltas, constructive and abandonment facies are not as vertically distinct due to appreciable reworking after abandonment(Fisher 1969).

Studies of the Mississippi delta illustrate a distinction between lobate and elongate (birdfoot) river-dominated deltas, primarily on the basis of sand distribution patterns (Fisk 1955, 1961, Fisk *et al.* 1954). In lobate deltas numerous distributary channels with closely spaced mouth bars and connecting barrier spits at the shoreline produce a sheet-like distribution of sand. Elongate deltas have fewer distributaries and sand facies are confined to narrow "bar-fingers" of distributary channel and mouth bar sediments, separated by thick "wedges" of pro-delta and interdistributary bay silts and muds.

In the Bideford Group, distributary channel sands and proximal mouth bar facies are restricted to the progradational sequences of Cycles 1, 3 and 9 which may suggest that these facies were impersistently developed lateral within individual delta complexes. This implies an elongate delta model in which these sequences represent passages into the axial part of bar fingers, whilst the remainder of the progradational sequences, which are dominated by fine sediments, represent sites laterally or distally removed from bar-finger axes (Fig. 3). The restriction of abandonment facies to sites above major (axial bar finger) sandstones implies marked differential compactional subsidence in the abandoned delta. This supports the elongate delta model as bar fingers persist as elevated areas after abandonment (e.g. Balize Bayou, modern Mississippi), whereas adjacent mud-dominated pro-delta–interdistributary bay depressions subside rapidly. The bar fingers are therefore relatively persistent sites of non-deposition, permitting biogenic reworking in the near shore area (Cycles 1, 3) and prolonged peat accumulation in the upstream area (Cycle 9). Adjacent areas experience only a brief cessation in deposition and the contact between cycles in these cases is therefore sharp, but with no lithological marker horizon. This situation contrasts with that in abandoned lobate deltas where the sheet-like sand distribution produces uniform subsidence, and the abandonment phase is represented by a marker horizon of variable facies which extends uniformly over the entire lobe (Elliott 1974a, 1975).

DISCUSSION

Three types of deltaic cycle are apparent in the Bideford Group with respect to proximal–distal and axial–lateral locations in bar finger sands (Fig. 3):

(1) *Axial cycles:* Constructive facies consist solely of progradational sequences, which include well developed mouth bar and/or distributary channel facies. Abandonment facies are restricted to this type. Examples—Cycles 1, 3 and 9.

255

(2) *Lateral cycles:* Progradational and aggradational sequences are both included in the constructive facies. The progradational sequences are usually dominated by fine grained sediments, with sand content in the upper part of the sequences determined by the distance from the bar finger axis. Aggradational sequences are confined to this type, being preferentially developed on shallow

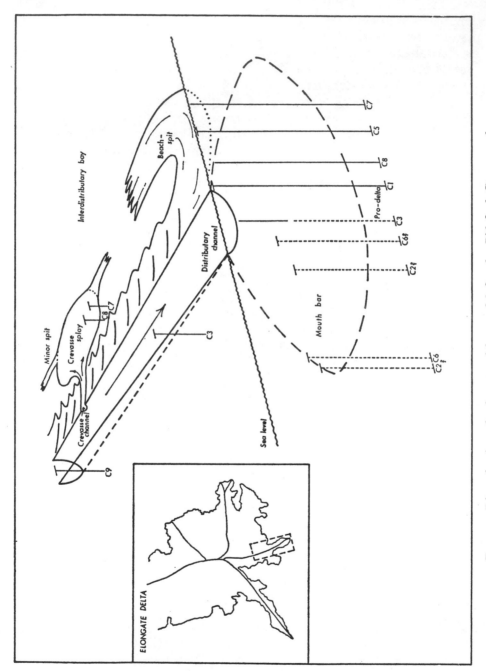

FIG. 3. River-dominated, elongate delta model for the Bideford Group cycles.

platforms provided by the margins of the bar fingers (Fig. 3). Examples—
Cycles 5, 7 and 8.

(3) *Distal cycles* (?): Cycles 2 and 6 resemble the progradational sequences
of the lateral cycles, but do not include aggradational sequences. They may
therefore represent lateral cycles deposited a considerable distance from the bar
finger axis, but alternatively they may be distal cycles reflecting limited delta
progradation into the basin. Only Cycle 4 cannot be considered in terms of
these types. This small-scale coarsening upwards sequence may represent a
minor river delta or crevasse sub-delta which was active in a shallow area of the
basin whilst the main depocentre was located in an adjacent, deeper part of the
basin. The temporary and localised shallowing of the basin indicated by this
sequence is presumably related to the presence of the underlying bar finger
sand (Raleigh Sandstone distributary channel in Cycle 3). Primary facies
distribution and subsequent compaction variations may account for other
instances of thickness variation between cycles in terms of shifting location of
the maximum depocentre, but this cannot be tested in the present study due to
a lack of lateral exposure on a regional scale.

Existing palaeocurrent data can also be re-considered in the light of the model.
Vector means of the major sandstones are consistently to the south or southwest (Fig. 2), and interpretation of these sandstones as bar finger sands
confirms that they indicate the general palaeoslope direction, at least within the
limits of spread exhibited by modern Mississippi bar fingers. In describing
turbidites in association with agitated water siltstones, Walker (1969 p. 138–9)
noted that whilst the turbidites indicated consistent flow to the south-east, the
cross laminated (agitated) siltstones demonstrated flow either to the east to
north-east or south-west. Here these "turbidites" are regarded as river flood
incursions into an intermediate depth mouth bar environment, frequently with
wave modification of the upper part of the beds which may account for the
anomalous palaeocurrent directions observed in the cross laminated siltstones.

Unfortunately it is rare to observe sedimentary structures in three-dimensions
in the polished coastal sections (except in the major sandstones) which precludes
the possibility of determining statistically valid vector means for sub-environ-
ments within the cycles.

4. Conclusions

In ancient successions delta regime and morphology can be deduced from the
general characteristics of the constructive and abandonment facies, and this is
important in a number of respects:

1. it provides a framework in which complex lateral and vertical facies
   variations, so characteristic of deltaic sediments, may be considered.
2. it permits interpretations of individual cycles to be refined. For example,
   the precise part of the delta represented in a cycle may be deduced, and
   predictions of sand body characteristics (shape, trend and size) may
   follow from this. Predictions of this nature are often imperative as
   deltaic sediments are important as oil, gas and ground-water reservoirs, and
   also constitute the "host" sediments in numerous coal-bearing successions.

3. as delta morphology is a response (via regime) to a variety of sedimentological, climatic and tectonic factors it may be considered as reflecting these factors. Delta morphology may therefore provide significant information on the characteristics of a sedimentary basin and its hinterland, and in thick deltaic successions it may be possible to detect temporal changes in factors.

ACKNOWLEDGEMENTS. The author is grateful to Drs. H. G. Reading and B. W. Sellwood for criticism; for a N.E.R.C. Research Studentship held at the University of Oxford, and the University of Reading for a Research Fellowship.

5. References

ALLEN, J. R. L. 1965. Late Quaternary Niger delta, and adjacent areas; sedimentary environments and lithofacies. *Bull. Am. Ass. petrol. Geol.* **49**, 41–61.

VAN ANDEL, TJ. H. & CURRAY, J. R. 1960. Regional aspects of modern sedimentation in northern Gulf of Mexico and similar basins, and palaeogeographic significance. *In* F. P. Shepard, F. B. Phleger & Tj. H. van Andel (eds.) *Recent sediments, northwest Gulf of Mexico* Am. Ass. petrol. Geol., Tulsa, Oklahoma, 345–64.

COLEMAN, J. M. & WRIGHT, L. D. 1973. Variability of modern river deltas. *Trans. Gulf-Cst Ass. geol. Socs* **23**, 33–6.

COLLINSON, J. D. 1970. Deep channels, massive beds and turbidity current genesis in the Central Pennine Basin. *Proc. Yorks. geol. Soc.* **37**, 495–519.

ELLIOTT, T. 1974a. Abandonment facies of high-constructive lobate deltas, with an example from the Yoredale Series. *Proc. Geol. Ass.* **85**, 259–65.

—— 1974b. Interdistributary bay sequences and their genesis. *Sedimentology* **21**, 611–22.

—— 1975 (in press.) The sedimentary history of a delta lobe from a Yoredale (Carboniferous) cyclothem. *Proc. Yorks. geol. Soc.* **40**.

FISHER, W. L. 1969. Facies characterisation of Gulf Coast Basin delta systems, with some Holocene analogues. *Trans. Gulf-Cst Ass. geol. Socs* **19**, 239–61.

—— BROWN, L. F., SCOTT, A. J. & McGOWEN, J. H. 1969. *Delta systems in the exploration for oil and gas.* Bur. Econ. Geol., Univ. Texas, Austin, 78pp.

FISK, H. N. 1955. Sand facies of Recent Mississippi delta deposits. *Proc. World Petrol. Cong.*, 4th, Rome sec. 1, 377–98.

—— 1961. Bar finger sands of the Mississippi delta. *In* J. A. Peterson & J. C. Osmond (eds.) *Geometry of Sandstone Bodies—a symposium.* Am. Ass. petrol. Geol., Tulsa, Oklahoma, 29–52.

—— KOLB, C. R. & WILBERT, L. J. JR. 1954. Sedimentary framework of the modern Mississippi delta. *J. sedium. Petrol.* **42**, 558–71.

MONEY, N. J. 1966. Sedimentary petrology of sandstones in the Upper Carboniferous of north Devon. *B.Sc. thesis Univ. Oxford* (unpubl.) 107pp.

OOMKENS, E. 1967. Depositional sequences and sand distribution in a deltaic complex. *Geol. Mijnb.* **46**, 265–78.

—— 1974. Lithofacies relations in the Late Quaternary Niger delta complex. *Sedimentology* **21**, 195–222.

PRENTICE, J. E. 1960. The stratigraphy of the Upper Carboniferous rocks of the Bideford region, north Devon. *Q. Jl geol. Soc. Lond.* **116**, 397–408.

—— 1962. The sedimentation history of the Carboniferous in Devon. *In* K.Coe(ed.) *Some aspects of the Variscan fold belt.* Manchester, 93–108.

DE RAAF, J. F. M., READING, H. G. & WALKER, R. G. 1965. Cyclic sedimentation in the Lower Westphalian of north Devon, England. *Sedimentology* **4**, 1–52.

READ, W. A., DEAN, J. M. & COLE, A. F. 1971. Some Namurian (E₂) paralic sediments in central Scotland: an investigation of depositional environment and facies changes using iterative fit trend surface analysis. *Q. Jl geol. Soc. Lond.* **127**, 137–76.

READING, H. G. 1965. Recent finds in the Upper Carboniferous of south-west England and their significance. *Nature Lond.* **208**, 745–8.

—— 1967. Sedimentation sequences in the Upper Carboniferous of northwest Europe. *C.r. 6e Congr. Strat. Géol. Carbonif.*, Sheffield 1967, IV, 1401–E2.

SCRUTON, P. C. 1960. Delta building and the delta sequence. *In* F. P. Shepard, F. B. Phleger & Tj. H. van Andel (ed.) *Recent Sediments, northwest Gulf of Mexico* Am. Ass. petrol. Geol., Tulsa, Oklahoma, 82–102.

WALKER, R. G. 1969. The juxtaposition of turbidite and shallow water sediments: a study of a regressive sequence in the Pennsylvanian of north Devon. *J. geol.* **77**, 125–43.

—— 1970. Deposition of turbidites and agitated-water siltstones: a study of the Upper Carboniferous Westward Ho! Formation, north Devon. *Proc. Geol. Ass.* **81**, 43–67.

WEBER, K. J. 1971. Sedimentological aspects of oilfields in the Niger delta. *Geol. Mijn.* **50**, 569–76.

WILLIAMS, D. F. 1968. The sedimentation of Westphalian Measures in the Little Haven–Amroth coalfield. *J. sedim. Petrol.* **38**, 332–63.

WRIGHT, L. D. & COLEMAN, J. M. 1973. Variations in morphology of major river deltas as functions of ocean wave and river discharge regimes. *Bull. Am. Ass. petrol. Geol.* **57**, 370–98.

—— 1974. Mississippi river mouth processes: effluent dynamics and morphologic development. *J. geol.* **82**, 751–78.

Received 26 May 1975; revised typescript received 18 July 1975.

TREVOR ELLIOTT, Sedimentology Research Laboratory, The University, Reading RG6 2AB

20

Reprinted from pp. 246, 247–255 of *Deltaic Sedimentation, Modern and Ancient*, J. P. Morgan, ed., Soc. Econ. Paleontologists and Mineralogists Spec. Pub. 15, 1970, 312 pp.

ALLEGHENY DELTAIC DEPOSITS

JOHN C. FERM[1,2]
Louisiana State University, Baton Rouge

ABSTRACT

Results of Recent delta studies provide a lithogenetic model which is applicable to Middle Pennsylvanian rocks of the northern Appalachian Plateau. The model is a three-dimensional lenticular body composed mainly of detrital rocks which grade upward and landward from clays (shales) to sand and which is the result of deltaic progradation. These detrital sediments are completely or partially enclosed in a veneer of chemically precipitated or indigenously formed peats (coals), root-penetrated clays (seat rocks) or carbonate sediments (limestone or ironstones) which were deposited on the offshore front or on marginal portions of the delta or accumulated on the delta-plain surface when it was no longer receiving appreciable detrital influx. Lateral shift of sites of major detrital deposition, a phenomenon common in Recent deltas, results in an *en echelon* arrangement of progradational detrital wedges, a distinctive feature of the ancient Appalachian sediments. The primary differences between these ancient deltaic sediments and their modern equivalents seem to arise from limited detrital supply for ancient deltas and from the site of ancient delta building in shallow, narrow embayments on the relatively stable continental plate. This contrasts with modern deltas, most of which are prograding toward oceanic basins.

[*Editor's Note:* Material, including figure 1, has been omitted at this point.]

A MODEL FOR RECENT DELTAS

Although the processes and resulting products of post-Pleistocene delta formation differ to some degree, available information seems sufficient to generate a generalized delta model. [For excellent examples, see Allen (1965), Coleman and Gagliano (1964), Scruton (1960) and van Andel, Postma and others, (1954), as well as contributions in the first part of this volume]. Recent marine deltas form when sediments, carried by rivers into relatively large bodies of open water, accumulate at the river mouth until the surface of the sediment pile reaches sea level. The emergence portion comprises the subaerial expression of the delta and

[1] A very substantial body of data was obtained from the open file of stratigraphic sections at the Ohio Geological Survey, and the most helpful cooperation of this group is gratefully acknowledged. In addition, I wish to thank J. M. Coleman and S. M. Gagliano of the Louisiana State University Coastal Studies Institute for their stimulating and informative discussions on modern delta processes which provided the basic links between ancient and Recent deltaic deposits. J. R. Conolly and J. P. Morgan both have read the manuscript and made many constructive criticisms. Photographs are by J. M. Coleman and R. S. Saxena, and other illustrations were prepared by the Cartographic Section of the Louisiana State University Department of Geology. Field work has been supported by National Science Foundation Grant G-18816 and by the Louisiana State University Graduate Research Council and the Department of Geology.

[2] Present address: Department of Geology, University of South Carolina, Columbia, S.C. 29208.

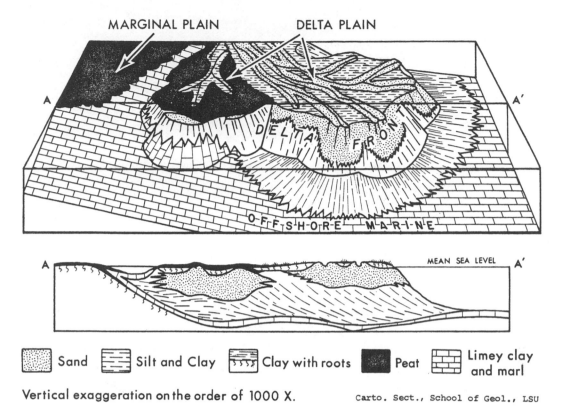

Vertical exaggeration on the order of 1000 X.   Carto. Sect., School of Geol., LSU

Fig. 2.—Lithic characteristics of an idealized prograding delta. Block diagram shows three dimensional characteristics of a single delta wedge, one portion of which is already abandoned. Section A-A′ shows the rock cross section resulting from progradation. The rock symbols are not typical of modern deltas but are those rock types common in the Pennsylvanian.

is the site of most Recent delta studies. Delta growth continues as sediment-laden streams pass over the emergent surfaces and deposit sand, silt, and clay over the frontal delta slope. As this process of building new land at the delta margin continues, the delta is said to prograde, and the product of progradation can be thought of as the typical delta sequence. Such a sequence consists of silty clay and silt deposited at the distal margins of the deltaic wedge, gradationally overlain by silty sand and sand which are laid down on the upper part of delta slope. (See figure 2.) These silty *foreset* or delta-front sediments are one of the most uniform and widespread components of Recent delta sequences. In contrast, the overlying *topset* or delta-plain beds, formed on the emerging or emergent portions of the delta, are often lithically complex. Overall, they are more coarse grained than the underlying sediments, but, because most of the available hydraulic energy is confined to linear channels, coarser sizes may be spatially isolated in discrete sedimentary units.

The areal distribution of these coarse-grained units depends on the geomorphic character of the delta surface. Where distributaries are well established and widely spaced, sand bodies occur as isolated linear units in the midst of fine-grained interdistributary deposits, for example, the *bar-finger* sands of Fisk (1961). Where distributaries shift frequently or are numerous, sand units tend to crosscut, overlap, and merge into single sheetlike sand bodies. Similar effects are found where wave energy is high and longshore drift transfers large quantities of coarser sediment laterally across the prograding delta front.

Because progradation is one type of marine regression, fauna and flora vary accordingly though the delta sequence. Silts and silty clay near the bottom of the sequence contain faunas indicative of normal or somewhat diminished salinity. Topset beds commonly incorporate fresh or brackish water fauna and flora, but, where interdistributary bays are large, marine or brackish forms may also be included. Organisms are not only static components but often contribute actively to delta process. In submarine portions, where the rate of sedimentation is not great, burrowers modify or destroy original sedimentary structures. On subaerial surfaces, vascular plants not only produce characteristic root structures but, where abundant, tend to trap sediment and fix distributary positions, thereby limiting the distribution of coarse detritus.

Laterally continuous with and marginal to the typical prograding delta are relatively thin deposits formed essentially by indigenous processes beyond the realm of major detrital influx. Seaward of the littoral zone and peripheral to the delta front these deposits are clays and marls, and, in the shore line area, these clayey and limey sediments intertongue with coarse-grained beach or tidal channel deposits, which in turn grade landward into root-penetrated marsh and swamp clays and peats. (See figure 2.) These *soil-swamp* sediments also tend to overlap the emergent portions of the delta wedge, particularly in those places where sediment influx is localized in well-organized stable channels. Under such circumstances, major detrital depositon is limited to areas directly adjoining the channels and to the delta front itself. Finally, as a result of progradation, offshore clays and marls are overridden by the advancing mass of detritus, and the resulting deltaic detrital wedge is almost completely surrounded by a variety of chemically or biochemically precipitated or modified materials.

If progradation is one major aspect of modern delta sedimentation, lateral shift or offset of prograded lobes is another. This process, which occurs to some degree in many delta complexes and is very pronounced in the Mississippi, consists of the gradual or abrupt abandonment of sediment-laden stream flow from one delta lobe and its diversion into an adjoining open water area. The exact factors governing this process are complex but usually can be associated with gradient advantages of the new stream course and subsidence rates. The first outcome of this process is the formation of a new prograded lobe on the flank or seaward of the older one. The second is the decay of the old lobe which proceeds at rates which vary with the degree to which the old delta is isolated from detrital supply, with the rate of subsidence and compaction, and with effectiveness of wave and tidal action. (See figure 3.) The overall result is, however, that the abandoned delta will be completely overlapped by deposits formerly restricted to peripheral portions of the delta wedge. Clays and marly clays previously restricted to subma-

rine areas beyond the delta front accumulate on the old foreset slopes while static marsh and swamp deposits overlap the subaerial surface and beach-barrier components develop along the margin. Subsidence of the old subaerial surface results in marine transgression and overlap of swamp and marsh beds by brackish or marine clays and marls.

The combined processes of progradation and lateral shift produce, in time, a deposit which can be referred to as a delta system or complex. The internal characteristics of such a system are only scantly known in most Recent deposits but in the example of the Mississippi may be reasonably inferred by extension from known data. This complex basically would consist of a series of overlapping and offset, prograded, mainly detrital wedges which are to some degree separated from one another by various kinds of deposits that usually are formed on the marginal portions of the delta. (See figure 3.) It is to this model deltaic complex that the Allegheny Formation of the northern Appalachian Plateau may be compared.

ALLEGHENY DELTAIC DEPOSITS

Before describing those features of Allegheny rocks that are analogous to the Recent delta model, certain significant differences between these ancient and modern sediments should be noted. Allegheny coals and limestones, for example, are more abundant, thicker, and occur in beds of greater lateral continuity than any known in recent deltas. Moreover, coal beds and other nonmarine strata of the Allegheny have much greater areal distribution than rocks indicating marine deposition, a characteristic which is the reverse of what is known of modern nonmarine deltaic deposits and their contemporary equivalent marine environments. These differences, however, suggest more a modification rather than a fundamental alteration of the basic delta pattern. Because coal and limestone are mainly biochemical sediments, their relatively large volume in the Allegheny implies that at least certain depositional sites were almost completely free of detrital influx for significant periods of time. This absence of detritus may be related in part to a low overall rate of detrital supply or deposition of all available detritus in localized areas. The great lateral extent of Allegheny coal and limestone beds, probably related in part to the general overall low level of detrital influx, can also be attributed to specific morphologic characteristics of Allegheny depositional sites. The distribution of known Allegheny marine beds (fig. 4) shows that potential areas for marine delta building during Allegheny time

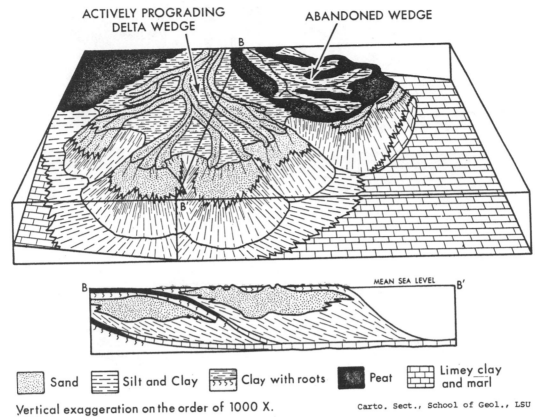

ACTIVELY PROGRADING
DELTA WEDGE   ABANDONED WEDGE

MEAN SEA LEVEL

Sand   Silt and Clay   Clay with roots   Peat   Limey clay and marl

Vertical exaggeration on the order of 1000 X.

Carto. Sect., School of Geol., LSU

Fig. 3.—Results of lateral shift of delta lobes concomitant with continued progradation. This diagram represents a continuation from figure 2, but the former area of active sedimentation has been abandoned and a new wedge has developed on the flank of the older one. The cross section B-B', showing the older wedge enveloped in coal, seat rock, and limestone and partially overlapped by a younger wedge, is typical of Allegheny lithic patterns on figure 5.

were relatively small and, in any event, are very much smaller than those of modern deltas which now build into oceanic basins. This small areal extent coupled with maximum estimated water depths on the order of 60 to 100 feet (Ferm and Williams, 1965) produces a setting in which a relatively small amount of detritus could completely fill an embayed area, producing a widespread surface where peat swamps could develop and where, upon minor subsidence, shallow water marine carbonates could form.

Marine embayments of such modest proportion relative to those of Recent deltas could also account for the greater areal distribution of coal-bearing nonmarine beds than of marine Allegheny facies. In modern deltas, nonmarine delta-plain deposits can form only after a very broad platform of essentially marine strata has been constructed, whereas in the Allegheny, nonmarine facies, continuously in existence on the margins of the embayments, could readily be extended across former marine areas with only a relatively minor amount of detrital fill.

Although some characteristics of Allegheny sediments have been influenced by the nature of their overall background, most depositional environments of Recent deltas can be recognized. Perhaps the most obvious are the ancient equivalents of deposits adjoining sites of deltaic detrital influx; coal beds almost certainly represent peats, whereas seat-rock units are probably soils which range from weakly leached marsh clays to well-drained soils with reasonably well-developed profiles. Allegheny marine limestones and calcareous clays seem to be correlative with clays and marls deposited offshore from the delta front. Almost as obvious are those Allegheny deposits of the prograding delta wedge which grade upward from fine-grained silty shale through coarse siltstone and sandstone at the top. Several of these wedges are illustrated on a generalized Allegheny cross section (fig.

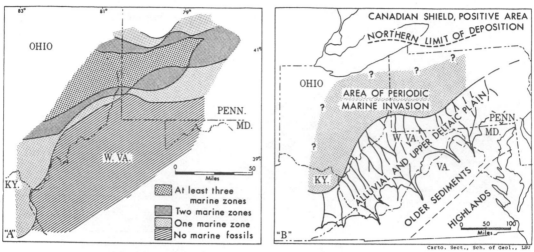

Fig. 4.—Distribution of Allegheny marine zones and generalized paleogeography. "A", map showing distribution of areas having one, two, and at least three marine zones and an area without marine fossils, which indicate that, although rather elongate, sites of Allegheny marine delta building were generally much less than 100 miles wide. "B", map based mainly on distribution of marine and nonmarine fossils and diagnostic lithic patterns (Ferm and Cavaroc, 1968) showing probable areas of marine and nonmarine deposition as well as adjoining source terranes.

5). One, labeled "Clarion α," is located in the lower left corner of the section and another, "Clarion β," is in the lower right. Overlying these are two more, designated "Lower Kittanning" and "Middle Kittanning," and in the far left center still another is partially shown between the Vanport "I" limestone and the Middle Kittanning coal bed. Each of the wedges portray characteristics of the Recent delta model cross section (fig. 3) in that they consist of a detrital sequence grading from fine to coarse upward and enclosed or enveloped by coals, seat rocks (soils), and marine limestones or clays—those sediments which are formed in sites peripheral to major detrital influx areas in Recent deltas. The sections on figure 5, of course, do not appear exactly like those on figure 3, but detailed mapping of each of the wedges (shown on figures 6, 7, and 8) indicates facies distributions quite similar to the model with cross section differences arising from the fortuitous position at which the section crosses the wedges.

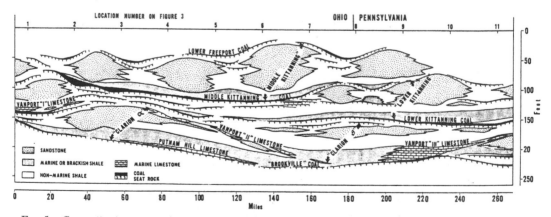

Fig. 5.—Generalized cross section of lower and middle Allegheny strata in eastern Ohio and western Pennsylvania. Location of section on figure 1. Thicknesses of some limestones and coal beds are slightly exaggerated and lateral extent of all rock units is greatly exaggerated. Relative lateral extent, however, is approximately in proportion; coal beds and marine limestones and shales are much more widespread than sandstones. This section is based on about 600 control points; thus, numbers at top of the section do not represent individual pieces of data but general locations. Stratigraphic names in quotations are not formal designations and are used here only as a matter of convenience.

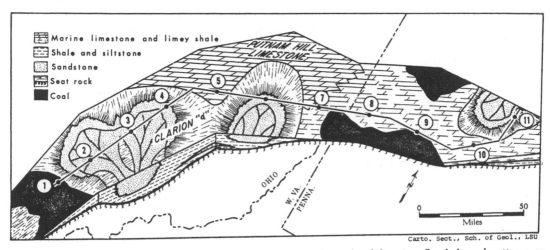

FIG. 6.—Paleogeography of the "Clarion α" delta wedge and associated features. Symbols and patterns on this and following figures 7 and 8 show only general rock characteristics but are comparable with the general cross section, figure 5, and with delta model on figures 2 and 3.

Among the most obvious differences between the sectioned wedges on figure 5 is the relative amount and position of sandstone within a wedge. Thus the "Clarion β" wedge on figure 5 and figure 9A shows the smallest amount of sandstone, and this is located at the top of the detrital unit. "Clarion α" and "Lower Kittanning" contain more sand, and, in some places, sandstone occupies almost all the vertical interval through the wedge. In the "Middle Kittanning," the proportion of sandstone is very much

greater and in many places occupies almost the entire interval between the Middle Kittanning and Lower Freeport coal beds. Figure 9 shows a typical exposure of this wedge and contrasts it with the sand-deficient wedge of "Clarion β." These differences between wedges seem to reflect the relative landward or seaward position at which the wedges are cut. (See figure 3.) Thus, some subsurface data indicate that about 40 miles south of locations 7 and 8 on figure 5 the "Clarion β" wedge takes on characteritistics

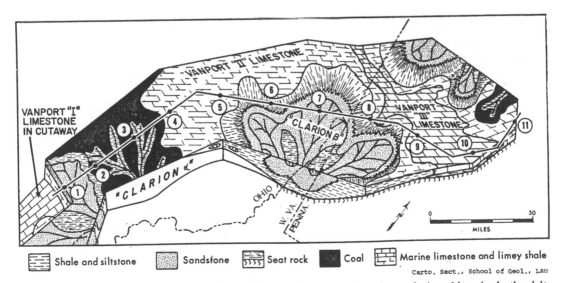

FIG. 7.—Paleogeography during deposition of the "Clarion β" wedge. Later during this episode the delta continued to prograde northeast, north, and westward, overlapping its offshore carbonate facies as well as part of the older "Clarion α" wedge. The small wedge on the southern flank of "Clarion α" is shown in cutaway fashion to indicate the distribution of the underlying limestone.

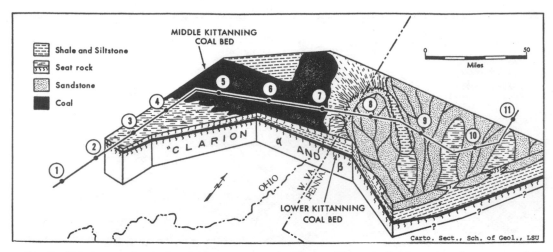

FIG. 8.—Paleogeography during the late stages of the "Lower Kittanning" delta and earliest stages of the Middle Kittanning unit. As progradation proceeds, the site of major influx shifts westward to the former site of marginal deposition.

of "Clarion α" at locations 2 and 3 and "Lower Kittanning" at location 9. Similarly the "Clarion β," "Lower Kittanning," and "Middle Kittanning" wedges are known to pass south and southeastward into strata containing a much greater proportion of sandstone and a relatively larger proportion of seat rock, with all rock types showing a much lower order of lateral continuity than those on figure 5. This southeastward facies has been described previously (Ferm and Cavaroc, 1968) as an alluviated portion of an upper deltaic plain. Therefore, if the differences between wedges shown on figure 5 can be attributed to progressive landward sectioning of delta wedges, then the Allegheny section between the "Brookville" and Lower Freeport coal beds in Ohio and western Pennsylvania can be interpreted not only as individual northward prograding wedges but also as a large deltaic complex which is also prograding from south to north.

The mechanics of northward progradation of

FIG. 9.—Two contrasting aspects of deltaic wedges in the Allegheny Formation. Part A, at Fallston, Pennsylvania, shows the seaward silty portion of the "Clarion β" delta wedge. The "Brookville" coal ("B" on figure) at the base of the wedge is overlain by silty shale and marly limestone with marine fossils that grade upward through siltstones and finally sandstones at the top. The lower Kittanning Coal at the top of the wedge is concealed in the trees at the top of the cut. Part B shows a landward equivalent of the type of sequence shown on part A. The interval illustrated here is between the Middle Kittanning and Lower Freeport Coals at Yellow Creek, Ohio. The Middle Kittanning Coal is concealed at the base of the picture and is overlain by a few feet of dark shales with marine fossils below the sandstone ledge. Sandstones and coarse siltstones dominate the remainder of the interval to the Lower Freeport Coal, which is concealed in the bench at the top of the cut. One mile to the right, this interval is composed almost entirely of sandstone, whereas within 1 mile in the opposite direction the same interval is mostly marine shale with a few beds of siltstone.

the delta complex appears to have been by forward and lateral shift of deltaic wedges, a characteristic well known in the Recent Mississippi delta. Thus, figure 6 shows the position of the "Clarion α" lobe on the flanks of two marginal seat rock and coal areas which are known to be older abandoned delta wedges. The following episode on figure 7 shows abandonment of the "Clarion α" lobe and development of two detrital wedges on its flanks. Subsequently the "Lower Kittanning" delta formed on the eastern margin of the abandoned "Clarion β" wedge with its thickest portion approximately overlying the thinned edges of "Clarion" deposits. Similarly the major part of the "Middle Kittanning" unit lies on the marginal peats and seat rocks which formed on the western flank of the preceding "Lower Kittanning" lobe (fig. 5).

Precise time relationships between the wedges is uncertain. Although it is reasonably clear that the upper portion of "Clarion β" at locations 7 to 9 on figure 5 is younger than equivalent portion of "Clarion α," the lowermost beds of "Clarion β" at location 8, could actually be contemporary with the upper beds of "Clarion α" at locations 3 and 4. Likewise the marine to brackish shale in the lower part of the "Lower Kittanning" wedge is certainly older than the "Middle Kittanning" deltaic detritus, but the upper sandy segment of the "Lower Kittanning" between locations 8, 9 and 10 (fig. 5) could very well be contemporary with the Middle Kittanning coal bed between locations 2 through 5. In any event, given some uncertainty about time relationships, the entire early to middle Allegheny episode seems to reflect northward progradation by lateral shift of individual deltaic wedges.

Similar patterns of lateral shift in the "Clarion" deltas of the northern margin of the embayment have been described elsewhere (Williams and Ferm, 1964) and are also illustrated on figures 6 and 7. These northern lobate deposits differ from those of the south in that lateral shift was the dominant process and was not accompanied by any significant southward progradation.

SUMMARY AND DISCUSSION

Sequences of rock types in the Allegheny Formation of the Appalachian Plateau are analogous to those of modern deltas. Clay shales and limestones of wide areal extent and containing marine fauna seem to be equivalent to marine clays and marls offshore from the delta front. Silty sequences, almost equally widespread, grading upward from clay shale to

coarse silt and sand are similar to the typical prograded sequences of modern deltas. Allegheny coal beds and seat rocks seem to be correlative with peats and root-penetrated clays occurring on the margins of modern deltas and on the emergent surfaces of Recent delta lobes. The geometric arrangement of offshore marine limestones and shales and of seat rocks and coals above, below, and lateral to deltaic detrital wedges indicate progradation over offshore marine strata, abandonment, stagnation, and eventual submergence of the delta plain, with both marine and nonmarine indigenous deposits forming contemporaneously on the periphery of the delta wedge. Location of these deltaic wedges, enclosed in marine limestone and shale, seat rock, and coal and *en echelon* to other similar wedges, indicates lateral migration of delta lobes similar to that of some Recent deposits. Progressive landward facies in the vertical succession of wedges suggest overall progradation of the entire delta complex.

Most differences between Allegheny and Recent deltaic deposits can be attributed to background factors which modified the basic delta process. Apparently a somewhat lower overall rate of detrital supply resulted in relatively larger volumes of marine limestone and coal during Allegheny time, whereas the small size of the embayments permitted even limited volumes of detritus to produce very widespread prograded areas. Some of these areas at least locally seem to have extended completely across the embayment, and many, after cessation of detrital deposition, provided depositional sites for areally extensive coal and/or marine limestone sequences.

Although not entirely within the scope of this paper, a comparison of these Pennsylvanian and Recent deltaic deposits would be incomplete without some exploration of possible effects of general sea level oscillation. Most modern marine deltas seem to have accumulated in a setting of post-Pleistocene sea level rise or stillstand. Wanless and Shepard (1936) have suggested that sequential recurrence of marine and nonmarine Carboniferous strata could be related to waning and waxing of continental glaciers. In contrast, Weller (1930), among others, has hypothesised a tectonic mechanism linking general emergence and increased sedimentation with tectonic uplift and linking submergence and diminished sedimentation to tectonic stability or subsidence. The present study indicates that much of the rock record of Pennsylvanian transgression and regression may be best attributed to deltaic progradation and decay, with deposition of available detritus governed by lateral delta migration. Such an explanation is in accord with that of Moore (1959) and Goodlet (1959) for the British Carboniferous rocks. This explanation does not, however, exclude the possibility of relative changes in sea level, whatever the cause. Indeed, over the total span of Pennsylvanian time such oscillations seem more likely than not. The problem is, thus, not one of alternatives between delta building and sea level control as an explanation for alternating marine and nonmarine Carboniferous rocks, but one of determining the degree to which the delta process has been effected by eustatic or isostatic changes.

REFERENCES

ALLEN, J. R. L., 1965, Late Quaternary Niger delta and adjacent areas: Am. Assoc. Petroleum Geologists Bull., v. 49, p. 547–600.

CAVAROC, V. V., JR., 1969, Geology of some Carboniferous deltaic plain sediments of western Pennsylvania (Ph.D. thesis): Baton Rouge, Louisiana State Univ., 200 p. (available from University Microfilms, Ann Arbor, Michigan.)

COLEMAN, J. M., AND GAGLIANO, S. M., 1964, Cyclic sedimentation in the Mississippi River deltaic plain: Gulf Coast Assoc. Geol. Soc. Trans., v. 14, p. 67–80.

FERM, J. C., AND CAVAROC, V. V., JR., 1968, A non-marine model for the Allegheny of West Virginia, *in* Late Paleozoic and Mesozoic continental sedimentation, northeastern North America: Geol. Soc. America Special Paper 106, p. 1–19.

——, AND WILLIAMS, E. G., 1965, Characteristics of a Carboniferous marine invasion in western Pennsylvania: Jour. Sedimentary Petrology, v. 35, p. 319–330.

FISK, H. N., 1961, Bar-finger sands of the Mississippi delta; *in* Geometry of sandstone bodies: Tulsa, Oklahoma, Am. Assoc. Petroleum Geologists, p. 29–52.

FLORES, R. M., 1966, Middle Allegheny paleogeography in eastern Ohio (Ph.D. thesis): Baton Rouge, Louisiana State Univ., 136 p. (available from University Microfilms, Ann Arbor, Michigan).

GOODLET, G. A., 1959, Mid-Carboniferous sedimentation in the Midland Valley of Scotland: Edinburgh Geol. Soc. Trans., v. 17, p. 217–240.

MOORE, D., 1959, Role of deltas in the formation of some British Lower Carboniferous cyclothems: Jour. Geology, v. 67, p. 522–539.

SCRUTON, P. C., 1960, Delta building and the deltaic sequence, *in* Recent sediments, northwestern Gulf of Mexico: Tulsa Oklahoma, Am. Assoc. Petroleum Geologists, p. 82–102.

VAN ANDEL, TJ. H., POSTMA, H., AND OTHERS, 1954, Recent sediments of the Gulf of Paria, *in* Reports of the Orinoco shelf expedition, Volume I: K. Nederlandsch Akad., Wetensch., Verh, v. 20, no. 5.

WANLESS, H. R., AND SHEPARD, F. P., 1936, Sea level and climatic changes related to late Paleozoic cycles: Geol. Soc. America Bull., v. 47, p. 1177–1206.

WEBB, J. E., 1963, Allegheny sedimentary geology in the vicinity of Ashland, Ky. (Ph.D. thesis): Baton Rouge, Louisiana State Univ. 168 p. (available from University Microfilms, Ann Arbor, Michigan).

WELLER, J. M., 1930, Cyclical sedimentation of the Pennsylvanian Period and its significance: Jour. Geology v. 38, p. 97–135.

WILLIAMS, E. G., AND FERM, J. C., 1964, Sedimentary facies in the lower Allegheny rocks of western Pennsylvania: Jour. Sedimentary Petrology, v. 34, 610–614.

ZIMMERMAN, R. K., 1966, Aspects of early Allegheny depositional environments in eastern Ohio (Ph.D. thesis): Baton Rouge, Louisiana State Univ. 123 p. (available from University Microfilms, Ann Arbor, Michigan).

21

Reprinted from pp. 1649, 1651, 1653–1664, 1665, 1666, 1669–1670 of *Geol. Soc. America Bull.* **83**(6)1649–1670 (1972)

Sedimentology of Upper Cretaceous Cody-Parkman Delta, Southwestern Powder River Basin, Wyoming

JOHN F. HUBERT   *University of Massachusetts, Amherst, Massachusetts 01002*
JOSEPH G. BUTERA   *Texaco, Inc., Midland, Texas 79701*
ROGER F. RICE   *Mobil Oil Corporation, New Orleans, Louisiana 70160*

ABSTRACT

The Upper Cretaceous Cody Shale and Parkman Sandstone in the southwestern Powder River basin, Wyoming, were deposited during progradation southeastward of a wave-dominated, high-destructive type of delta. The Cretaceous cordillera in eastern Idaho and southwestern Montana was the source of the homogeneous group of litharenite and feldspathic litharenite sandstones. The source rocks were pre-Belt crystalline rocks, the Belt Supergroup of argillite and quartzite, Paleozoic and Mesozoic sedimentary rocks, and Cretaceous volcanic rocks and ash falls.

The upper several hundred feet of the 2,500-ft Cody Shale comprise a flyschlike sequence deposited on the upper prodelta slope. The lower 220 ft of the overlying Parkman Sandstone accumulated on the delta platform and along the strandline. Abundant pillow structures occur in sandstone beds deposited at the top of the prodelta slope and on the outer delta platform. Mapping of structural elements in pillow horizons shows that the subaqueous delta surface dipped southeast with southwest-northeast contours. The Cody graded beds were deposited by ocean-bottom currents that flowed northeast, perpendicular to the paleoslope, rather than by downslope turbidity currents.

The sandstone beds on the delta platform were deposited by strong waves and subordinate tidal currents, both of which were highly variable in direction. Cross-bedding consistently dips landward and along shore in the strandline sandstones; they were deposited in a complex of beaches, swash bars, spits, and subtidal areas swept by alongshore drift. The upper 200 ft of the Parkman Sandstone are deltaplain carbonaceous mudstones, lensing channel sandstones, and a few thin lignite beds. The region was one of low relief and southeast gradient, with floodplains, meandering stream courses, lakes, swamps, and a coastal marsh with tidal creeks.

Lowering of the delta surface by compaction of the thick sequence of prodelta mud resulted in local deposition of 12 to 60 ft of marine sandstone at the top of the Parkman Sandstone which represents the destructional phase of the delta. Subsequent transgression of an unnamed marine tongue of the Cody Shale terminated the delta cycle.

[*Editor's Note:* Material, including table 1 and figures 1 and 2, has been omitted at this point.]

269

SEDIMENTARY FACIES

Prodelta Slope Facies

At Teapot Dome, the Cody Shale above the Sussex Sandstone Member is about 700 ft thick. Only the upper 300 ft of sandstone and siltstone beds, mostly graded and interstratified with highly bioturbated shale, are exposed. These strata are a flyschlike sequence deposited on the upper prodelta slope of a large delta lobe (Figs. 2, 3). Typical of prodelta sequences (Fisk and others, 1954; Scruton, 1960), the beds are progressively coarser grained and thicker upward in the section. The Cody sequence grades into outer delta-platform sandstone of the overlying Parkman Sandstone (Figs. 4, 5, 6).

Many of the graded beds have groove, flute, and longitudinal-ridge markings which, along with parting-step lineations and ripple marks, show that the currents consistently flowed northeast (Fig. 7).

The Cody graded beds contain a partial (or rarely, complete) sequence of internal structures as described by Bouma (1962). The complete sequence of internal divisions is (A) graded bedding, (B) lower division of horizontal lamination, (C) ripple cross-lamination, (D) upper division of horizontal lamination, and (E) pelite. This sequence reflects continuous deposition of each bed from a waning current. In a random, continuous sequence of 100 beds, 89 percent show only the B, C, D, DE, or E divisions (Fig. 8). The B-division beds are considerably coarser grained than are the D-division beds. The 11 percent BC, BDE, and BCDE beds tend to be among the thickest

beds, as is common in flyschlike sequences.

The frequency distribution of the thickness of the graded beds is approximately log normal (Fig. 9), which is characteristic of flyschlike sequences (Dott, 1963; Scott, 1966; Hubert, 1967). The sedimentary processes responsible for sequences of graded beds tend to produce many more thin beds than thick. The thin graded beds of the Cody Shale are chiefly very fine sandstone and siltstone, implying that most of the sequence was deposited by relatively weak currents.

The submarine topography of the Cody-Parkman delta was evidently similar to that of the lower Cody Shale between the Shannon Sandstone Member and the Niobrara Formation investigated by Asquith (1970) in the southwestern Powder River Basin near Teapot Dome. Asquith established inclined time-stratigraphic horizons by correlation of bentonite-rich shale beds on electric-resistivity logs in closely spaced oil wells. The time-stratigraphic horizons suggest that the prodelta slope for the Cody-Parkman delta probably dipped seaward from 1 to 2 degrees. Water depth at the edge of the delta platform was perhaps 50 to 100 ft, increasing to 500 to 700 ft at the base of the slope. These values are our best estimates; there is no conclusive evidence for water depths.

The Cody flyschlike sequence closely matches the definition of "distal turbidites" of Walker (1967). Important similarities are thickness of the Cody Shale graded beds where the median thickness is 3 cm (Fig. 9); fine grain size of the graded beds; predominance of B, C, D, DE, and E beds with relatively few BDE and BCDE beds (Fig. 8); grooves more common than flutes; and lack of scour channels and

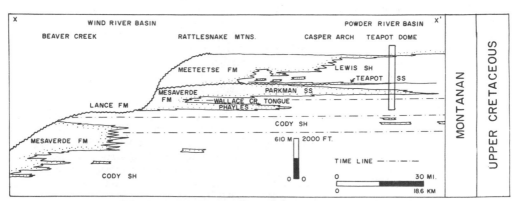

Figure 3. Age and correlation of the Mesaverde Formation in east-central Wyoming (*after* Barwin, 1961; J. A. Barlow, Jr., and J. D. Haun, 1969, written commun.). The time lines are based on ammonite zonation.

[*Editor's Note:* Figure 4 has been omitted.]

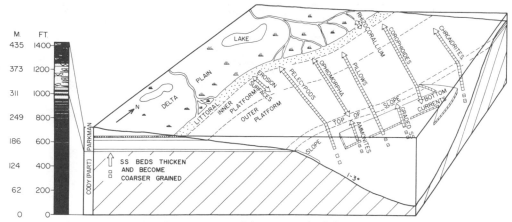

Figure 5.   Depositional environments of the upper Cody Shale and Parkman Sandstone.

amalgamated beds. The sequence is distal in its position on the upper prodelta slope seaward of the delta platform. It is not distal, however, in that mostly mud was deposited farther down the prodelta slope to a water depth of about 500 to 700 ft and out into the basin.

Top of the Prodelta Slope Facies

A 40-ft sequence of sandstone, gray siltstone,

and mudstone was deposited at the break in slope at the outer edge of the delta platform. The sandstone beds are thicker and more numerous than in the underlying prodelta slope facies of the Cody Shale and form the basal unit of the Parkman Sandstone. The sandstone beds are horizontally laminated or festoon cross-bedded; graded bedding is absent. Fragments of marine pelecypods occur in sandstone

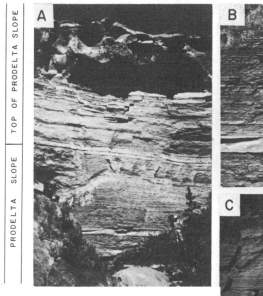

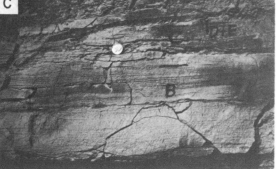

Figure 6.   A. Transition between the upper Cody Shale and lower Parkman Sandstone, section 1. Note pillow horizon at top of picture. B. Part of Cody flyschlike sequence in Figure 6A. Pick gives scale. C. Cody graded bed with BCDE internal structures.

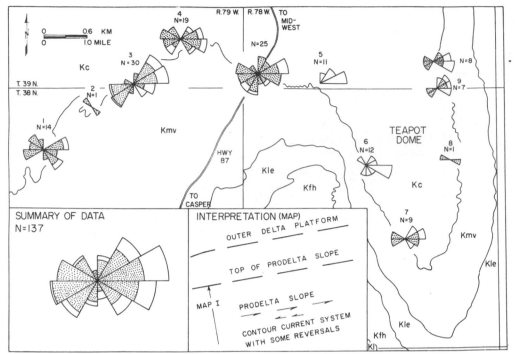

Figure 7. Prodelta slope paleocurrents in the upper Cody Shale based on data from 137 graded beds. Flutes and ripple marks are blank pattern in the roses; grooves and parting-step lineations are stippled. The "equal-area" roses have the frequency in each 30° interval plotted on a geometric scale proportional to the square root of the number of readings. Explanations of symbols: Kc, Cody Shale; Kmv, Mesaverde Formation; Kle, Lewis Shale; Kfh, Fox Hills Sandstone; Kl, Lance Formation.

at section 8 (Fig. 4). The currents that deposited the sandstone beds had no consistent orientation (Fig. 10).

Outer Delta-Platform Facies

The outer delta-platform facies of sandstone with subordinate gray siltstone and shale is about 130 ft thick west of Teapot Dome. The interval thins southeastward in the direction of delta progradation to about 60 ft on the southeast flank of the dome. Siltstone and shale beds become thicker and more numerous to the southeast (Fig. 4). The currents that deposited the sandstone beds on the outer delta platform were highly variable due to complex tidal and wave patterns (Fig. 11).

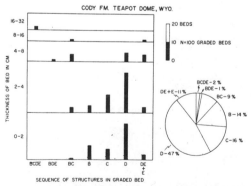

Figure 8. Sequence of internal structures in 100 graded beds in the upper Cody Shale, Teapot Dome.

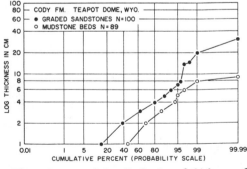

Figure 9. Cumulative distribution of thickness of 100 graded beds in the upper Cody Shale plotted on logarithmic-probability paper.

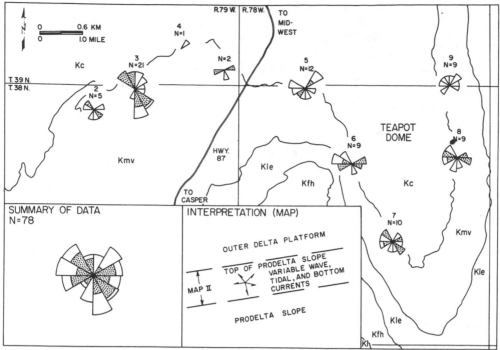

Figure 10. Paleocurrents for sandstones at the top of the prodelta slope facies, Parkman Sandstone. The roses are "equal-area" plots. Symbols as in Figure 7.

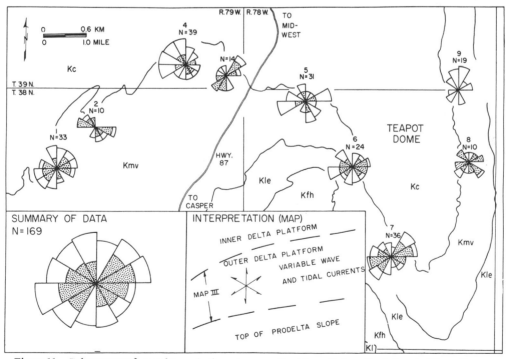

Figure 11. Paleocurrents for sandstones in the outer delta-platform facies, Parkman Sandstone. The roses are "equal-area" plots. Symbols as in Figure 7.

Figure 12. Elongate sandstone pillow in outer delta-platform facies, Parkman Sandstone, section 8.

U-shaped burrows of *Corophioides* are most common in sandstone beds of the outer delta platform, but also occur rarely on the prodelta slope and inner delta platform. Feeding trails of *Chrondrites* are present in sandstone from the prodelta slope through the outer delta platform. Fragments of marine pelecypods occur in sandstone from sections 1, 5, and 7 (Fig. 4).

Pillow structures are abundant in sandstone deposited at the top of the prodelta slope and on the outer delta platform (Fig. 4). Most pillows are elongate flattened cylinders (Figs. 12, 13). The average size of the pillows is about 1 ft thick, 2 ft wide, and 8 ft long. The largest pillows are 7 ft thick and 10 ft wide. Although a few pillow horizons are continuous in outcrop for 0.5 mi, they are too limited in areal extent for use as stratigraphic markers between sections.

The laminae within each pillow structure are concave upward, subparallel to the oval exterior (Fig. 14). The pillows formed at the sand/water interface because the sand layers above and below the pillows are undeformed. The upper surfaces of most pillow horizons were bevelled by erosion before deposition of the next overlying sand layer. Isoclinal folds with

nearly horizontal axial planes occur in many pillows, demonstrating the importance of lateral movement in the formation of the pillows (Fig. 15).

Each pillow horizon formed by slight to substantial downslope gravitational sliding of the sand layers on an underlying thin bed of mud. Water-saturated mud, with its low viscosity and great fluidity, began to move because of the stress gradient created by the weight of the overlying sand layer on the slight slope, about 1°, of the delta surface. Downslope gliding of the sand layer over the moving mud deformed the sand plastically into pillows. Water-saturated lower density mud moved upward between the more dense sand pillows to compensate for shortening of the sand layer. The pillows occur in fine and very fine sand, grain sizes particularly susceptible to failure because of their tendency to be deposited with metastable packing and consequently weak cohesion (Dott, 1963).

Compared to slowly deposited marine clay, the more rapid flocculation of deltaic mud produces larger clay flocs, more loosely packed and randomly oriented clay particles, and more entrapped pore water (Dott, 1963). The large volume of pore water is primarily responsible for the instability of the intrinsically cohesive clay particles. The water saturated clay beds are very sensitive to forcible expulsion of the pore water which physically disrupts the particle packing, thus decreasing cohesion.

The presence of loosely packed water-saturated mud on a sloping surface is not the only factor that controlled formation of the pillows because pillows are rare in the mud-rich prodelta slope facies, but abundant in sands deposited at the top of the prodelta slope and on the outer delta platform. It is evidently the loading of a thick sand bed onto a water-saturated mud layer that triggers downslope movement and

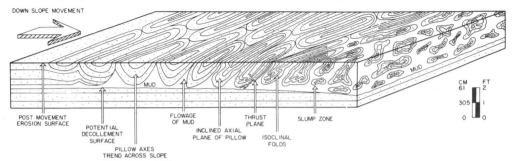

Figure 13. Schematic geometry of pillow horizons and its relation to paleoslope direction. Horizontal and vertical scales are the same.

Figure 14.   Sandstone pillows in the outer delta-platform facies, Parkman Sandstone, section 6.

formation of pillows. The chief role of pore fluid in the mud is the slight buoying effect which reduces shear resistance to critical levels to permit flowage (Terzaghi, 1956; Hubbert and Rubey, 1959).

Some pillows moved downslope to pile up like box cars behind a stalled locomotive so that an occasional pillow slid over its neighbor along a small reverse fault (Fig. 16). The horizontal displacement along these faults is commonly about 3 to 6 ft. Only rarely did a sand layer disintegrate into a jumbled mass of fragments imbedded in mud. The downslope movement evidently did not produce large piles of displaced sediment, although any thickened areas would tend to be bevelled by postmovement erosion.

At ten localities across a 12-mi-wide area, 541 structural elements were measured (Fig. 17). The axes of pillows are consistently aligned parallel to the southwest-northeast contours of the delta surface, and perpendicular to the southeast-slope direction. The strike of the axial planes of the pillows, and of areas of mud upwelling between pillows, are also consistently parallel to the contours of the delta surface. The strike of the thrust planes between pillows is

parallel to the slope contours, and the planes dip upslope. The poles to the axial planes of the isoclinal folds within pillows tend to form a girdle passing through the center of each Schmidt net, bifurcating the plots of the pillow axes. The strike of the axial planes of the folds is parallel to the slope contours. The axial planes of the folds are overturned either upslope or downslope. The axes of the folds tend to lie

Figure 15.   Cross-section of isoclinal folds with nearly horizontal axial planes within sandstone pillow on outer delta-platform facies, Parkman Sandstone, section 6. Coin (quarter) gives scale.

Figure 16. Thrust plane between pillows in the outer delta-platform facies, Parkman Sandstone, section 6. Mud (arrow) has flowed along the thrust between the pillows. Pick is above arrow.

parallel to the slope contours, but show considerable scatter and steep plunges.

A clear boundary exists between the northern area where pillows are absent (solid line contact on Fig. 17) and the area to the south where pillows are abundant (dashed line contact). The gradient of the delta surface in the north was evidently too low to allow formation of pillow horizons. The southern area, in the direction of delta progradation, was sufficiently inclined to allow sliding of some sand layers. Furthermore, pillow axes just south of the demarcation line at sections 3 and 5, and north of section 9, show superior orientation due to less lateral movement and deformation of the sand layers than do those farther southeast in the direction of downslope movement. To the south, the pillow axes show greater variability in their orientation.

The slope of the delta surface could be mapped because the pillows formed by a combination of loading and downslope movement. Similar pillow horizons have been described in the literature, commonly from deltaic sequences. The descriptions suggest that gravity has been effective in formation of the pillows. Studies of such structures probably could be valuable for mapping paleoslopes. A few examples are the Mississippian Bedford Shale and Berea Sandstone of Ohio (Pepper and others, 1954), Upper Devonian Chemung rocks of south-central New York (Sorauf, 1965), Upper Cretaceous Panther Sandstone of central Utah (Howard and Lohrengel, 1969), Middle Devonian Marcellus Formation of eastern New York (Wolff, 1969), Carboniferous rocks, Pembrokeshire, England (Kuenen, 1949), and Precambrian Torridonian Sandstone, Scotland (Stewart, 1963).

The geometry of the pillow horizons demonstrates that the delta slope dipped to the southeast, and the depth contours trended southwest to northeast (Fig. 18). It is thus possible to test the hypothesis that the Cody graded beds were transported and deposited in the downslope direction by gravity-driven turbidity currents. The mapping criteria are independent of the current direction indicated by the sole marks, parting-step lineations, and ripple marks of the graded beds which, under the turbidity current hypothesis, are assumed to indicate the slope direction. The currents that deposited the graded beds flowed northeast, perpendicular and across the southeast-dipping delta slope. The graded beds were deposited by a system of transitory or shifting ocean-bottom currents in

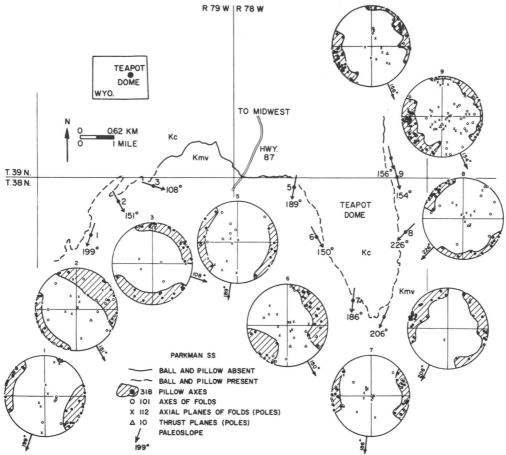

Figure 17.   Map of 541 structural elements in pillow horizons in the lower Parkman Sandstone. The data at each outcrop are plotted on the lower hemisphere of a Schmidt net.

a prodelta environment, rather than by turbidity currents.

Mapping of slopes by criteria independent of current measurements from graded beds, especially slump fold axes, has made it increasingly clear that the current systems that deposited graded beds in many flyschlike sequences flowed across slope (Murphy and Schlanger, 1962; Scott, 1966; Hubert, 1966; Klein, 1966; Hubert, 1967). In other basins, the currents flowed in the downslope direction (Walker, 1970), providing permissive evidence for turbidity currents, but also for ocean-bottom currents. If graded beds, in general, are deposited by ocean-bottom currents, then paleogeographic reconstructions such as slope direction, location of source areas, and configuration of bottom topography must be reevaluated. The entire "turbidite problem" has been reviewed by Van der Lingen (1969).

Inner Delta Platform Facies

The 50-ft sequence deposited on the inner delta platform is mostly sandstone with a few thin siltstone layers. About half of the sandstone is festoon cross-bedded and half is horizontally laminated. At most outcrops, several sandstone layers contain abundant burrows of *Ophiomorpha* which aid in correlation of the measured sections (Figs. 4, 5). The complex burrows are 0.5 to 1 inch in diameter and as much as 5 ft in vertical extent. The burrows were constructed by a decapod crustacean similar to the modern *Callianassa maior* Say (Weimer and Hoyt, 1964). *Ophiomorpha* burrows occur only in horizontally laminated sandstone because the

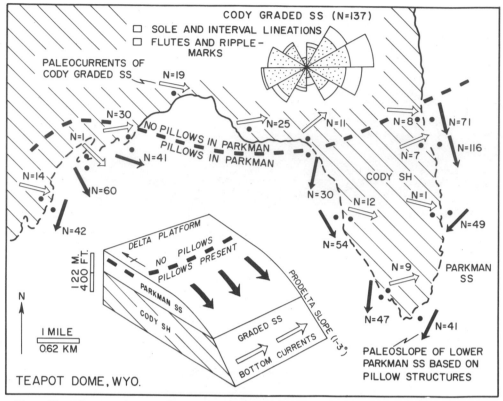

Figure 18. Divergence between the current system that deposited the Cody graded beds and the paleoslope as determined by pillow structures in the lower Parkman Sandstone.

crustacean avoided areas of actively migrating megaripples, where moving sand could clog its entrance and exit holes.

Although *Ophiomorpha* horizons are most abundant in inner delta-platform sandstone, there is one occurrence in strandline sandstone at section 5 and another at the top of the prodelta slope at section 8. *Ophiomorpha* thus lived in littoral and neritic water depths, perhaps up to 50 to 100 ft.

U-shaped *spreiten* burrows of *Corophioides* occur in inner delta-platform sandstone at sections 6 and 9 and in outer delta-platform sandstone at section 6. The U-shaped burrows are 3 in. wide and 1 ft high with the tubes 0.5-inch in diameter. The tubes are always oriented perpendicular to the bedding surfaces.

Several large erosion surfaces occur in the Parkman Sandstone on the west flank of Teapot Dome. The erosion surfaces, two at section 6 and four at section 7, are restricted to the inner delta-platform facies, stratigraphically just below the littoral-beach sandstone (Fig. 4). Local

relief on the erosion surfaces is 5 to 10 ft (Fig. 19). The surfaces are veneered by discoidal pebbles and cobbles of intraformational mudstone. Conglomerate lenses 1 to 2 ft in thickness lie on many low areas.

In plan view, each erosion surface is elongate with the long axis oriented perpendicular to the southwest-northeast trend of the strandline. The two erosion surfaces at section 6 can be traced southward on the outcrop for about 3 mi before they disappear in the subsurface at the end of the dome. The erosion surfaces are absent both northwest of section 6, and on the east flank of the dome, so that their east-west width is about 0.5 to 1 mi. When the contact between the carbonaceous brown mudstone of the coastal-plain marsh and the strandline sandstone is used as a marker horizon, the two erosion surfaces at section 6 are seen to descend to a lower stratigraphic position at section 7 (Fig. 4). The drop is about 50 ft vertically in a horizontal distance of about 2 mi.

Several lines of evidence suggest that the

Figure 19. Erosion surface (arrows) cut into sandstone of the inner delta-platform facies, Parkman Sandstone, section 7.

erosion surfaces are the offshore extensions of open-mouth river estuaries (Fig. 20). *Ophiomorpha* horizons occur in sandstone deposited between two erosion surfaces at section 7, implying relatively shallow water. A barrier island-lagoon system was not present along the strandline at Teapot Dome because of the absence of lagoonal mud above the strandline sandstone. The surfaces are oriented perpendicular to the strandline and deepen seaward. The spatial separation of flow and ebb currents in modern estuaries produces complicated megaripple patterns. The currents on the Parkman inner delta platform lacked a consistent orientation (Fig. 21), evidently due to complicated wave and tidal currents. The erosion surfaces were not formed by storm erosion, for they would be more numerous in the direction of the strandline, and they would be more widely distributed across the dome.

It is unlikely that the Parkman erosion surfaces were formed by tidal scour between tidal current sand ridges, although the size and shape of the erosion surfaces are compatible with this hypothesis. Modern tidal current ridges are elongate sand bodies that form where strong tides, commonly 8 to 10 ft in vertical range, are

confined either by shallowing of the bottom or at the entrance to a river estuary. Their internal structure is not well known. Most modern tidal ridges are 25 to 100 ft high, 5 to 40 mi long, 1 to 2 mi apart, and oriented parallel to the tidal currents (Off, 1963; Houbolt, 1968). The Late Cretaceous inland sea probably had a relatively low vertical tidal range implying that tidal current sand ridges were not prominent features of the Parkman delta platform.

Littoral-Beach Facies

Horizontally laminated and festoon crossbedded sandstone comprises the strandline facies. The facies overlies the *Ophiomorpha*-bearing sandstone of the inner delta platform and underlies the carbonaceous, brown mudstone deposited in coastal marshes of the delta plain (Fig. 22). Most of the paleocurrents at each outcrop flowed directly toward or along the strandline (Fig. 23). The current rose at each strandline outcrop commonly has a mode oriented opposite to the direction of flow of the rivers in the overlying delta plain (Fig. 24). The onshore and alongshore current pattern for the Parkman littoral sandstones suggests that they were deposited in a complex of

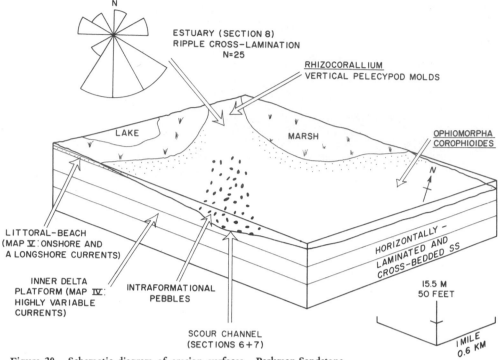

Figure 20. Schematic diagram of erosion surfaces Parkman Sandstone.
cut into sandstone of the inner delta-platform facies,

beaches, swash bars, spits, and subtidal areas swept by alongshore drift.

The foreshores of modern beaches are commonly underlain by planar cross-bed sets with cross-laminae that dip landward. After a severe storm erodes a beach, a new equilibrium beach profile is constructed by landward migration of an intertidal "ridge-and-runnel system" (Hayes and others, 1969). The ridge forms landward-dipping, planar cross-bed sets as it migrates on-shore over the eroded beach surface. Megaripples and ripples form in the runnel between the foreshore and the ridge as the ebb tide flows parallel to the beach until reaching a break in the ridge where it can drain seaward. The resulting cross-beds at any foreshore locality dip landward, due to the migrating ridge, and along the strandline, due to water flowing in the runnel. The process is repeated until several ridge-and-runnel systems coalesce to form a new equilibrium beach profile. The great predominance of festoon rather than planar cross-bed sets in the Parkman littoral sandstones implies that ridge-and-runnel systems were relatively unimportant, and the vertical tidal range was small.

The trend of the shoreline was mapped across an area 12 mi wide based on the cross-beds that dip landward and alongshore in the strandline sandstones, and the downslope, seaward-dipping cross-beds of fluvial sandstones in the overlying delta-plain facies. The arcuate strandline trended northeast, east, northeast, and then north (Fig. 23).

The delta-plain marshes joined the beaches without an intermediate belt of coastal sand dunes. Modern coastal dunes are characterized by "planar, cuniform" cross-bed sets and cross-bed dip angles up to 42° due to wetting and drying by salt spray (Land, 1964; McBride and Hayes, 1962). Neither of these criteria is present in the Parkman strandline sandstone.

Delta-Plain Facies

A 200-ft delta-plain sequence of nonmarine carbonaceous brown and gray mudstone, lensing channel sandstone, and a few thin lignite beds comprise the upper Parkman Sandstone. The region was one of low relief with flood plains, meandering stream courses, lakes, swamps, and a coastal marsh with tidal creeks.

Three tidal channel sandstones occur near the

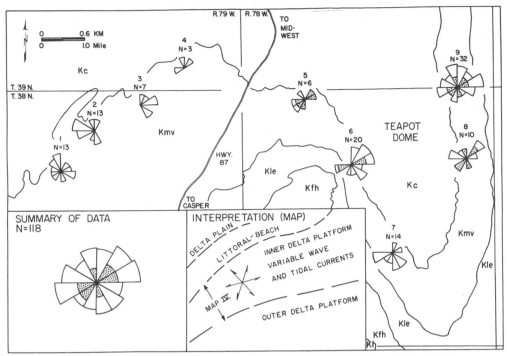

Figure 21.   Paleocurrents for sandstones of the inner delta-platform facies, Parkman Sandstone.   The roses are "equal-area" plots. Symbols as in Figure 7.

base of the sequence at section 8 (Fig. 22). The thickest sandstone is about 5 ft and has an erosional base; the top grades into carbonaceous brown mudstone. Charcoal laminations are common in the sandstone, and there are a few 3-in. silty coal layers. Current ripples with wave lengths of about 1.5 in. are very abundant throughout the sandstone. The ripple cross-lamination is severely disturbed by numerous U-shaped burrows of *Rhizocorrallium* with a diameter of 1/16 in. and height of 1.5 in. Vertical molds of an unknown pelecypod genus in living position are 0.5 in. wide and 1.3 in. high. The currents that deposited the ripple marks flowed dominantly southeast toward the strandline, but numerous current reversals to the northwest indicate tidal rhythms (Fig. 20).

The fine to very fine-grained sandstones form a multistory complex of lensing channel sandstones. Individual sandstone bodies vary from a feather edge to about 50 ft in thickness. The channel bodies have erosional lower contacts, overlain locally by intraformational mudstone conglomerate eroded from adjacent floodplains during lateral stream migration. Most of the channel sandstone bodies fine upward. A few

show the point-bar accretionary sequence of festoon cross-beds followed by horizontal lamination and then current ripples (Visher, 1965). Other common structures are climbing ripples and planar cross-bedding. The cross-beds show that the delta plain sloped southeast (Fig. 24).

The palms *Sabalites montana* and *Ficus* sp., conifer *Araucarites longifolia*, and a sedge-like plant similar to *Typha* or *Cyperacites* suggest a warm moist climate and a low coastal swampland (Heydenburg, 1966). The dinosaur *Trachodon*, crocodile *Deinosuchus*, turtle shell fragments, and fairly common small bone fragments imply luxurious vegetation to support a large vertebrate population (Wegemann, 1911). The fresh-water gastropod *Viviparus* and fresh-water pelecypod *Unio* occur in a sandstone lens near the top of the delta-plain sequence (Heydenburg, 1966).

Destructional Marine Sandstone Facies

The destructional phase of a delta cycle is initiated when the sediment supplied to the area of active progradation is greatly decreased, commonly as a result of a shift in the main river course due to an upstream river diversion

Figure 22. Beach-littoral, coastal marsh, tidal creek, and stream channel deposits in the Parkman Sandstone, section 8.

lar unconformity in the middle of the sequence (Fig. 25). The cross-bed azimuths imply wave and tidal currents of highly variable direction (Fig. 26).

[Editor's Note: Material has been omitted at this point.]

(Scruton, 1960). The delta surface subsides due to compaction of the thick prodelta sequence of water-saturated mud. Marine waves and tidal currents commonly form a thin blanket of sand by selective sorting of the surface sediments.

The destructional phase of the Cody-Parkman delta is represented by 12 ft of marine sandstone with subordinate mudstone that transitionally overlie the delta-plain sediments on the east side of the dome (Fig. 4). Compaction of the delta surface is demonstrated by an angu-

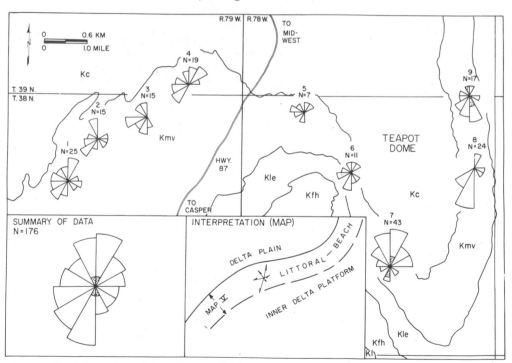

Figure 23. Paleocurrents for sandstones of the littoral-beach facies, Parkman Sandstone. The roses are "equal-area" plots. Symbols as in Figure 7.

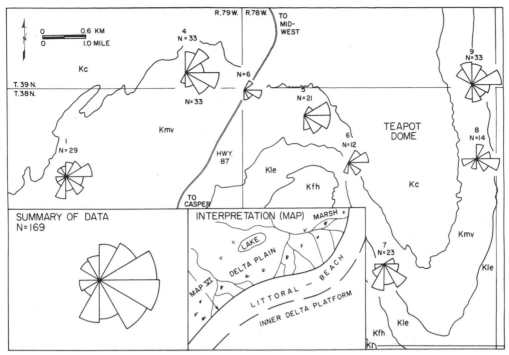

Figure 24. Paleocurrents of fluvial sandstones of the delta-plain facies, Parkman Sandstone. The roses are "equal-area" plots. Symbols as in Figure 7.

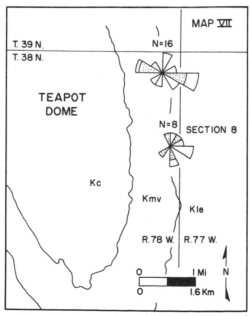

Figure 26. Paleocurrents of marine sandstones of the delta-destructive facies, Parkman Sandstone. The roses are "equal-area" plots. Symbols as in Figure 7.

REFERENCES CITED

Asquith, D. O., 1970, Depositional topography and major marine environments, Late Cretaceous, Wyoming: Am. Assoc. Petroleum Geologists Bull., v. 54, p. 1184–1224.

Barwin, J. R., 1961, Stratigraphy of the Mesaverde Formation in the southern part of the Wind River basin, Wyoming: Wyoming Geol. Assoc. 16th Field Conf. Guidebook, p 171–179.

Bouma, A. H., 1962, Sedimentology of some flysch deposits: Amsterdam, Elsevier Pub. Co., 168 p.

Dott, R. H., Jr., 1963, Dynamics of subaqueous gravity depositional processes: Am. Assoc. Petroleum Geologists Bull., v. 47, p. 104–128.

Fisk, H. N., McFarlan, E., Jr., and Kolb, C. R., 1954, Sedimentary framework of the modern Mississippi Delta: Jour. Sed. Petrology, v. 24, p. 76–99.

Folk, R. L., 1968, Petrology of sedimentary rocks: Austin, Hemphill's Book Store, 170 p.

Folk, R. L., and Ward, W. C., 1957, Brazos River bar; a study in the significance of grain size parameters: Jour. Sed. Petrology, v. 27, p. 3–26.

Friedman, G. M., 1961, Distinction between dune, beach, and river sands from their textural characteristics: Jour. Sed. Petrology, v. 31, p. 514–529.

Gill, J. R., and Cobban, W. A., 1966, Regional unconformity in Late Cretaceous, Wyoming: U.S. Geol. Survey Prof. Paper 550-B, p. 20–27.

Hayes, M. O., Anan, F. S., and Bozeman, R. N., 1969, Sediment dispersal trends in the littoral zone; a problem in paleogeographic reconstruction, in Coastal environments of northeastern Massachusetts and New Hampshire: Amherst, Univ. Massachusetts Pub., no. 1, p. 290–315.

Heydenburg, R. J., 1966, Stratigraphy and depositional environments of the Parkman Sandstone Member, Mesaverde Formation, near Midwest, Natrona County, Wyoming [M.S. thesis]: Laramie, Univ. Wyoming, 81 p.

Houbolt, J.J.H.C., 1968, Recent sediments in the southern bight of the North Sea: Geologie en Mijnbouw, v. 47, p. 245–273.

Howard, J. D., and Lohrengel, C. F., Jr., 1969, Large non-tectonic deformational structures from Upper Cretaceous rocks of Utah: Jour. Sed. Petrology, v. 39, p. 1032–1039.

Hubbert, M. K., and Rubey, W. W., 1959, Role of fluid pressure in mechanics of overthrust faulting; I. Mechanics of fluid-filled porous solids and its application to overthrust faulting: Geol. Soc. America Bull., v. 70, p. 115–166.

Hubert, J. F., 1966, Sedimentary history of Upper Ordovician geosynclinal rocks, Girvan, Scotland: Jour. Sed. Petrology, v. 36, p. 677–699.

—— 1967, Sedimentology of prealpine flysch sequences, Switzerland: Jour. Sed. Petrology, v. 37, p. 885–907.

Kane, W. T., and Hubert, J. F., 1963, Fortran program for calculation of grain-size textural parameters on the IBM 1620 computer: Sedimentology, v. 2, p. 87–90.

Klein, G. deV., 1966, Dispersal and petrology of sandstones of Stanley-Jackford boundary, Ouachita fold belt, Arkansas and Oklahoma: Am. Assoc. Petroleum Geologists Bull., v. 50, p. 308–326.

Kuenen, Ph. H., 1949, Slumping in the Carboniferous rocks of Pembrokeshire: Geol. Soc. London Quart. Jour., v. 104, p. 365–385.

Land, L. S., 1964, Eolian cross-bedding in the beach dune environment, Sapelo Island, Georgia: Jour. Sed. Petrology, v. 34, p. 389–394.

McBride, E. F., and Hayes, M. O., 1962, Dune cross-bedding on Mustang Island, Texas: Am.

Assoc. Petroleum Geologists Bull., v. 46, p. 546–551.

Murphy, M. A., and Schlanger, S. O., 1962, Sedimentary structures in Ilhas São Sabastião formations (Cretaceous), Reconcavo Basin, Brazil: Am. Assoc. Petroleum Geologists Bull., v. 46, p. 457–477.

Off, Theodore, 1963, Rhythmic linear sand bodies caused by tidal currents: Am. Assoc. Petroleum Geologists Bull., v. 47, p. 324–341.

Parker, J. M., 1958, Stratigraphy of the Shannon Member of the Eagle Formation and its relationship to other units in the Montana Group in the Powder River Basin, Wyoming and Montana: Wyoming Geol. Assoc. 13th Ann. Field Conf. Guidebook, p. 90–102.

Pepper, J. F., De Witt, Wallace, Jr., and Demarest, D. F., 1954, Geology of the Bedford Shale and Berea Sandstone in the Appalachian Basin: U.S. Geol. Survey Prof. Paper 259, p. 1–109.

Pryor, W. A., 1961, Petrography of Mesaverde sandstones in Wyoming: Wyoming Geol. Assoc. 16th Field Conf. Guidebook, p. 34–52.

Rich, E. I., 1958, Stratigraphic relation of Cretaceous rocks in parts of Powder River, Wind River, and Big Horn basins, Wyoming: Am. Assoc. Petroleum Geologists Bull., v. 42, p. 2424–2443.

Sabins, F. F., 1962, Grains of detrital, secondary, and primary dolomite from Cretaceous strata of Western Interior: Geol. Soc. America Bull., v. 73, p. 1183–1196.

Scott, A. J., and Fisher, W. L., 1969, Delta systems and deltaic deposition, in Fisher, W. L., and others, eds., Delta systems in the exploration for oil and gas: Austin, Texas Univ., Bur. Econ. Geology, p. 10–29.

Scott, K. M., 1966, Sedimentology and dispersal patterns of a Cretaceous flysch sequence, Patagonian Andes, southern Chile: Am. Assoc. Petroleum Geologists Bull., v. 50, p. 72–107.

Scruton, P. C., 1960, Delta building and the deltaic sequence, in Shepard, F. P., Phleger, F. B., and van Andel, Tj. H., eds., Recent sediments, northwest Gulf of Mexico: Tulsa, Am. Assoc. Petroleum Geologists, p. 82–102.

Sorauf, J. E., 1965, Flow rolls of Upper Devonian rocks of south-central New York: Jour. Sed. Petrology, v. 35, p. 553–563.

Stewart, A. D., 1963, On certain slump structures in the Torridonian sandstones of Applecross: Geol. Mag., v. 100, p. 205–218.

Terzaghi, Karl, 1956, Varieties of submarine slope failures: Austin, Texas Soil Mechanics and Foundation Eng. Conf., 8th, Proc., 41 p.

Van der Lingen, G. J., 1969, The turbidite problem: New Zealand Jour. Geology and Geophysics, v. 12, p. 7–50.

Visher, G. S., 1965, Fluvial processes as interpreted from ancient and Recent fluvial deposits, in Middleton, G. V., Primary sedimentary structures and their hydrodynamic interpretation: Soc. Econ. Paleontologists and Mineralogists Spec. Pub. 12, p. 116–132.

Walker, R. G., 1967, Turbidite sedimentary structures and their relationship to proximal and distal depositional environments: Jour. Sed. Petrology, v. 37, p. 25–43.

—— 1970, Review of the geometry and facies organization of turbidites and turbidite-bearing basins, in Lajoie, J., ed., Flysch sedimentology of North America: Geol. Assoc. Canada Spec. Paper 7, p. 219–252.

Wegemann, C. H., 1911, The Salt Creek oil field, Wyoming: U.S. Geol. Survey Bull. 670, 52 p.

Weimer, R. J., 1961, Uppermost Cretaceous rocks in central and southern Wyoming and northwest Colorado: Wyoming Geol. Assoc. 16th Ann. Field Conf. Guidebook, p. 17–28.

—— 1970, Rates of deltaic sedimentation and intrabasin deformation, Upper Cretaceous of Rocky Mountain region, in Morgan, J. P., Deltaic sedimentation, modern and ancient: Soc. Econ. Paleontologists and Mineralogists Spec. Pub. 15, p. 270–292.

Weimer, R. J., and Hoyt, J. H., 1964, Burrows of *Callianassa major* Say, geological indicators of littoral and shallow neritic environments: Jour. Paleontology, v. 38, p. 761–767.

Wilson, M. D., 1970, Upper Cretaceous-Paleocene synorogenic conglomerates of southwestern Montana: Am. Assoc. Petroleum Geologists Bull., v. 54, p. 1843–1867.

Wolff, M. P., 1969, The Catskill deltaic complex-deltaic phases and correlations of the Middle Devonian Marcellus Formation in the Albany region, in New England Intercollegiate Geol. Conf. Guidebook, sec. 20, p. 1–41.

MANUSCRIPT RECEIVED BY THE SOCIETY JULY 22, 1971
REVISED MANUSCRIPT RECEIVED DECEMBER 13, 1971

Editor's Comments
on Papers 22 Through 27

PARALIC DEPOSITS

In the paralic environments sand and mud come to their last resting place before reaching the open sea. Although deltaic complexes commonly include beaches, bays and lagoons, tidal flats, and barrier islands, these facies also occur along nondeltaic coasts. Just as fluvial channel-fill and delta distributary-bar sandbodies form stratigraphic traps, so do deposits of beaches and offshore bars. For this reason considerable effort has been directed toward distinguishing among them (LeBlanc 1972; MacKenzie 1972; Potter 1969; Busch 1974). Tidal-flat and lagoonal sediments, on the other hand, comprise finer-grained sequences rather like floodplain and lacustrine deposits.

Ancient tidal deposits, both detrital and chemical, are more numerous than formerly supposed. Reineck (1972) summarized their characteristic features, and Ginsburg (1975) has assembled a useful volume of recent studies. Some of the papers that describe detrital successions involve Ordovician deposits in eastern California and adjacent Nevada (Klein 1975) and in northeastern British Columbia (Jansa 1975), Devonian sediments in southern Tunisia (Rizzini 1975), Pennsylvanian strata in Oklahoma (Visher 1975), and Early Cretaceous estuarine sandstone filling tidal scours on tidal flats in northeastern Wyoming (Campbell and Oaks 1973). Additional information about the repeated sequences of tidal-channel to supratidal deposits suggests that the thickness of cycles may record prehistoric tidal ranges. For example, a 12- to 15-m paleotidal range has been inferred from cycles in Early Precambrian rocks of South Africa (Von Brunn and Hobday 1976).

More varied coastal complexes commonly involve interbedded tidal-flat and beach facies. Johnson (1975) has published a well-illustrated study of this association. His detailed analysis of two fining-upward late Precambrian sequences in northern Norway provides criteria for discriminating between tide- and wave-dominated shoreline deposits by their grain size, bed thickness, and sedimentary structures. Similar complexes are recorded in late Precambrian rocks in Scotland (Anderson 1976), in Ordovician deposits in the central Appalachians (Thompson 1975), in an Early Cretaceous sequence in eastern Colorado (Reinert and Davies 1975), in Late Cretaceous deposits in Utah (Cotter 1975), and in late Cenozoic sediments in Switzerland (Van der Linden 1963) and in southern New Jersey (Carter 1975). Dickinson et al. (1972) have reviewed the criteria for recognizing ancient barrier coastlines, and Davies (1976) has constructed a model for barrier-island sedimentation based on both lithology and sedimentary structures. Lawson (1976) has described an example of late Precambrian erosion and local accumulation of conglomerate along a cliffed shoreline in northwestern Scotland.

The six papers selected illustrate both the variety of paralic deposits and the different modes of analyzing them. They are based largely on stratigraphic framework and association of sedimentary features, supplemented by lithologic and paleontologic data. Walker and Harms (1975) have described a muddy shoreline along the fringe of the Devonian Catskill delta in central Pennsylvania (Paper 22). Here a succession of muddy regressive sequences prograded rapidly into a cratonic sea with a low tidal range and weak activity. Kuijpers (1971) found that the late Devonian deposits in southern Ireland record a transition from fluvial facies of a coastal plain to transgressive tidal-marine ones (Paper 23). The great thickness of the succession (1000 m) and the slow vertical transition from fluvial to tidal deposits suggest that aggradation almost kept pace with basin subsidence (or sea-level rise).

MacKenzie (1975) has described an Early Cretaceous tidal-sand-flat facies in Colorado (Paper 24) that is the upper part of a shoal-water deltaic complex. His interpretation is based on critical observations of ripple-modified tabular sets of cross-strata, partially mud-filled channels, symmetrical and current ripples, mud cracks, nested U-shaped burrows, dinosaur footprints, and sedimentary dispersal patterns. Klein (1972) reviewed the characteristics of ancient fining-upward tidal sequences and made a paleotidal range model based on data from Recent intertidal sediments (Paper 25).

Campbell (1971) has constructed a cross-sectional depositional model of a Late Cretaceous sand-beach shoreline in northwestern New Mexico (Paper 26). The succession includes imbricate patterns of parallel-bedded, regressive-beach sandstone overlain unconformably by transgressive offshore-bar sandstones. Landward the beach deposits interfinger with coal-swamp sediments or are truncated and overlain by dune sandstone. Seaward extensively burrowed offshore siltstone and mudstone overlie and locally grade laterally into the offshore-bar sandstone.

Hobday's (1974) three-dimensional model and plan of a late Carboniferous sandstone complex in northern Alabama (Paper 27) consists mainly of beach and barrier-island systems that exhibit four distinct varieties of sedimentary structures. These were built synchronously by two directions of sediment influx, partly across a platform of deltaic coal measures and partly across marine facies with no prior deltaic phase.

REFERENCES

Anderson, R. 1976. Tidal-shelf sedimentation: An example from the Scottish Dalradian. *Sedimentology* 23:429–458.

Busch, D. A. 1974. Stratigraphic traps in sandstones—exploration techniques. *Am. Assoc. Petroleum Geologists Mem. 21*, pp. 1–174.

Campbell, C. V., and Oaks, K. Q. 1973. Estuarine sandstone filling tidal scours, Lower Cretaceous Fall River Formation, Wyoming. *Jour. Sed. Petrology* **43**:765–788.

Carter, C. H. 1975. Miocene-Pliocene beach and tidal deposits, southern New Jersey. In *Tidal Deposits,* ed. R. N. Ginsburg. New York: Springer-Verlag, pp. 109–111.

Cotter, E. 1975. Late Cretaceous sedimentation in a low-energy coastal zone: The Ferron Sandstone, Utah. *Jour. Sed. Petrology* **45**:669–685.

Davies, D. K. 1976. Models and concepts for exploration in barrier islands. In *Sedimentary Environments and Hydrocarbons,* ed. R. S. Saxena. New Orleans, Louisiana: Amer. Assoc. Petrol. Geologists and New Orleans Geol. Soc. Short Course, pp. 79–115.

Dickinson, K. A.; Berryhill, H. L.; and Holmes, C. W. 1972. Criteria for recognizing ancient barrier coastlines. *Soc. Econ. Paleontologists and Mineralogists Spec. Pub. 16,* pp. 192–214.

Ginsburg, R. N., ed. 1975. *Tidal Deposits.* New York: Springer-Verlag, pp. 1–428.

Jansa, L. F. 1975. Tidal deposits in the Monkman Quartzite (Lower Ordovician), northeastern British Columbia. In *Tidal Deposits,* ed. R. N. Ginsburg. New York: Springer-Verlag, pp. 153–161.

Johnson, H. D. 1975. Tide- and wave-dominated inshore and shoreline sequences from the late Precambrian, Finnmark, North Norway. *Sedimentology* **22:**45–74.

Klein, G. deV. 1975. Tidalites in the Eureka Quartzite (Ordovician), eastern California and Nevada. In *Tidal Deposits,* ed. R. N. Ginsburg. New York: Springer-Verlag, pp. 145–151.

Lawson, D. E. 1976. Sandstone-boulder conglomerates and a Torridonian cliffed shoreline between Gairloch and Stoer, northwest Scotland. *Scottish Jour. Geol.* **12:**67–88.

LeBlanc, R. J. 1972. Geometry of sandstone reservoir bodies. *Am. Assoc. Petroleum Geologists Mem. 18,* pp. 133–190.

MacKenzie, D. B. 1972. Primary stratigraphic traps in sandstone. *Am. Assoc. Petroleum Geologists Mem. 16,* pp. 47–63.

Potter, P. E. 1967. Sandbodies and sedimentary environments: A review. *Am. Assoc. Petroleum Geologists Bull.* **51:**337–365.

Reineck, H. E. 1972. Tidal flats. *Soc. Econ. Paleontologists and Mineralogists Spec. Pub. 16,* pp. 146–159.

Reinert, S. G., and Davies, D. K. 1976. Third Creek field, Colorado: A study of sandstone environments and diagenesis. *Mtn. Geologist* 47–60.

Rizzini, A. 1975. Sedimentary sequences in Lower Devonian sediments (Uan Caza Formation), southern Tunisia. In *Tidal Deposits,* ed. R. N. Ginsburg. New York: Springer-Verlag, pp. 187–195.

Thompson, A. M. 1975. Clastic coastal environments in Ordovician molasse, central Appalachians. *Tidal Deposits,* ed. R. N. Ginsburg. New York: Springer-Verlag, pp. 135–143.

Van der Linden, W. J. M. 1963. Sedimentary structures and facies interpretation of some molasse deposits. *Geologica Ultraiectina* **12:**1–42.

Visher, G. S. 1975. A Pennsylvanian interdistributary tidal-flat deposit. In *Tidal Deposits,* ed. R. N. Ginsburg. New York: Springer-Verlag, pp. 179–185.

Von Brunn, W., and Hobday, D. K. 1976. Early Precambrian tidal sedimentation in the Pongola Supergroup of South Africa. *Jour. Sed. Petrology* **46:**670–679.

22

Reprinted from pp. 103–108 of *Tidal Deposits*, R. N. Ginsburg, ed., Springer-Verlag,
New York, 1975, 428 pp.

Shorelines of Weak Tidal Activity:
Upper Devonian Catskill Formation,
Central Pennsylvania

Roger G. Walker and John C. Harms

A common expectation among stratigraphers is to find tidal deposits where marine strata merge with nonmarine strata, either in laterally or vertically exposed sequences. This expectation would seem especially well founded where the marine to nonmarine transition occurs in mud-rich sediments. However, we see no evidence of significant thicknesses of intertidal deposits in a part of the Upper Devonian Catskill Formation, which is dominantly muddy and contains more than 20 marine-nonmarine regressive sequences (Walker and Harms, 1971). Although the regional setting would appear favorable for extensive tidal-flat development (and these sediments have been interpreted as tidal deposits) (Allen and Friend, 1968), no features of these sediments suggest periodic flood and ebb of tides.

We are indebted to the following for their assistance during this study: D. M. Hoskins and his colleagues at the Pennsylvania Geological Survey; R. John Knight, C. H. Carter, G. V. Middleton, and R. Goldring. For funds, Walker is indebted to NATO and the National Research Council of Canada. Permission to publish has been given by the Marathon Oil Company.

We believe that a muddy shoreline prograded rapidly into a sea of low tidal range, and thus any intertidal fringe was restricted in area and migrated too rapidly to form a significantly thick record.

Two aspects of these rocks should be stressed: first, the sedimentary features that define the regressive sequences; and second, the predicted tidally induced features that appear to be consistently absent.

It is not our intention to reiterate work already published. Full, illustrated descriptions have been given by Walker and Harms (1971) and Walker (1971), and a field guidebook with full measured sections has recently been published (Walker, 1972).

Occurrence

The lower 600 meters of the Catskill Formation (the Irish Valley Member) contains a series of repeated marine-nonmarine lithofacies sequences (termed "motifs" by Walker and Harms, 1971), each of which contains low tidal activity shoreline deposits. There are about 20 to 25 repeated motifs in the Susquehanna Valley area, ranging in thickness from 4 to 45 meters, and averaging about 25 meters. They overlie turbidities and slope shales (Trimmers Rock Formation), and pass upward into nonmarine alluvial fining-upward cyclothems (Walker and Harms, 1971; Walker, 1971, 1972). The motifs extend along the depositional strike for at least 50 mi, and each one may have prograded as much as 20 mi into the basin.

Lithofacies Sequences in the Motifs

Individual examples of the Irish Valley motifs vary in thickness, facies type, and lithologic type and proportion. All motifs, however, contain elements that reflect, by their character and organization, related events and sedimentation processes. Consequently, an idealized motif can be described here (Fig. 12–1), and readers are referred to completed measured sections (Walker, 1972) for details of individual motifs. The thicknesses given below refer to Figure 12–1, but motif thicknesses vary from 4 to 45 meters.

There are five parts:

1. 0 to 0.5 meters. A basal bioturbate sandstone with scattered quartz granules, brachiopod and crinoid fragments, and some phosphatic nodules and bone fragments. The lower surface of the sand is commonly sharp and planar, but in places is burrowed and irregular. The maximum thickness is about 50

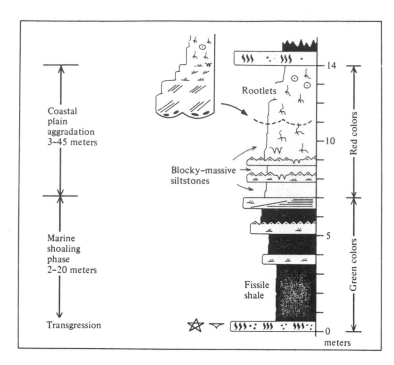

Figure 12-1. An idealized Irish Valley motif, here shown 14 meters thick. In the Susquehanna Valley area, individual motifs range from 4 to 45 meters thick. The five main parts of the motif are discussed in the text. In less than 10 percent of the motifs observed, an alluvial fining-upward sequence was observed, shown diagrammatically in the top center of the figure.

cm. All sedimentary structures have been destroyed by the bioturbation. This sand is followed rather abruptly by

2. 0.5 to 3.5 meters. Olive green fissile shales, with scattered brachiopod and crinoid fragments. Rarely, this unit may form the base of the motif where the bioturbate sand is absent. There are no siltstone or sandstone laminae in these fissile shales. They grade upward into

3. 3.5 to 7.0 meters. Green siltstones and mudstones, with very fine grain sandstone beds between 10 and 50 cm thick. The sandstones show low-angle cross-bedding, horizontal lamination, or ripple cross-lamination. A marine fauna is rare, but symmetrical ripples with 10 to 15 cm wavelengths are common on the top of sandstones. Burrows and trails are common. Conspicuously absent are flaser and lenticular bedding, as well as scouring or channeling. This unit grades upward into

4. 7.0 to 12.0 meters. Drab red siltstones characterized by rootlet traces and desiccation cracks. Thin sandstone beds (2 to 10 cm thick) commonly have symmetrical ripples of 2 to 10 cm wavelength, and vertical and horizontal trails and burrows are common. This unit grades upward into

5. 12.0 to 14.0 meters. Massive, blocky, red mudstones characterized by rootlet traces, desiccation cracks, and tan calcareous concretions. Calcite is disseminated throughout the mudstones. Sandstones are rare, unless occurring in a fining-upward alluvial facies sequence. The sandstones in such sequences are 2 to 9 meters thick, have an intraformational "lag" conglomerate at the base, and are characterized by parallel lamination, cross-bedding, and, at the top, ripple cross-lamination.

"Tidal" Features Notable by Their Absence

Despite the dominantly silty and muddy nature of the motifs, the flaser and lenticular bedding so characteristic of the North Sea tidal flats (Straaten, 1954; Reineck, 1960) was never observed. Small channels, with low-angle muddy and silty lateral accretion surfaces, also characteristic of the lower parts of tidal flats, were never observed. In the rare sand bodies thicker than about 1 meter, there were no reactivation surfaces, or features indicating tidal emergence runoff (Klein, 1970). Despite the prograding nature of the shoreline in these motifs, the tidal "fining-upward" facies sequence (Klein, 1971) was never developed. Our interpretation of the motifs is therefore based upon the idea of weak, low flux tidal activity.

Basic Interpretation of the Motifs

The motifs represent prograding shorelines. The marine fissile shales (0.5 to 3.5 meters) represent offshore clay deposition, and the drab red siltstones (7.0 to 12.0 meters), with rootlets and desiccation cracks, represent vegetated mud flats at the distal margin of an alluvial plain. The portion from 3.5 to 7.0 meters, by its position in the sequence, must be nearshore marine/coastal margin. However, winnowed sand bodies are conspicuously absent, and there is no evidence of offshore bars or barriers, nor of beaches, nor of lower tidal-flat sheet or channel sands. The uppermost unit (12.0 to 14.0 meters) is also interpreted as distal floodplain. The tan calcareous concretions are calichelike, and the fining-upward sandstones probably represent meandering channels draining the alluvial plain.

After maximum progradation, cut off of sediment supply, coupled with subsidence, resulted in marine transgression and initiation of a new motif. The bioturbate sand (0 to 0.5 meter) contains quartz granules winnowed from the underlying motif. The very slow deposition during transgression results in total bioturbation, and as fully marine conditions are reestablished, brachiopods, and crinoids reappear.

In essence, therefore, each motif represents a prograding, low-energy, muddy shoreline, where tidal and other longshore or semipermanent currents are incapable of winnowing sediment into clean sand bodies during progradation. The closest modern equivalent is the prograding muddy shoreline of southwestern Louisiana (Beall, 1968), and detailed comparisons have been made by Walker and Harms (1971). The extent of the motifs and lack of associated sand bodies probably eliminate an estuarine or bay environment.

Tidal Range

Three lines of evidence suggest a low tidal range:

1. Absence of channels (excluding the fining-upward alluvial sequences). In modern areas of moderate to high tidal range, water is carried off the inundated tidal flats in a network of tidal channels. In the Irish Valley motifs, the absence of such channels in the zone of marine to nonmarine transition suggests that large quantities of water were *not* carried off tidal flats following each high tide, implying a low tidal range.

2. Close vertical association of marine and nonmarine elements. In some of the motifs, rootlets and desiccation cracks occur less than 2 meters stratigraphically above marine faunas. The tidal range is unlikely to have exceeded 2 meters; otherwise, brachiopods would have been exposed at low tide and the plants inundated at high tide. Even allowing for compaction and subsidence during progradation, approximately 2 meters can be proposed for the *maximum* possible tidal range. In this critical transitional zone, there are no small-scale directional or textural alternations that would suggest ebb and flood, nor are there any overall textural gradients that would suggest vertical succession of a lower, slightly coarser tidal flat below an upper slightly finer tidal flat.

3. Absence of winnowed sand bodies. The nearshore and shoreline area can be readily identified in the motifs, as it must occur between the green marine facies and the red siltstones with rootlets and desiccation cracks. However, clean winnowed sand bodies thicker than 50 cm are absent in over 90 percent of the motifs studied, implying no formation or preservation during shoreline progradation of beaches and beach ridges, cheniers, surf zone sandbars, or barrier island complexes (it is not clear in Allen and Friend's (1968) publication why barrier island-lagoon complexes were suggested for this part of the section). The winnowing process is primarily controlled by wave action and the rate of supply of mud. If mud is supplied faster than the waves can winnow sand, a muddy shoreline will prograde. However, the problem remaining in the context of this volume is the importance of tidal currents and tidal range, rather than wave action alone. A high tidal range, and hence a higher tidal flux, should result in more powerful currents in the nearshore area. These, in turn, should result in more channeling and winnowing, perhaps more than is recorded in the motifs. The absence of winnowed sand bodies can therefore be related primarily to a high rate of mud input and low activity, and, secondarily, to low tidal flux.

References

ALLEN, J. R. L., and FRIEND, P. F. 1968. Deposition of the Catskill Facies, Appalachian region: With notes on some other Old Red Sandstone basins. *In* Klein, G. deV., ed., Late Paleozoic and Mesozoic continental sedimentation, northeastern North America. *Geol. Soc. Am. Special Paper 106,* 21–74.

BEALL, A. O. 1968. Sedimentary processes operative along the western Louisiana shoreline. *J. Sed. Petrol. 38,* 869–877.

KLEIN, G. DEV. 1970. Depositional and dispersal dynamics of intertidal sand bars. *J. Sed. Petrol. 40,* 1095–1127.

——— 1971. A sedimentary model for determining paleotidal range. *Geol. Soc. Am. Bull. 82,* 2585–2592.

REINECK, H.-E. 1960. Über die Entstehung von Linsen- und Flaserchichten. *Abhandl. Deut. Akad. Wiss. Berlin 1*(3), 369–374.

STRAATEN, L. M. J. U. VAN. 1954. Composition and structure of recent marine sediments in the Netherlands. *Leidse Geol. Mededel. 19,* 1–110.

WALKER, R. G. 1971. Nondeltaic depositional environments in the Catskill clastic wedge (Upper Devonian) of central Pennsylvania. *Geol. Soc. Am. Bull. 82,* 1305–1326.

——— 1972. Upper Devonian marine-nonmarine transition, Southern Pennsylvania. *Pennsylvania Geol. Surv. Bull. G.62,* 25 pp.

WALKER, R. G., and HARMS, J. C. 1971. The "Catskill Delta": A prograding muddy shoreline in central Pennsylvania. *J. Geol. 79,* 381–399.

23

Reprinted from *Geol. en Mijnbouw* **50**(3):443–450 (1971)

TRANSITION FROM FLUVIATILE TO TIDAL-MARINE SEDIMENTS IN THE UPPER DEVONIAN OF SEVEN HEADS PENINSULA (SOUTH COUNTY CORK, IRELAND)

E.P. KUIJPERS [1]

ABSTRACT

Along the east coast of Seven Heads peninsula (southern Ireland) a more than 1000 m thick north dipping succession of Upper Devonian strata is investigated. The lithology and sedimentary structures of the facies types are described and briefly compared with recent sediments. They indicate a gradual transition from a continental "Old Red Sandstone" facies along a coastal plain facies into overlying (tidal) marine facies.

INTRODUCTION

The southeastern part of Seven Heads peninsula (southern Ireland) consists of relatively fine-grained Upper Devonian strata, forming a broad syncline (see fig. 1a and 1b): The Seven Heads bay syncline (N a y l o r, 1964; N a y l o r et al., 1969). Its southern flank is described in the present paper. The upper part of the strata belonging to the northern flank is dealt with in a separate paper by d e R a a f and B o e r s m a (1971). The beds belonging to the southern flank are almost entirely accessible and form a section comprising more than 1000 m of sediment, cut off near the synclinal axis by a fault of moderate importance (see fig. 1b). As has already been pointed out by Naylor the lower part of this section is of "Old Red Sandstone" facies, whereas the upper part belongs to the lower portion of his marine "Cork beds", namely the Old Head Sandstone Group. It is intended that the relationship between this study and the local litho-stratigraphy as outlined by N a y l o r (1964, 1969), M a c C a r t h y et al. (1971, in press) will be dealt with in detail in a subsequent publication. He mentioned furthermore that the top of the Old Head Sandstone Group exposed on the Old Head of Kinsale peninsula is not an outcrop on the east coast of Seven Heads peninsula. Naylor did not examine in detail the lithology and the sedimentary structures in this part of his large area of investigation. The present paper is a preliminary note on a part of a more extensive research work in the South Cork area (d e R a a f, 1970).

LITHOLOGY, SEDIMENTARY STRUCTURES AND PRESUMED ENVIRONMENTS OF DEPOSITION

GENERAL

The upward replacement of the Old Red Sandstone facies b y (tidal) marine beds occurs throughout a transitional unit. Therefore the succession of strata has been subdivided from bottom to top in 3 major facies types (see fig. 1b and fig. 2).

3. "Old Head Sandstone Group" facies (450 m): Upper subdivision.
2. Transitional facies (185 m): Middle subdivision.
1. Upper Old Red Sandstone facies (420 m): Lower subdivision.

---

[1] Department of Sedimentology, Geological Institute of the University, Utrecht, The Netherlands.

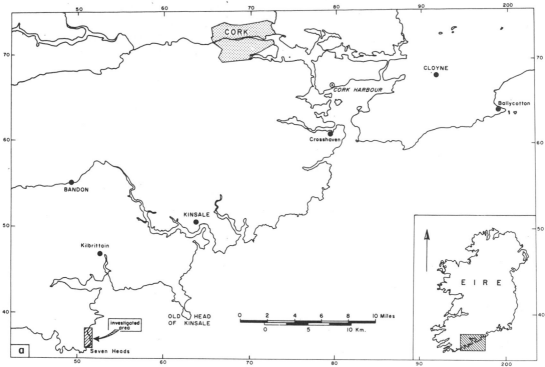

Fig. 1a
Location map.

DESCRIPTION AND ENVIRONMENTAL INTER-PRETATION OF THE SUBDIVISIONS

1. *The lower subdivision (Upper Old Sandstone facies), 420 m thick.*

Description

Intensive cleavage usually hampers the analysis of the lithologic and sedimentary features of the lower subdivision, especially in deposits of a fine-grained character. Red shades are predominant; green shades, where present, chiefly occur in the coarser-grained parts.

The bulk of this subdivision consists of reddish

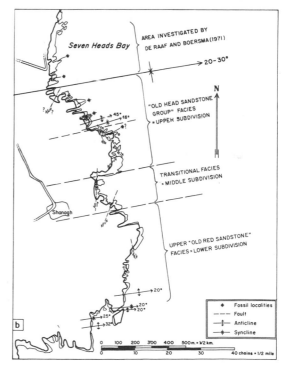

Fig. 1b
Map of the investigated area showing general tectonics and the distribution of main sedimentary complexes (partly after N a y l o r, 1964). Offset along faults do usually not exceed 10 m.

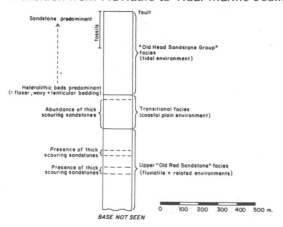

Fig. 2
Diagrammatic picture of the investigated Upper Devonian stratigraphical section.

mudrock (term. I n g r a m, 1953), which is mostly so strongly bioturbated that often only remnants of the original stratification are preserved (see photo 1). The latter show up as silty to sandy laminae in the mudrock or, exceptionally, as wave-built ripple surfaces.

The next important component consists of sandstones of a generally very fine grain-size, which are regularly intercalated in the mudrock. Their thickness usually varies from a few mm to 40 cm; occasionally beds of more than one metre thick occur. Because of the cleavage the quantitative importance of the thinner intercalations could not be ascertained. The bulk of these very fine-grained sandstones appear to be of a rather sheet-like character. Though their lower boundary is sharp, scouring features and lag deposits are rarely encountered. Generally these sandy units pass upward gradually into the overlying mudrock. Where their internal structures can be observed they show unidirectional, small-scale cross-lamination in places merging laterally and vertically into horizontal lamination. The cross-lamination generally shows a slightly climbing ripple pattern with here and there a preserved ripple stoss-side. Sporadically a single mega cross-stratified set occurs in these fine-grained sandstones.

The coarsest and generally thickest (up to $3\frac{1}{2}$ m) beds of the lower subdivision are represented by fine to medium-grained sandstones, which have been found only in a limited number in two zones (see

fig. 2). Their lower boundary scours into the underlying beds. Locally some lag deposits were observed. The sandstones show either a sharp or a gradual upper boundary. In the latter case the fining upward goes with a vertical change in sedimentary structures as follows: In the lower, coarser part the sandstones, if not massive, show horizontal to slightly inclined bedding, which is repeatedly interrupted by irregular cross-stratified intervals connected with erosional truncation and scour and fill (C o l e m a n and G a g l i a n o, 1965). Cross-stratification due to ripple-migration is rare. The overlying fine-grained sandstones show horizontal lamination and especially towards the top unimodal small-scale cross-lamination, usually of a slightly climbing pattern. Here and there scouring sandstones are structurally more complex and show a repetition of the structural intervals mentioned above. The scouring sandstones are sometimes wedging out, which appears in some cases to be due to original depositional conditions, in other cases to tectonics.

Environmental interpretation

The scouring sandstone bodies are structurally and sequentially very closely comparable to channel sediments of modern fluviatile streams as described by several authors (A l l e n, 1970a, p. 311). The predominance of even bedding over cross-stratified units built up by migrating ripples pleads for deposition in low-sinuosity streams (A l l e n, 1970a).

Photo 1
Red burrowed mud with remnants of the original stratification. Photo taken a few metres above the first scouring sandstone in the lower subdivision (see fig. 2), looking north.

Photo 2
Thick scouring sandstones with a fining upward tendency.
Photo taken of a section situated a few metres above the base
of the upper subdivision, looking east.

As to the mudrock, when directly overlying the scouring sandstone units it may correspond with deposits present in alluvial floodplains (A l l e n, 1970a). However, the mudrock units of excessive thickness (more than 60 m, see fig. 2) suggest deposits in a more permanent and extensive alluvial basin. It is thought that these alluvial basins – and floodplains – are generally unimpeded, because of the regular occurrence of usually sheet-like sandstones, which probably are sudden influxes of coarser material in the mudrock.

2. *The middle subdivision (transitional facies), 185 m thick.*

Description

The lower boundary is taken at the occurrence of the first of a large number of scouring sandstones (see fig. 2). The top is formed by the base of the first bed showing "herringbone structures" (v a n S t r a a t e n, 1951); current ripples remain unidirectional in character throughout the middle subdivision.

Green and grey shades are here present in all types of lithology, whereas red and pale-purple shades only occur in strongly bioturbated mudrock in the lower part of this subdivision; the uppermost red bed was observed 50 m above its base and the uppermost pale-purple bed 70 m below its top.

The lithology is characterized by thick sandstones with scouring bases (see photo 2), which are on the whole slightly predominant in quantity compared with the mudrock. The granulometric, structural and sequentional aspects of these scouring sandstones are closely comparable to those of the scouring sandstones present in the lower subdivision: The only difference is their greater individual thickness (up to 12 m). Compared with the lower subdivision these scouring sandstones usually pass upward more gradually, and along silt-laminated sand with small scale cross-laminated intercalations, into mudrock with sandy and silty laminae, which locally display erosional truncations (see photo 3). Towards the top of the subdivision the sandy laminae are here and there replaced by sand lenticles with internal unidirectional cross-lamination. The mudrock is generally slightly bioturbated; however, minor intervals are churned especially where red and pale-purple shades are found.

Compared to the lower subdivision fewer minor sandstone bodies are intercalated in the fine-grained deposits; they comprise beds of a channel as well as of a sheet-like character. The lithologic aspect of this subdivision changes from bottom to top in the following way:

a. A gradual increase of the sand admixture in the mudrock takes place. This is manifested by a transition from mudrock with scattered intercalations of silty to sandy laminae to mudrock with regular intercalations of sandy laminae and sand lenticles.

b. A decreasing bioturbation was observed, which shows up as a gradual replacement of almost homogenized mudrock by less burrowed fine-grained deposits.

c. A gradual replacement from red to green, and from green to grey rocks takes place.

d. A relative decrease in thickness, number and scouring features of the sandstones characterizes the uppermost part of this subdivision.

e. Vertical and lateral transitions between rock types become more common towards the top layers of the middle subdivision.

Environmental interpretation

The scouring sandstones in this subdivision are similar to the scouring sandstones in the Upper Old Red Sandstone facies and are accordingly again interpreted as beds deposited by fluviatile streams of presumed low-sinuosity character. The thickness of some of the fining upward scouring sandstones points to distributaries with an important discharge, characteristic for the lower reaches of a fluvial system.

The finer grained deposits with occasionally intercalations of thin sandstone bodies are interpreted as stream topstrata, generally incorporating levees. The gradual upward change of the character of the sandstones, the mudrock, and the rock colours point to a modification in the depositional conditions similar to the Old Red Sandstone facies and suggest already the approach of a marine (tidal) environment (see upper subdivision), as will be shown in more detail in a later publication. On these considerations a coastal plain environment is assumed for this subdivision.

3. The upper subdivision (= Old Head Sandstone Group facies), 450 m thick.

Description

As already mentioned by N a y l o r (1964; et al. 1969) the lower 270 m of this upper subdivision are closely comparable in lithology and sedimentary structures to the Breamrock formation of the Old Head of Kinsale (see fig. 1), which has been dealt with briefly in a previous paper by the author (K u i j p e r s, 1971). The upper 180 m correspond to Naylor's Holeopen formation, (highest formation

Photo 4
Detail of steep side of a small channelling sandstone lense. Photo taken just above the base of the upper subdivision, looking north.

in the Old Head Sandstone Group), which, as remarked previously, is not entirely exposed.

The bulk of the lower 270 m contains lenticular bedding, especially in the lower part associated with sand-laminated mudrock, wavy and flaser bedding (R e i n e c k & W u n d e r l i c h, 1968). These bedding types occur in alternation and form complexes of a varying sand/mud ratio, having the same character as in the Breamrock formation on the Old Head of Kinsale (lithologic type 2 of K u i j p e r s, 1971). In these "heterolithic" beds current-ripple structures — usually of herringbone type — are a common feature; wave-built ripples were rarely observed. Locally thin-bedded sandstone intercalations — commonly horizontally laminated — occur. Minor intervals of fine-grained material are occasionally completely bioturbated especially in the lowest part of this subdivision.

Fine to medium-grained sandstones having a sharp base with little scouring or even none (lithologic type 1 of K u i j p e r s, 1971 are in places intercalated in the lower 270 m of this subdivision. Their upward junction to the mud-containing sediments is frequently sharp. In these sandstones horizontal lamination with slight bedding-plane divergencies is characteristic; also small-scale cross-lamination, usually of herringbone type, is common and is generally found more towards the top of these sandstones. To a less extent irregular cross-stratified intervals connected

Photo 3
Sand/mud interlaminations displaying erosional truncations as locally present in this type of lithology. Photo taken in the top layers of the middle subdivision, looking north.

with erosional truncation and scour and fill were observed.

Throughout the lower 270 m of this succession small as well as large-scale occasionally strongly downcutting channels occur generally with a fine-grained infill. A few small channelling sandstone lenses with a steep side or even an overhanging one were observed near the base of this subdivision (see photo 4). Other structures present include slump structures, convolutions and load casts.

Poorly preserved casts of marine fossils (usually concentrated in wash-outs) are fairly common in the uppermost 180 m of the studied section. Together with the appearance of these fossils a few lithologic and structural changes were observed (see d e R a a f and B o e r s m a, 1971):

a. The series show a gradual increasing amount of sandstones (see fig. 2); here and there some of these sandstones are completely bioturbated. More often than in the lower part of the tidal succession the sandstones show a gradual upward junction to the heterolithic beds, which keep their same particular character; locally the mud-rich hetero-lithic beds are more common towards the top of a heterolithic sequence, thus yielding a type of fining upward cyclothem. T e r w i n d t (1971) describes a comparable feature in subtidal channels in the southwestern Netherlands, when dealing with his litho-facies II.

b. It was often observed in the uppermost 200 m that the horizontally laminated sandstones with slight bedding-plane divergencies alternate laterally and vertically with mega cross-stratified sandstones with herringbone structures due to oppositely migrating mega-ripples (on an average 15 cm high), thus pointing to a higher streampower of the currents (A l l e n, 1970a).

c. Wave-built and interference ripples are common; they occur frequently in the top centimetres of sandstone beds internally built up by current-produced structures.

Environmental interpretation

These deposits are interpreted as tidal, just as the Breamrock formation of the Old Head of Kinsale (K u i j p e r s, 1971). The small channelling sand-stone lenses with steep sides may have originated in an intertidal environment; however, real proof for intertidal sedimentation could not be found.

H ä n t z s c h e l and R e i n e c k (1968) mention that fossil concentrations in washouts as found in the uppermost 180 m of the sequence have so far been found only in shallow water.

Modern areas, as the delta of the Klang and Lancat (C o l e m a n et al., 1970) and the lower deltaic plain of the Niger (A l l e n, 1970b) show, when going through an intertidal zone seaward into a subtidal zone, amongst other phenomena, an increasing sand content caused by a higher frequency of relatively more powerful tidal streams as well as by increasing wave-action. In this respect comparable features were observed upward in the ancient tidal succession, but differences, which probably are also due to dissimi-larities in vegetation, remain between the modern and ancient depositional areas mentioned.

PALEO-CURRENT DIRECTIONS

Paleo-current directions were based on measure-ments of the foreset slopes of small-scale cross-laminated units. Measurements of the foreset slope of

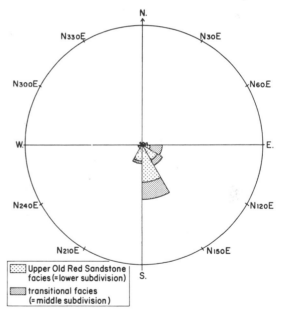

Fig. 3
Current rose based on 52 readings of foreset slopes of small-scale cross-laminated units occurring throughout the upper Old Red Sandstone facies and the transitional facies. Readings are plotted as current directions.

mega-ripple structures were taken too; their quantity, however, is small. The readings were made throughout the succession and, if possible, at regular distances.

The paleo-current directions prevailing in the Old Red Sandstone facies and the transitional facies suggest that the source area for this continental deposits had been situated in the north-northwest (see fig. 3).

Paleo-current directions based on measurements in the tidal facies of the Old Head Sandstone Group clearly show a bimodal current distribution (see fig. 4). North-northwesterly current directions (flood-currents?) and south-southeasterly current-directions (ebb-currents?) are a common feature especially in the lower 300 m of these tidal deposits. In the top of the sequence the current distribution has a more random character.

CONCLUSION

A gradual, vertical transition from strata formed in fluviatile and related environments along coastal-plain

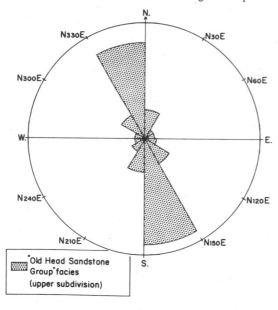

Fig. 4
Current rose based on 133 readings of foreset slopes of small-scale cross-laminated units occurring throughout the tidal facies of the Old Head Sandstone Group. Readings are plotted as current-directions.

deposits to strata formed in a marine tidal environment was observed. The great thickness of these fluviatile, coastal plain and tidal strata occurring in superposition and the slow vertical transition between these environments suggest that the transgression took place in a basin, in which sedimentation almost kept place with subsidence.

ACKNOWLEDGEMENTS

The author gratefully acknowledges financial assistence from The Netherlands Organization for the Advancement of Pure Research (Z.W.O.) for this project, which is still in progress, and also wishes to thank Prof.Dr. J.F.M. de Raaf and Dr. J.R. Boersma (University of Utrecht) for assistence in the field and for helpful criticism in preparing the manuscript.

REFERENCES

Allen, J.R.L. (1970a) – Studies in fluviatile sedimentation: A Comparison of fining-upward cyclothems, with special reference to coarse-member composition and interpretation. J. Sediment. Petrol. 40, p. 298-324.
—— (1970b) – Sediments of the modern Niger delta: A summary and review. Soc. of Econ. Paleontol. and Mineralog., Spec. Publ. 15, Deltaic sedimentation, modern and ancient, p. 138-151.
Coleman, J.M. and S.M. Gagliano (1965) – Sedimentary structures: Mississippi River deltaic plain. Soc. of Econ. Paleontol. and Mineralog., Spec. Publ. 12, Primary sedimentary structures and their hydrodynamic interpretation, p. 133-148.
Coleman, J.M., S.M. Gagliano and W.G. Smith (1970) – Sedimentation in a Malaysian high tide tropical delta. Soc. of Econ. Paleontol. and Mineralog., Spec. Publ. 15, Deltaic sedimentation, modern and ancient, p. 185-197.
Häntzschel, W. and H.E. Reineck (1968) – Fazies-Untersuchungen in Hettangium von Helmstedt (Niedersachsen). Mitt. aus dem Geol. Staatsinstitut in Hamburg. 37, p. 5-39.
Ingram, R.L. (1953) – Fissility of mudrocks. Geol. Soc. Amer. Bull., 64, p. 869-878.
Kuijpers, E.P. (1971) – Preliminary note on the Breamrock formation (Upper Devonian), County Cork, Ireland – An Ancient Tidal Deposit. Sediment. Geol., 5, p. 83-89.
MacCarthy, J.A.J., P.R.R. Gardiner and R.R. Horne (1971) – The Garryvoe Conglomerate Formation (Tournaisien) of southern Ireland. Bull. of the Geol. Survey of Ireland, 2 (in press).
Naylor, D. (1964) – The Upper Devonian and Lower

Carboniferous stratigraphy of South County Cork, Ireland. Unpublished thesis, Trinity College, Dublin, 308 p.

Naylor, D., P.C. Jones and M.J. Clarke (1969) — The Stratigraphy of the Cork beds (Upper Devonian and Carboniferous) in Southwest Ireland. Sci. Proc. Roy. Dublin, Soc. A3, p. 171-191.

Raaf, J.F.M. de (1970) — Environmental studies concerning Upper Devonian (and Lower Carboniferous) coastal sections of South County Cork. Geol. en Mijnb., 49, p. 247-248.

Raaf, J.F.M. de and J.R. Boersma (1971) — Tidal deposits and their sedimentary structures (seven examples of western Europe). Geol. en Mijnb., 50, p. 479.

Reineck, H.E. and F. Wunderlich (1968) — Classification and origin of flaser and lenticular bedding. Sedimentol., 11, p. 99-104.

Straaten, L.M.J.U. van (1951) — Texture and genesis of Dutch Wadden Sea sediments. Proc. 3rd Intern. Congr. Sedimentol., Groningen-Wageningen, p. 225-244.

Terwindt, J.H.J. (1971) — Litho-facies of inshore estuarine and tidal inlet deposits. Geol. en Mijnb., 50, p. 515.

24

Reprinted from pp. 117–125 of *Tidal Deposits,* R. N. Ginsburg, ed., Springer-Verlag,
New York, 1975, 428 pp.

Tidal Sand Flat Deposits
in Lower Cretaceous Dakota Group
Near Denver, Colorado

David B. MacKenzie

Occurrence

West of Denver, on the east side of the Dakota hogback,
large bedding plane surfaces dipping 35° to 40° east expose
the upper 30 meters of the 100-meter thick Lower Cretaceous
Dakota Group (Fig. 14–1). The Dakota Group of this area is
primarily a shoal-water deltaic assemblage within which many
different subenvironments are recognized (MacKenzie, 1971,
1972; Weimer and Land, 1972).

Facies

The section (Fig. 14–2) is made up of well-sorted fine- to
very fine-grained quartzose sandstones (Table 14–1) with
less than 10 percent interbedded shales. The lower half consists

Discussion with those familiar with Holocene tidal flats—H. E. Reineck
and J. D. Howard, in particular—has been very helpful. Observations
made by my colleague J. C. Harms, as well as his helpful review of
this manuscript, are gratefully acknowledged.

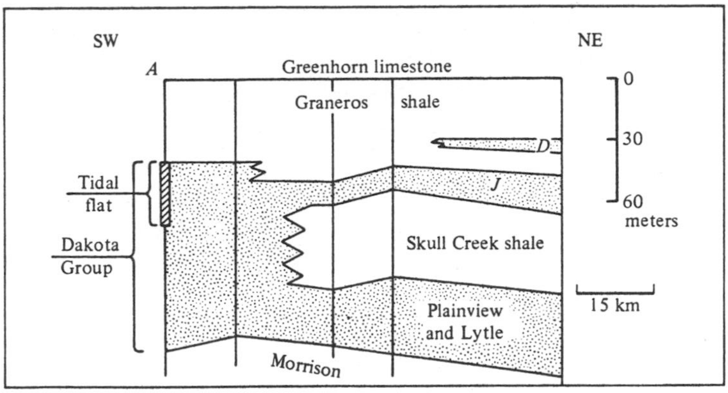

Figure 14-1. Southwest-northeast stratigraphic cross section showing position of tidal-flat deposits. *A* is The Alameda Avenue section located about 10 miles west-southwest of the state capitol building in Denver.

mostly of a fining-upward sequence of fine-grained, cross-stratified sandstone; the upper half consists of very fine-grained tabular-bedded sandstones, locally incised by sand- or partially mud-filled channels.

RIPPLE-MODIFIED TABULAR SETS OF CROSS-STRATA. A 9-meter thick unit in the lower-middle part of the section (*F* in Fig. 14–2) is a fine-to medium-grained sandstone with tabular sets of cross strata 0.1 to 0.6 meters thick. Carbonized wood fragments are abundant throughout. The cross-laminae, which commonly consist of alternating fine- to medium-grained sandstones, are concave up and have maximum dips (restored) of 25°. They dip uniformly north. In plan view, the traces of the

Table 14–1   *Grain size analyses*

Bed	Comments	Median φ	Median μm	5th percentile φ	Sorting (Folk and Ward), φ
U		3.25	105	2.56	2.01
B		3.93	66	3.08	3.20
N	Rippled pocket	3.51	88	2.85	1.45
	Unrippled surface	3.36	97	2.55	2.64
X		4.28	52	3.18	2.84
R		3.68	78	2.68	2.06
D		2.78	146	2.42	1.34
Z		3.45	92	2.63	2.06
P	ss above scour	2.63	162	2.24	0.51
	ss below scour	3.08	119	2.52	1.19
F		2.15	226	1.85	0.56
G		3.10	117	2.58	0.97

305

cross-laminae on boundaries between sets are remarkably straight—in some places for at least 6 meters

A distinctive feature is the presence of long-crested, asymmetric ripples on the boundaries between sets of cross-strata (Fig. 14–3D). The ripple crests trend at a large angle to the strike of the cross-strata; the crests trend 009° and are generally asymmetric toward the west. The ripples have wavelengths of 4 to 8 cm and amplitudes of 2 to 5 mm. Shorter wavelengths are commoner toward the top of the unit. Invertebrate tracks and trails are present locally on rippled surfaces in the lower part. Either low amplitude symmetric ripples or straie are also present locally on the surfaces of cross-strata.

Clearly, the dominant mode of sedimentary transport was unidirectional toward the north; equally clearly, the body of water was shallow enough and broad enough for closely spaced ripples to form intermittently on the bed as a result of waves moving transverse to average current direction. Repeated alternations of currents and wavy slack water are consistent with a

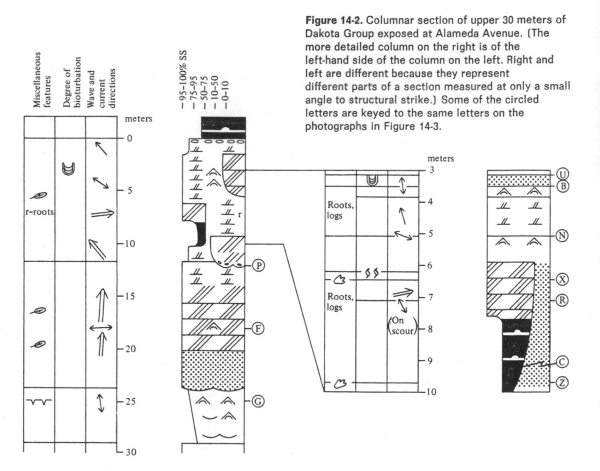

Figure 14-2. Columnar section of upper 30 meters of Dakota Group exposed at Alameda Avenue. (The more detailed column on the right is of the left-hand side of the column on the left. Right and left are different because they represent different parts of a section measured at only a small angle to structural strike.) Some of the circled letters are keyed to the same letters on the photographs in Figure 14-3.

A

3 meters

B

3 meters

C

R   C   D   C   C   Z

3 meters

3 meters

D

3 meters

307

Figure 14-3 (opposite). *A.* A surface with long-crested ripples at position *N* in Fig. 14.2. Width of photograph is 25 meters. *B.* A surface with dinosaur footprints at position *X* in Fig. 14.2. Width of photograph is 4 meters. *C.* Scour beneath mud-filled channel in the lower half of the right-hand column of Fig. 14-2. Width of photograph is 35 meters. *D.* A surface showing ripple-modified sets of cross strata at position *F* in Fig. 14-2. Width of photograph is 22 meters.

tidal environment. The preservation of wave ripples beneath advancing sand waves may require some sort of stabilization of the sand—perhaps binding by algal mat.

CHANNEL, PARTIALLY MUD-FILLED. From about 6 to 9 meters stratigraphically beneath the top of the Dakota Group (*C* in Fig. 14–2) is an erosional scour at least 60 meters broad, overlain first by shale and siltstone and then by sandstone. Preserved for observation are three fragments (each marked *C* in Fig. 14–3*C*) of a once continuous curved surface that truncates at least 2.5 meters of underlying beds. The upper part of the surface has ripples with long crests subparallel to the dip. In Figure 14–3*C* the scour can be seen to truncate the surface labeled *R*, characterized by rootlets and by imprints of driftwood in a very fine-grained sandstone. The surface of truncation is overlain by 2 meters of medium gray silty shale with interlaminated siltstone and then by 2 meters of fine-grained, trough cross-stratified sandstone. The axes of the troughs dip eastward (080°) (Fig. 14–3*C*).

RIPPLES. Some bedding plane surfaces have remnants of current ripples; others have conspicuous, long-crested, symmetrical, wave-generated ripples (Fig. 14–3*A*). The long-crested, symmetrical ripples on the very fine-grained sandstone at *B* have a wavelength of 5 cm, an amplitude of 5 mm, and an orientation of 082°. Those on the surface of the very fine-grained sandstone at *N* occur in irregular pockets on an otherwise unrippled surface. They much resemble features observed in intertidal zones at Sapelo Island, Georgia, and in the North Sea (H. E. Reineck, personal communication). They have an average wavelength of 12 cm, an amplitude of 1.5 cm, and an orientation of about 020°.

At the edges of the irregular patches at *N*, the ripples rise to the level of the unrippled surface. So the patches are not the result of recent selective removal of an overlying layer. How they do originate is not clear. One interpretation is that the unrippled part was flattened by swash action and then stabilized by algal mat. It is only where this algal mat was somehow ripped up that ripples were able to form. Although algal mat is common in the upper parts of some intertidal sand flats, once formed it is probably too tough to be breached later by relatively gentle wave motion in shallow water. Had the rippled depressions on surface *N* been formed by ripping up of algal mat, one might expect associated features indicating a more vigorous environment.

An alternative interpretation is that ripples are preserved only in gentle depressions, those once present on the slightly higher and much larger surface having been destroyed as the water shoaled and swash or shallow sheet flow flattened the bed. In either interpretation, short-period water depth changes are indicated.

MUD CRACKS. The rippled, very fine- grained sandstone, indicated by *G*, is mud cracked locally. The mud cracks are up to 3 mm wide and several centimeters long. The thin yellow film that cracked was deposited as a mud in the hollows between ripple crests. The sequence of mud deposition (slack water), mud cracks (emergence), and overlying rippled sand (re-submergence) is suggestive of short-period fluctuations in water level.

NESTED U-SHAPED BURROWS. Although tracks and trails are common on many of the bedding planes in the upper third of the section, bioturbation is conspicuous in only one bed (*U* in Fig. 14–2). It is a 34-cm thick, very fine-grained sandstone. Seen in vertical section, each burrow has the shape of many *U*s nested together (Fig. 14–4). A typical one is 4 cm wide and 18 cm deep. Their presence is often revealed only by closely spaced vertical planes of weakness along which the sandstone of the *U* bed breaks.

Burrows similar to these have been described earlier from the Dakota Group (Howell, 1957) and from the Devonian Baggy Beds in Devonshire, England: *Diplocraterion yoyo* (Goldring, 1962). The organisms that made this burrow cannot be identified. But a crustacean *Corophium* does make this kind of burrow in the upper parts of tidal flats of Jade Bay, Germany (Reineck, 1967), and some species of clams may make a similar burrow also.

PLANT ROOTLETS. Tiny outlines of plant rootlets are visible in at least two beds (Fig. 14–2). Sometimes the carbonaceous

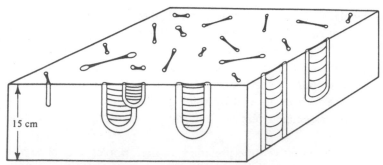

Figure 14-4. Block diagram showing nested U-shaped burrows in bed *U* of Fig. 14-2.

filaments are preserved (especially on freshly broken surfaces); sometimes they have been oxidized so that only the molds around them remain. Their presence in a bed means that before the bed was covered by other sediments, plants became rooted in it. The waters covering the surface of the bed must have been very shallow, or the bed may have been emergent. The size of the rootlets (1 to 2 mm in diameter and up to 8 cm long) suggests rushes or small bushes.

DINOSAUR FOOTPRINTS. Dinosaur footprints are found on beds *X* and *Z*. Bed *X* is a coarse siltstone with at least three different sets of footprints. Selected individual prints are outlined in black in Figure 14–3*B*. The footprints are three-toed and either 55 cm wide and long or 40 cm wide and long. Most of the prints are indistinct, but deep. This suggests that although the tracks were formed under water, most of the animals' weight was borne by the bottom; hence the water must have been quite shallow.

SEDIMENTARY TRANSPORT DIRECTIONS. As shown to the left of the columnar section (Fig. 14–2), the orientations of the primary sedimentary structures are variable, particularly the long-crested ripples. However, the cross-strata of the sandstones generally show a north component of sedimentary transport. The more marine facies of the Dakota Group also lie to the north and east.

Examples of local upward alternation in direction of sedimentary transport—herringbone cross-stratification (Klein, 1971)—are rare. A possible example occurs at the base of the sandstone channel fill at *P*.

Comparison with Holocene Deposits

Observation of modern tidal flats and their subtidal equivalents suggests that progradation would result in a 3- to 10-meter thick, fining-upward sequence (MacKenzie, 1972). The fills of migrating ebb gullies would be selectively preserved in the lower parts; the deposits of the higher parts of the tidal flat of one cycle would tend to be selectively removed by erosion associated with succeeding cycles.

Comparison of the · Alameda Avenue deposits with this Holocene model points up one obvious disparity—thickness. Either the 30-meter thick section contains several cycles, or subsidence was sufficiently pronounced for the thickness of one cycle to be greatly increased. The presence of mud cracks in the lower part of the section suggests that there are at least two and probably more cycles present.

The cross-stratified sandstones of the lower part of the Alameda Avenue sequence could well represent the fills of migrating ebb gullies. Many of the features associated with the higher parts of Holocene tidal flats are, in fact, found only in the upper 10 meters of the Alameda Avenue section: root beds, intense bioturbation, partially mud-filled channel, vertebrate tracks. Not found at Alameda Avenue but reported from Holocene tidal flats are "longitudinal oblique beds" as parts of channel fills, features attributable to "emergence runoff" (Klein, 1971), flaser bedding, or algal mats.

Any alternative to the tidal sand flat interpretation presented here should account for the following key features:

1. Overall fining-upward sequence
2. Presence of broad surfaces with abundant tracks and trails, interbedded with root beds indicating emergence
3. Widespread occurrence of wave-generated ripples
4. Presence of fairly numerous channels filled with sand deposited by currents flowing generally in a seaward direction.

To explain this association one might invoke other parts of a shoal water delta where tidal influences were minimal or nonexistent. In that case, the local scours would be interpreted as deltaic distributary channels; the broad, thin beds with abundant tracks and trails as interdistributary bay deposits; and the root beds as deposits of interdistributary swamps. One argument against such an interpretation is the scarcity of muds; present shoal water deltas of the world contain a much higher proportion of mud than does the upper 30 meters of the Dakota Group at Alameda Avenue.

Many features of these rocks indicate deposition in a marginal marine environment in which water level was undergoing short-period fluctuation. Neither period nor range is known. Of present-day processes, tidal movement, whether wind-driven or astronomic in origin, is the preferred explanation.

References

GOLDRING, R. 1962. The trace fossils of the Baggy Beds (Upper Devonian) of North Devon, England. *Palaontol. Z. 36*, 232–251.

HOWELL, B. F. 1957. New Cretaceous scoleciform annelid from Colorado. *J. Paleontol. Soc. India 2*, 149–152, pl. 16.

KLEIN, G. DEV. 1971. A sedimentary model for determining paleo-tidal range. *Bull. Geol. Soc. Am. 82*, 2585–2592.

MACKENZIE, D. B. 1971. Post-Lytle Dakota Group on west flank of Denver basin, Colorado. *Mountain Geol. 8*, 91–131.

————— 1972. Tidal sand flat deposits in Lower Cretaceous Dakota Group near Denver, Colorado. *Mountain Geol. 9*, 269–277.

REINECK, H.-E. 1967. Layered sediments of tidal flats, beaches, and shelf bottoms of the North Sea. *In* Lauff, G. H., ed., *Estuaries.* Am. Assoc. Advan. Sci., pp. 191–206.

WEIMER, R. J., and LAND, C. B., JR. 1972. Field guide to Dakota Group (Cretaceous) stratigraphy, Golden-Morrison area, Colorado. *Mountain Geol. 9*, 241–267.

25

Reprinted from *Internat. Geol. Congr., 24th, Ottawa, 1972, Rept. of Section 6,*
pp. 397–405

Determination of Paleotidal Range in Clastic Sedimentary Rocks

GEORGES DEVRIES KLEIN,
U.S.A.

ABSTRACT

Prograding clastic tidal flat coastlines generate a graded, fining-upward sequence of sediments in response to controlled zonation of sediment processes across intertidal flats. These transport zones are (1) tidal bedload transport combined with emergence runoff prior to exposure (producing low tidal flat sand), (2) alternation of bedload and suspension sedimentation (producing mid flat interbedded sand and mud) and (3) suspension sedimentation (producing high tidal flat mud and clay). The distinction between subtidal and low intertidal flat sands is achieved by recognizing combinations of sedimentary structures indicating late-stage emergence runoff prior to tidal flat exposure. The contact between these two sands, one with and the other without emergence runoff structures, coincides with the position of mean low water. The top of the graded, fining-upward sequence consists of clay-sized sediment and coincides with mean high tide.

Such prograded fining-upward sequences contain within them a preserved record of tidal range. In Holocene sediments, the thickness of the interval representing low tidal flat, mid flat and high tidal flat sediments coincides with mean tidal range. In fossil equivalents, the thickness of similar sequences gives a quantitative measurement of paleotidal range. Range determinations measured so far include: Late Precambrian: 0.3—13.0 m; Cambrian: 1.1—7.9 m; Silurian: 3.3—6.1 m; Devonian: 2.0—8.9 m; Jurassic: 0.8—4.1 m. In comparison to Holocene variation in tidal range (0—17.0 m), paleotidal range variation has not changed appreciably since Late Precambrian time.

INTRODUCTION

MEASURING VARIATION IN TIDAL RANGE through geological time has remained a problem because very few models based on Holocene sediment studies were available for determining paleotidal range. A model for determining paleotidal range of clastic sedimentary rocks was proposed recently by Klein (1971a) and it was based on Holocene sediment studies of intertidal sediments. This paper summarizes the results of some paleotidal range determinations using Klein's (1971a) model. The data were obtained from the Late Precambrian Uppermost Sterling Quartzite and Wood Canyon Formation, and the Cambrian Harkless Formation of eastern California and western Nevada. Additional data include those based on studies of the Silurian of Pennsylvania (Smith, 1967, 1968), the Devonian of New York and Pennsylvania (Johnson, 1968; Walker and Harms, 1971) and the Jurassic of England (Klein, 1965).

As detailed previously (Klein, 1971a), sediments deposited by tidal currents contain a distinct association of sediment parameters reflecting ten phases of tidal sediment transport and deposition in clastic and carbonate environments. For convenience, such sediments are termed *tidalites*. Tidalites are sediments deposited by tidal processes in both the intertidal zone and shallow, subtidal, tide-dominated environments less than 200 meters deep. A special subset is termed for convenience as *intertidalites* which are sediments and sedimentary rocks demonstrably deposited in the *intertidal zone*. It is mandatory to document sedimentary parameters indicating exposure, evaporation and late-stage emergence runoff prior to exposure to identify an intertidalite.

Many sedimentary features occurring in tidalites and intertidalites do occur in some sediments deposited in fluvial and other shallow marine environments by different processes. Klein (1970a, p. 982) and Swett, Klein and Smit (1971), however, have documented differences in sediment parameters that enable one to distinguish tidalites and intertidalites from fluvial, deltaic, eolian and beach sediments.

SUMMARY OF CLASTIC INTERTIDALITE PALEOTIDAL RANGE MODEL

Clastic intertidalite sedimentation is subdivided into a distinct zonation of sediment transport processes distributed parallel to shore from high tide level to low tide level (Van Straaten and Kuenen, 1958). These transport zones are (1) bedload tidal current transport and late-stage sheet-like runoff prior to exposure, (2) alternation of bedload tidal processes with suspension deposition and (3) suspension sedimentation (Fig. 1). Across the clastic tidal flat environment, a seaward-coarsening textural distribution occurs in response to such sediment transport zonation (Fig. 1). In prograding tidal flat environments along the North Sea Coast of the Netherlands and Germany (Van Straaten, 1954; Van Straaten and Kuenen, 1958; Reineck, 1963, 1967) and England (Evans, 1965, 1970), intertidalite sedimentation generates a graded, fining-upward sequence (Fig. 1). The base of the sequence consists of sands which are cross-stratified

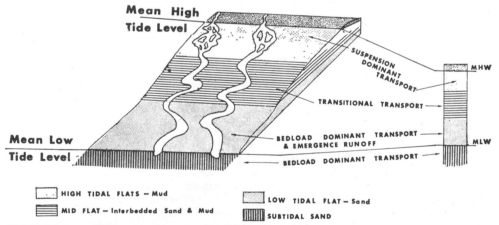

FIGURE 1 — Model of clastic intertidalite for North Sea Coast of the Netherlands (Van Straaten, 1954; Van Straaten and Kuenen, 1958) and Germany (Reineck, 1963, 1967) showing control of sediment distribution by sediment transport zones. Tidal flat progradation generates a graded, fining-upward sequence. Thickness of sediment interval representing low tidal flat, mid flat and high tidal flat coincides with mean tidal range. Abbreviations: MLW — Mean Low Water; MHW — Mean High Water. Coquille pattern indicates location of supratidal marsh. (Adapted from Klein, 1971a, Figure 1.)

TABLE 1 — Paleotidal Range Determinations in Uppermost Sterling Quartzite, Wood Canyon Formation and Harkless Formations, Eastern California and Western Nevada.

Formation and Locality	Thickness of Paleotidal Range Sequences (in meters)		
Uppermost Sterling Quartzite (Late Precambrian). Johnson Canyon, east side of Panamint Range, Inyo County, California. Common Corner of Secs. 19,30, T21S, R46E, and Secs. 25,29, T21S, R46E. (2 Measurements)	4.9. $\overline{X}$ = 8.95 meters	13.00.	
Lower Member, Wood Canyon Formation (Late Precambrian). Road Cut, U.S. Highway #95, Nye County, Nevada, SW¼, NE¼, Sec. 25 and NE¼, SE¼, Sec. 11, T16S, R52E. (8 Measurements).	2.35. 0.70. 1.47. C = 1.79 m.	1.72. 2.56. 2.20.	1.30. 2.01.
Middle Member, Wood Canyon Formation, (Late Precambrian). Tungsten Canyon, Bare Mountain, Nye County, Nevada, SW¼, NW¼, Sec. 18 (extended), T13S, R58E. (4 Measurements).	3.18. 1.88. $\overline{X}$ = 1.96 m	1.10.	1.68.
Striped Hills, Nye County, Nevada, SW¼, SW¼, Sec. 5, T15S, R50E. (18 Measurements)	1.78. 2.11. 0.49. 0.98. 2.72. 4.66. $\overline{X}$ = 2.29 m	1.15. 1.27. 0.89. 1.47. 4.57. 2.10.	1.28. 0.54. 3.26. 3.60. 3.51. 4.99.
South Nopah Range, Inyo County, California. N½, Sec. 10, T20N, R8E (6 Measurements).	2.62. 2.44. $\overline{X}$ = 2.04 m	2.69. 3.55.	1.13. 1.82.
Salt Spring Hills, San Bernardina County, California. N½, Secs. 17, 18, T.18N, R7E. (4 Measurements).	1.19. 1.93. $\overline{X}$ = 1.88 m	1.87. 2.92.	1.49.
Harkless Formation (Cambrian) Nevada Highway #3, Esmeraldo County, Nevada. SE¼, SW¼, Sec. 27, T3S, R.38E. (6 Measurements).	2.43. 1.12. $\overline{X}$ = 3.80 m	7.89. 3.13.	5.60. 2.63.
Lida Summit, Nevada Highway #3, Esmeraldo County, Nevada, SE¼, Sec. 7, T6S, R40E. (1 Measurement).	4.70.		

with bipolar-bimodal dip orientation produced by reversing tidal currents. Such sands may also extend into the shallow subtidal environment, but low tidal flat sands differ from the subtidal ones by containing sedimentary features indicating late-stage, sheet-like emergence runoff in water less than a half-meter immediately before tidal flat exposure (Klein, 1963a, 1970b, 1971a, his Table 1).

Low tidal flat sands grade upward into flaser-bedded sands and silts, lenticular-bedded sands and interbedded layers, 0.5 cm thick, of sand and clay comprising tidal bedding (Wunderlich, 1970, p. 106). Such mixed lithologies occur in the mid flat environment, which is dominated by alteration of bedload and suspension sedimentation. The mid flat sediments are overlain by high tidal flat clays deposited by suspension sedimentation (Reineck, 1967; Van Straaten and Kuenen, 1958). The top of the clay is overlain by supratidal marsh (Fig. 1).

The fining-upward sequence which is generated by prograding tidal coasts contains *within it* a preserved record of mean tidal range at the time of deposition and is here designated at *paleotidal range sequence*. The position of mean low water occurs within the basal sandstone member and coincides with the boundary of the sandstone containing features indicative of bedload transport only, and features indicating both bedload transport and late-stage emergence runoff prior to exposure (see Klein, 1971a, his Table 1). In the Lower Fine-grained Quartzite (Precambrian) of Scotland, Klein (1970a, p. 978) distinguished inter-

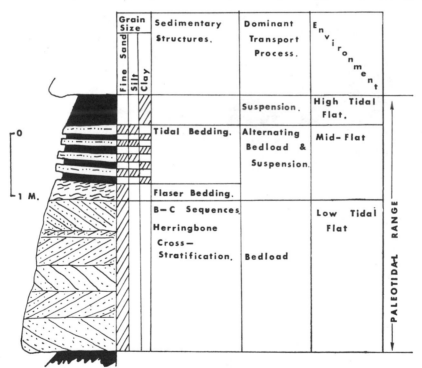

FIGURE 2 — Graphic log of intertidalite paleotidal range sequence in Middle Member, Wood Canyon Formation (Late Precambrian), Striped Hills, Nye County, Nevada (East Center, SW¼,SE¼, Sec. 15, T15N, R.50E). Log also shows interpretation of dominant sediment transport processes and depositional environments based on texture and sedimentary structures.

tidal from subtidal sandstones by application of criteria that segregated these two types of tidal sediment transport. The top of the paleotidal range sequence coincides with mean high water where the high tidal flat muds are in contact with overlying marsh deposits. In Holocene sediments, the thickness of the sediment interval between the two levels indicating mean high and mean low water coincide with Holocene mean tidal range (Fig. 1). This thickness, when preserved in fossil intertidalites, is equivalent to paleotidal range. An example of a sedimentological log of a Precambrian paleotidal range sequence is shown in Figure 2.

Application of the sedimentary model permits one to recognize fossil intertidalites and to determine paleotidal range by measuring the thicknesses of paleotidal range sequences as outlined above. Discrepancies from true range may be caused by basin subsidence (producing anomalously thick low tidal flat sands), compaction (producing thinner sequences) and erosion (producing recognizable truncated sequences). Compaction is a negligible factor, however, because observations by the present author of shear-strength of intertidal clays in the Minas Basin, Bay of Fundy, Nova Scotia average to 0.75 kg/cm² for surface clays and 1.71 kg/cm² for clays occurring 5 cm below the surface (Fig. 3). Therefore, compaction of intertidal clays is an extremely early diagenetic phenomenon which must precede progradation. Eroded sequences can be documented in the field by identifying missing elements of the fining-upward model.

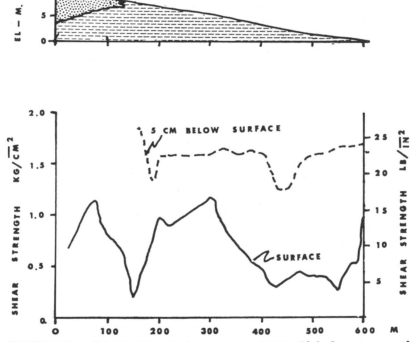

FIGURE 3 — Distribution of shear stress of intertidal clays measured on surface (solid line) and 5 cm below surface (dashed line) across a 600-m-wide intertidal mudflat at Five Islands, Minas Basin, Bay of Fundy, Nova Scotia, Canada. Stipled pattern represents reworked Pleistocene fluvial-glacial sandy gravels.

In Holocene tidal coasts, lateral variation in tidal ranges are controlled by coastline morphology, among other factors. On the North Sea Coast of the northern Netherlands and northwestern Germany, tidal ranges change laterally from 1.34 to 3.63 (U.S. Dept. of Commerce, 1970, p. 179-180). However, in the Middle Member of the Wood Canyon Formation (Late Precambrian) of Nevada and California, a minimal amount of lateral variation in mean thickness of paleotidal range sequences was found (Fig. 4). Over an outcrop belt 100 km long and 25 km wide, mean paleotidal ranges varied by no more than 41 cm (Fig. 4). Perhaps the factor of geographic variation in fossil intertidalites is less significant than predicted from Holocene studies. Therefore, the determination of paleotidal range through geological time appears now to be possible.

APPLICATION OF THE MODEL

The Late Precambrian uppermost Sterling Quartzite, and Lower and Middle members of the Wood Canyon Formation, and the Cambrian Harkless Formation of eastern California and western Nevada are characterized by intertidalites dominated with fining-upward sequences produced by tidal coastline progradation. Within these units, paleotidal range sequences are preserved. Figure 2 shows a typical paleotidal range sequence that comprises the dominant sedimentary motif in the Wood Canyon and Harkless Formations.

Thicknesses of Paleotidal range sequences in these three formations are summarized in Table 1 and are interpreted to represent paleotidal range for the late Precambrian and early Cambrian.

Paleotidal range sequences similar to the ones reported above also occur in other formations. Smith (1967, 1968) described identical sequences from the Silurian Clinton Formation of eastern Pennsylvania, whereas Johnson (1968) reported them from the Devonian Tully clastic equivalents of New York. Walker and Harms (1971) determined paleotidal ranges from the Devonian Catskill Formation of Pennsylvania by measuring the thicknesses of sedimentary intervals between rootlet zones and the nearest underlying bed containing nearshore marine fossils. The Middle Jurassic Great Oolite Series (Klein, 1963b, 1965) also contains fining-upward sequences developed by prograding tidal flats.

Paleotidal ranges (Table 1) measured from the late Precambrian and Cambrian are compared (Fig. 5) to the thicknesses of paleotidal range sequences from the intertidalites mentioned above and to Holocene tidal range variation 0 - 17.0). Although the data collected so far (Fig. 5) represent an initial investigation in time-stratigraphic variation of paleotidal range, they do indicate that paleotidal range variation is less than known Holocene variation. Furthermore, when examined time-stratigraphically, it would appear that there has not been any unusual changes in paleotidal range variation since late Precambrian time.

SOME IMPLICATIONS

The lack of appreciable variation in paleotidal ranges from Late Precambrian and younger clastic sediments raises important implications for early earth-moon history, models concerning the rotation of the earth, and supposed consequences for paleotidal sedimentation. A detailed statement of the problem is beyond the scope of this paper, but two points are of interest here.

(1) At least two models are available concerning the age of the earth-moon system. One calls for formation of the earth-moon system some time prior to 3.2 eons ago (e.g. Wise, 1969; Singer, 1970), whereas another calls for formation of

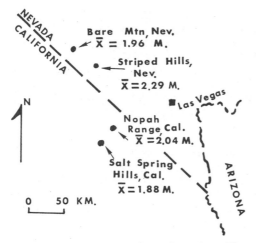

FIGURE 4 — Map of southwestern Nevada, eastern California and western Arizona, USA, showing mean paleotidal ranges measured in Middle Member, Wood Canyon Formation (Late Precambrian), at four localities (data in Table 1).

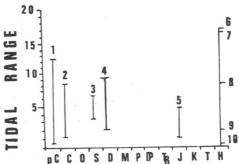

FIGURE 5 — Paleotidal ranges in meters for (1) Late Precambrian (Sterling Quartzite; Wood Canyon Formation); (2) Cambrian (Harkless Formation); (3) Silurian (Clinton Formation; Smith, 1967, 1968); (4) Devonian (Catskill Formation); Tully clastic equivalents; Walker and Harms, 1971; Johnson (1968); (5) Jurassic (Great Oolite Series; Klein, 1965). Holocene tidal ranges from U.S. Dept. of Commerce (1970) show variation from selected areas as follows: (6) Minas Basin, Bay of Fundy; (7) Severn Estuary, England; (8) The Wash, England; (9) Charleston, South Carolina; and (10) Gardner Island, Breton Sound, Louisiana.

the earth-moon system during Late Precambrian time (McDonald, 1964; Olson, 1966, 1970; Merifield and Lamar, 1968, 1970; Lamar and others, 1970). Advocates for a Late Precambrian age argued that the earth's surface would experience exaggerated tidal processes during early earth-moon history (Late Precambrian - Early Paleozoic) which would result both in unusually strong bottom current velocities (up to ten times present-day velocities) and unusually high tidal ranges in excess of six times present-day ranges (Olson, 1966, 1970; Merifield and Lamar, 1968, 1970, p. 33; Lamar and Merifield, 1967; Lamar and others, 1970). These high velocities and ranges were allegedly a consequence of a speedier rotation of the earth and shorter earth-moon distances than today. The paleotidal range measurements reported herein do not support the arguments by proponents of a Late Precambrian age for the earth-moon system, particularly because the extreme high tidal ranges suggested are absent. My paleotidal range determinations are consistent with hypotheses requiring a much older age.

(2) The data reported herein imply a contradiction with the interpretation of analysis of supposed daily growth bands (some tidally controlled) of Paleozoic corals and molluscs by Wells (1963, 1970), Scrutton (1964, 1970), and Pannella and McClintock (1968), whose work indicated that the number of days per year during the early and middle Paleozoic were far greater than the present 365 days per year. This increase in the number of days was in response to a speedier rotation of the earth (MacDonald, 1964) and, as stated above, requires supposedly more active tidal flow systems in shallow water environments during the late Precambrian and early Paleozoic.

At least two reasons probably account for the apparent contradiction between paleotidal range data and the coral and molluscan daily growth band counts. First, the daily growth band model is subject to several variations (Wells, 1970), including nutrient supply (Scrutton, 1970) and temperature variation (cf. Williams and Naylor, 1969) rather than tidal rythms *per se,* as maintained by Pannella and McClintock (1968). The daily and tidal growth band models require further calibration and consideration of problems such as animal shock. Second, it is noteworthy that in areas of extremely high tidal range such as the Bay of Fundy, bottom current velocities are of the same order of magnitude (Klein, 1970b, Tables 2, 9) as velocities in areas of moderate tidal ranges such as the tidal coast of the Netherlands (Postma, 1967). Although there may have been increased tidal intensity along late Precambrian and early Paleozoic coasts in response to a speedier rotation of the earth, the *rate of increase* of bottom current velocities and tidal ranges may not have responded so directly as previously suggested. A slight increase in tidal ranges and bottom velocities may have existed and enhanced the preservation potential of Late Precambrian and Early Paleozoic intertidalites. If the current velocities were as high as maintained by Olson (1966, 1970) and Merifield and Lamar (1968, 1970, p. 33), then using Klein's (1970b, Table 9) bottom current velocity and Froude Number data, one could predict supercritical flow for Late Precambrian and early Paleozoic bottom tidal currents. Such flow would of course generate upper flow regime sedimentary structures. However, Late Precambrian and early Paleozoic clastic tidalites and intertidalites are dominated by *lower flow regime* sedimentary structures (Klein, 1970a, 1971b; Swett, Klein and Smit, 1971) indicating that the current velocities could not be as high as maintained by Olson and Merifield and Lamar. The paleotidal range measurements reported herein are perhaps more consistent with a model calling for a slight increase (perhaps as much as 50 per cent) in tidal ranges and tidal current velocities during the Late Precambrian and early Paleozoic in response to a speedier rotation of the earth.

ACKNOWLEDGMENTS

Field research in eastern California and western Nevada was provided by funds from the University of Illinois (Urbana), the University of California, Berkeley (while serving as a Visiting Faculty Member) and the American Philosophical Society. Shear strength measurements from Nova Scotia were obtained under research support from the National Science Foundation (Grant GA-21141). I thank C. W. Collinson, C. J. Mann and M. W. Reynolds for helpful comments on an earlier manuscript version and R. H. Dott, Jr. and R. N. Ginsburg for discussions about the application of the model and its implications. The writer remains responsible for the ideas reported herein.

REFERENCES

Evans, G., 1965. Intertidal flat sediments and their environments of deposition in the Wash. Geol. Soc. Lond. Q. J., 121, p. 209-245.
————, 1970. Coastal and nearshore sedimentation: a comparison of clastic and carbonate deposition. Geol. Assoc. Proc. 81, p. 493-508.
Johnson, K. G., 1968. The Tully clastic correlatives (Upper Devonian) of New York State: model for recognition of alluvial. dune (?), tidal, nearshore (bar and lagoon) and offshore sedimentary environments in a tectonic delta complex. Unpub. PhD dissert., Rensselaer Polytechnic Inst., 122 p.
Klein, G. de V., 1963a. Bay of Fundy intertidal zone sediments. J. Sed. Petrol., 33, p. 844-854.
————, 1963b. Intertidal zone channel deposits in Middle Jurassic Great Oolite Series, southern England. Nature, 197, p. 1060-1062.
————, 1965. Dynamic significance of primary structures in Middle Jurassic Great Oolite Series, southern England. In Middleton, G. V. (*Editor*), Primary sedimentary structures and their hydrodynamic interpretation — a symposium. Soc. Econ. Mineral. Paleontol. Spec. Pub. No. 12, p. 173-191.

————, 1970a, Tidal origin of a Precambrian quartzite — the lower Fine-grained Quartzite (Dalradian) of Islay, Scotland. J. Sed. Petrol., 40, p. 973-985.

————, 1970b, Depositional and dispersal dynamics of intertidal sand bars. Jour. Sed. Petrol., 40, p. 1095-1127.

————,1971a, A sedimentary model for determining paleotidal range. Geol. Soc. Am. Bull., 82, p. 2585-2592.

————, 1971b, Environmental model for some sedimentary quartzites (Abs). Am. Assoc. Pet. Geol. Bull., 55, p. 347.

Lamar, D. L., and Merifield, P. M., 1967. Cambrian fossils and origin of the earth-moon system. Geol. Soc. Am. Bull., 78, p. 1359-1368.

Lamar, D. L., McGann-Lamar, J. V., and Merifield, P. M., 1970. Age and origin of the earth-moon system. *In* Runcorn, S. K. *(Editor)*, Paleogeophysics. Academic Press, New York, p. 41-52.

MacDonald, G. J. F., 1964. Tidal friction. Rev. of Geophysics., 2, p. 467-541.

Merifield, P. M., and Lamar, D. L., 1968. Sand waves and early earth-moon history. J. Geophy. Res., 73, p. 4767-4774.

————, 1970. Paleotides and the geologic record. *In* Runcorn, S. K. *(Editor)*, Paleogeophysics. Academic Press, New York, p. 31-40.

Olson, W. S., 1966. Origin of the Cambrian-Precambrian unconformity. Am. Sci., 54, p. 458-464.

————, 1970. Tidal amplitudes in geological history. N.Y. Acad Sci. Trans. Ser. 2, 32, p. 220-233.

Pannella, G., and MacClintock, C., 1968. Biological and environmental rhythms reflected in molluscan shell grow. *In* Macurda, B. D. Jr. *(Editor)*, Paleobiological aspects of growth and development — a symposium. Paleontol. Soc. Mem., 2, p. 64-80.

Postma, H., 1967. Sediment transport and sedimentation in the estuarine environment. *In* Lauff, G. H. *(Editor)*, Estuaries. Am. Assoc. Adv. Sci. Pub., 83, p. 158-179.

Reineck, H. E., 1963. Sedimentgefüge im Bereich der südlichen Nordsee. Abh. Senckenberg. Naturforsch. Ges., No. 505, p. 1-138.

————, 1967. Layered sediments of tidal flats, beaches and shelf bottoms. *In* Lauff, G. H. *(Editor)*, Estuaries. Am. Assoc. Adv. Sci. Pub., 83, p. 191-206.

Scrutton, C. T., 1964. Periodicity in Devonian coral growth. Paleontology. 7, p. 552-558.

————, 1970. Evidence for a monthly periodicity in the growth of some corals. *In* Runcorn, S. K. *(Editor)*, Paleogeophysics. Academic Press, New York, p. 11-16.

Singer, S. F., 1970. Origin of the moon by capture and its consequences. Am. Geophy. Union. Trans., 51, p. 637-641.

Smith, N. D., 1967. A stratigraphic and sedimentological analysis of some lower and middle Silurian clastic rocks of the north-central Appalachians. Unpubl. PhD Dissert., Brown Univ., 195 p.

————, 1968. Cyclic sedimentation in a Silurian intertidal sequence in eastern Pennsylvania. J. Sed. Petrol., 38, p. 1301-1304.

Swett, K., Klein, G. de V., and Smit, D. E., 1971. A Cambrian tidal sand body, the Eriboll Sandstone of northwest Scotland — An ancient-Recent analog. J. Geol., 79, p. 400-415.

U.S. Dept. of Commerce, 1970. Tide Tables, Europe and West Africa. Washington. Env. Sci. Services Admin., 211 p.

Van Straaten, L. M. J. U., 1954. Sedimentology of Recent tidal flat deposits and the Psammites du Condroz (Devonian), Geol. en. Mijnb., 16, p. 25-47.

Van Straaten, L. M. J. U. and Kuenen, Ph.H., 1958. Tidal action as a cause of clay accumulation. J. Sed. Petrol., 28, p. 406-413.

Walker, R. G., and Harms, J. C., 1971. The Catskill "Delta" — a prograding muddy shoreline in central Pennsylvania. J. Geol., 79, p. 381-399.

Wells, J. W., 1963. Coral growth and geochronomentry. Nature., 197, p. 948-950.

————, 1970. Problems of annual and daily growth-rings in corals. *In* Runcorn, S. K. *(Editor)*, Paleogeophysics. New York, Academic Press, p. 3-10.

Williams, B. G. and Naylor, E., 1969. Synchronization of the locomotor tidal rhythm of *Carcinus*. Jour. Exp. Biol., 51, p. 715-725.

Wise, D. U., 1969. Origin of the moon from the earth: some new mechanisms and comparisons. J. Geophys. Res., 74, p. 6035-6045.

Wunderlich, F., 1970. Genesis and development of the "Nellenköpfchenschichten" (Lower Emsian, Rheinian Devonian) at locus typicus in comparison with modern coastal environments of the German Bay. J. Sed. Petrol., 40, p. 102-130.

26

Reprinted from pp. 395, 396–397, 399–409 of *Jour. Sed. Petrology* **41**(2):395–409
(1971)

DEPOSITIONAL MODEL—UPPER CRETACEOUS GALLUP BEACH SHORELINE, SHIP ROCK AREA, NORTHWESTERN NEW MEXICO[1]

CHARLES V. CAMPBELL
Esso Production Research Company, Houston, Texas

ABSTRACT

To provide an example of ancient deposits across a sand-beach shoreline for comparison with both Recent and other ancient deposits, a cross-sectional model was constructed from surface exposures across the Upper Cretaceous Gallup shoreline in the Ship Rock area of northwestern New Mexico. The model shows lower Gallup regressive-beach sandstone overlain unconformably by upper Gallup transgressive offshore-bar sandstones. Landward, the beach sandstone interfingers with coal-swamp deposits or is truncated and overlain by dune standstone. Seaward, the beach sandstone grades through shoreface sandstone into a transition that gives way to offshore siltstone and mudstone. Offshore siltstone and mudstone overlie and in part grade laterally into offshore-bar sandstone, which has no contiguous landward equivalents and transgresses all beach and related facies.

Within their particular stratigraphic frameworks, sedimentary structures, supported by lithologic and paleontologic data, identify both the beach and the offshore-bar sandstones as well as their related facies. Both the beaches and the offshore bars consist of imbricate patterns of parallel beds. Beds in the backshore beach consist of cross laminae that dip either in diverse directions or landward and that compose laminasets, which are swale fillings; cross laminae in the foreshore beach dip uniformly seaward at low angles. Beds of shoreface sandstone are either churned by burrowing organisms or composed of large-scale truncated wave-ripple laminae. In the transition between shoreface and offshore siltstone and mudstone, beds of mudstone interfinger with beds of sandstone and siltstone, which show small-scale truncated wave-ripple laminae and churned structure. Siltstone and mudstone beds alternate in offshore rocks, and the siltstones contain both current-ripple and truncated wave-ripple laminae, except where burrowing organisms have destroyed the laminae. Most beds of offshore-bar sandstone are churned, but some show high-angle cross laminae that dip parallel with the length of the bar and approximately parallel with the distant shoreline. Burrowing organisms have completely churned the muddy sandstone beds of the offshore-bar transition.

[*Editor's Note:* Material, including figure 1, has been omitted at this point.]

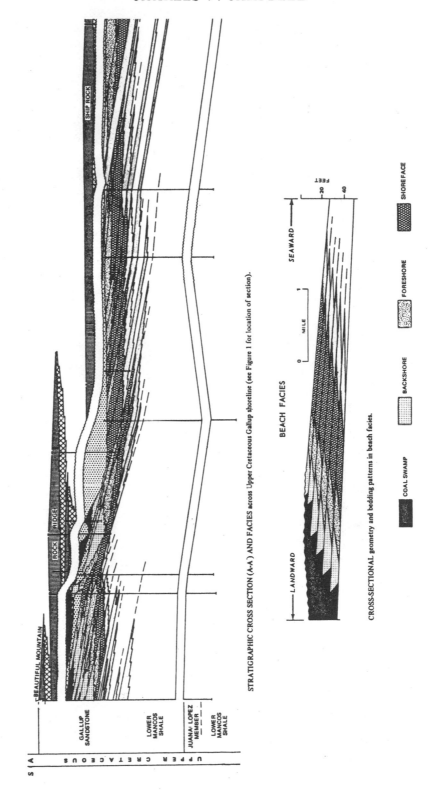

STRATIGRAPHIC CROSS SECTION (A-A) AND FACIES across Upper Cretaceous Gallup shoreline (see Figure 1 for location of section).

CROSS-SECTIONAL geometry and bedding patterns in beach facies.

323

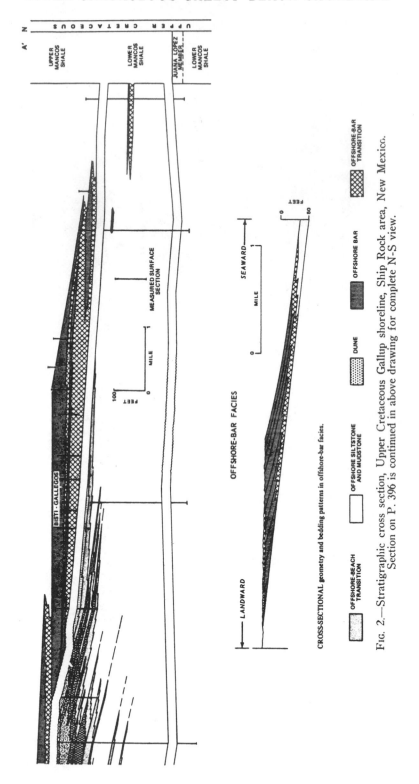

CROSS-SECTIONAL geometry and bedding patterns in offshore-bar facies.

Fig. 2.—Stratigraphic cross section, Upper Cretaceous Gallup shoreline, Ship Rock area, New Mexico. Section on P. 396 is continued in above drawing for complete N–S view.

[*Editor's Note:* Material has been omitted at this point.]

DESCRIPTION OF CROSS-SECTIONAL MODEL

The cross section of Upper Cretaceous rocks extends 21 miles across the Gallup shoreline and shows three distinct depositional episodes (fig. 2), which are, from base to top, (1) the Juana Lopez Member of the Lower Mancos Shale, (2) the lower Gallup regressive-beach complex, and (3) the upper Gallup offshore bars.

(1) The Juana Lopez is remarkably consistent in thickness (about 40 feet) for more than 100 miles north to south and more than 150 miles east to west. Throughout its extent, the Juana Lopez consists of alternating thin beds of siltstone, fossiliferous sandstone, and mudstone. [Dane (1960, p. 50–52) thinks that the top of this consistent unit is approximately synchronous, but Lamb (1968, p. 847) suggests that the top may be diachronous.] The siltstone beds internally show both current-ripple and truncated wave-ripple laminae; but their surfaces are often covered by interference types of wave-ripple patterns. *Inoceramus*, oyster fragments, shark's teeth, and shallow-water types of foraminifers are abundant. All of these features suggest slow deposition on a stable shelf; siltstone was spread over the depositional area by wave and unidirectional currents, while waves planed the bottom to a uniform depth. This slightly wavy bottom formed the base for accumulation of the Gallup regressive-beach complex.

(2) The Gallup beach complex consists of an imbricate or shingled system of individual beach sandstone bodies (fig. 2) which represent separate cycles or episodes of beach building. The overall relationship between bodies is regressive; each younger beach lies slightly seaward of the underlying one, although occasional cycles transgress over more landward facies. Each beach also rests upon the seaward-sloping face of the underlying one. This imbricate arrangement forms a complex of beach sandstone bodies more than 100 miles wide perpendicular to the average shoreline trend, and this complex extends at least 200 miles along depositional strike.

Each beach cycle consists of the following seaward succession of gradational facies (fig. 2): (1) coal swamp, (2) beach sandstone, including backshore and foreshore, (3) shoreface sandstone, and (4) offshorebeach transition, which grades seaward into offshore siltstone and mudstone. However, facies in the overlying beach are usually displaced seaward from the underlying facies. Consequently, the common vertical sequence through a beach complex corresponds to a succession of repeated, partially complete, lateral sequences of beach and associated facies.

Of these beach-cycle facies, the three most landward may be overlain unconformably by dune sandstone. Both the dune sandstone and all of the beach-cycle facies are transgressed and unconformably overlain by Gallup offshore bars and associated facies.

(3) Upper Gallup offshore bars are the basal sandstones of the Upper Mancos transgression (fig. 2). Successively younger bars transgress older ones and rest progressively upon older sediments of the beach complex and then upon the swamp and fluvial deposits. The facies composing each offshore bar is offshore-bar sandstone, and it grades seaward through offshore-bar transition into offshore siltstone and mudstone. Furthermore, the offshore-bar sandstone wedges out landward; and no landward equivalent facies is found contiguous to the bars. Either offshore siltstone and mudstone or another offshore bar and associated facies overlie offshore bars.

DESCRIPTION OF GALLUP FACIES

The groups of Gallup facies are (1) dune sandstone; (2) facies of the beach cycle—coal-swamp deposits (characterized by the presence of coal beds and not further described), backshore, foreshore, shoreface, and offshore-beach transition; and (3) offshore facies—offshore siltstone and mudstone and offshore-bar sandstone along with its transition. Each group of facies appears within a discrete geometric form that has a distinctive internal bedding pattern. Each facies is characterized by diagnostic sedimentary structures. Texture and composition may be similar in adjacent facies, and fossils may overlap facies boundaries.

Dune Sandstone

External Geometry. Gallup dune sandstone forms either pods or sheets (fig. 2). The pods rest upon beach or coal-swamp deposits or are enclosed by swamp deposits, and the sheets veneer the truncated landward ends of beaches.

Internal Bedding Pattern. Dune sandstone consists of a complex of trough-shaped beds, from 6 inches to 5 feet thick, which are bounded by curved, nonparallel bedding surfaces.

Sedimentary Structures. Cross laminae compose the trough-shaped beds of dune sandstone. (This structure is commonly termed "festoon crossbedding.") Laminae in adjacent beds form a crisscross pattern in exposures perpendicular to the wind direction (fig. 4-A), and laminae dip in one direction between nearly parallel bedding surfaces in exposures parallel with the wind direction (fig. 4-A). Occasional structures in Gallup dune sandstone include contorted laminae, root casts, and rare burrows, tracks, and trails. Furthermore, where dunes are transgressed by offshore bars, as much as 3 feet of sandstone at the top of the dune may be churned by burrowing organisms.

Texture, Composition, and Fossils. Gallup dune sandstone is well sorted but contains somewhat more "fines" than associated beach sandstone. The dune sandstone consists of quartz grains with minor feldspar. Dolomite and collophane grains are very rare. No fossils were recovered from samples of dune sandstone.

Facies of Beach Cycle

External Geometry. Gallup beach and associated shoreface sandstones of each beach-building cycle form elongated sandstone prisms that range from 3 to 20 feet in thickness and 1 to 3 miles in width (fig. 2). Their length parallel with the shoreline cannot be measured from the outcrops studied, but it may compare with that of some modern beaches which extend uninterrupted a distance of 100 miles along the shoreline. In addition, the seaward-sloping uppermost depositional surface of each cycle generally parallels its base, although examples of both seaward and landward convergence of these surfaces appear locally on Figure 2. Finally, a succession of these beach cycles, each younger one building on the seaward face of the underlying one, forms an extensive seaward-imbricated system of sandstone bodies that have the general geometric form of a sheet sandstone.

Internal Bedding Pattern. Beds composing a beach cycle are bounded by even, parallel bedding surfaces. These surfaces parallel the uppermost seaward-sloping surface of the particular beach episode, and they continue from the

beach through successive lateral facies (fig. 2). Occasionally, the bedding surfaces converge or diverge slightly seaward where the beach cycle as a whole, repsectively, thins or thickens. Beds range in thickness from 4 inches to 2 feet.

Each younger bed in the vertical sequence of a beach lies slightly seaward from the underlying one; the youngest bed extends farthest seaward; and the oldest, farthest landward (fig. 2). Because of this imbrication, beach sandstone bodies commonly are drawn with flat bases and convex-upward tops, even though these representations ignore the internal layered structure and facies changes in beach and laterally equivalent deposits.

Backshore Sandstone. The even, parallel beds of backshore sandstone (fig. 4-B) consist of a complex of laminasets, each of which is a partially preserved swale filling. The swales trend generally parallel with the shoreline. Laminae usually dip from both sides into the swales, but some swales are filled predominantly with cross lamine that dip landward. Locally, laminae in some beds parallel bedding surfaces. Less common sedimentary structures in backshore sandstone are ripple marks, ripple laminae, rib-and-furrow structure, air-heave structures, and locally profuse burrows.

Foreshore Sandstone. Cross laminae in the even, parallel beds of foreshore sandstone dip uniformly in a common direction (fig. 4-C). This direction is seaward, and the strike of the cross laminae is thus parallel with the shoreline. The dip of the laminae is usually less than 6 degrees. Additional uncommon sedimentary structures in foreshore sandstone include both wave- and current-ripple marks, both wave- and current-ripple laminae, parting lineation, swash and rill marks, and vertical burrows.

Shoreface Sandstone. Either one or the other of two intergradational structures compose the even, parallel beds of shoreface sandstone. First is churned structure, where burrowing organisms have churned the sandstone and destroyed all original laminae. If compositional differences of the sandstone are slight, this structure is described objectively as mottled or homogenous (structureless). Second is the structure termed truncated wave-ripple laminae (Campbell, 1966), which consists of truncated sets of wave-ripple laminae that form laminasets within beds (fig. 4-D). Crest-to-crest distances (wavelengths) of wave ripples of Gallup shoreface beds commonly range from 5 feet to 15 feet, but lesser wavelengths are found. Heights (amplitudes) of the wave ripples average 10 inches. Although crests of the individual wave ripples show diverging orientations, their overall trend is parallel with the shoreline. Other shoreface

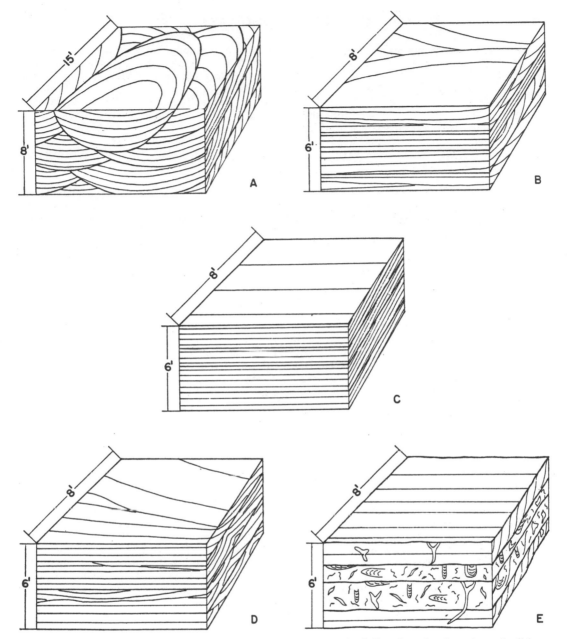

Fig. 4.—Diagrams illustrating bedding and lamination patterns in Gallup dune, beach cycle, and offshore-bar sandstones.

A. Dune structure. In exposures perpendicular to the wind direction, dune sandstone shows a crosscross pattern of laminae that compose the trough-shaped beds. In exposures parallel with the wind direction, dune sandstone consists of sets of cross laminae that dip in a single direction and are contained between nearly parallel bedding surfaces.

B. Backshore structure. Cross laminae dip in diverse directions between parallel bedding surfaces or are parallel with the bedding surfaces. Seaward is toward the front of the block.

C. Foreshore structure. Cross laminae dip uniformly seaward between parallel bedding surfaces. Seaward is toward the front of the block.

D. Shoreface structure. Large-scale truncated wave-ripple laminae form laminasets between parallel bedding surfaces. However, churning by burrowing organisms commonly destroys all evidence of original laminations in shoreface sandstone. Seaward is toward the front of the block.

E. Offshore-bar structure. Cross laminae, in offshore-bar sandstone dip uniformly in a single direction and at a high angle between wavy-parallel bedding surfaces. Cross laminae may be partly to completely destroyed by the churning of burrowing organisms.

structures are wave-ripple marks, burrows, and contorted laminae, commonly of the pattern termed convolute laminae.

Offshore-Beach Transition. The even, parallel beds of the transition between shoreface sandstone and offshore siltstone and mudstone show structures similar to shoreface sandstone. However, truncated wave ripple laminae appear only in sandstone and siltstone, and the wave-lengths of the ripples are almost always less than 1 foot. Fine siltstone and mudstone, as well as much of the coarse siltstone and sandstone, are churned by burrowing organisms. Also present are burrows, wave-ripple marks, and rare current-ripple and contorted laminae.

Texture and Composition. Combined sieve and pipette analyses of samples from Gallup beach cycles show a gradual seaward and downward decrease in grain size. With the decrease in grain size, mud and silt-size matrix becomes increasingly abundant, and the offshore-beach transition consists of alternating sandstone and siltstone with interfingering mudstone beds. Backshore and foreshore sandstones are slightly better sorted than shoreface sandstones, but statistical moment measures (sorting, skewness, and kurtosis) do not clearly separate these sandstone facies.

Sandstone facies within Gallup beach episodes vary little in composition. All consist predominantly of quartz. Feldspar, although minor, is the second most abundant mineral, except in some shoreface sandstones and offshore-beach transitions where rock fragments predominate over feldspar. Dolomite, glauconite, and collophane grains are most common in shoreface and offshore-beach transition samples. Carbonaceous particles exceed 10 percent of the total grains in some offshore-beach transitions and are sparsely present in all facies. Compositions of the siltstones and mudstones were not determined.

Fossils. Megafossils are rare in Gallup beach facies. *Inoceramus* prisms and fragments, along with oyster fragments, are present in all facies (except the coal-swamp facies). Plant-fragment imprints are especially abundant in some offshore-beach transitions, as well as in the coal-swamp deposits. However, trace-fossil associations combine with foraminiferal faunas to provide a continuous sequence of diagnostic fossils across the Gallup shoreline. From the beach seaward the sequence of trace fossils is (1) *Ophiomorpha* in both backshore and foreshore, (2) *Ophiomorpha* plus sigmoidal *Teichichnus* and *Hallimondia* in the shoreface, and (3) the shallow-marine association (listed in Table 1) in the offshore-beach transition and offshore environments. Continuing seaward through the offshore siltstone and mudstone, an arenaceous

benthonic foraminiferal assemblage, which appears first in the offshore-beach transition, gives way to calcareous benthos that grade into a predominantly planktonic assemblage (Table 1). Spores and pollen are usually most abundant along with the planktonic foraminifers.

Offshore Facies

External Geometry. Offshore siltstone and mudstone, which form sheet deposits, overlie or enclose the elongated lenses of offshore bars and their transitions. An average offshore-bar lens is less than 20 feet thick by 2 miles wide by 4 to 40 miles long. Where two or more lenses are superposed in regressive order, widths and thicknesses of the total sandstone body are appropriate multiples of the above figures. In addition, offshore bars in cross section show (1) a flat landward top that diverges seaward from the plane base and (2) a sloping seaward top that converges with the base (fig. 2).

Internal Bedding Pattern. Offshore siltstone and mudstone consist of even and wavy, parallel (and sometimes nonparallel) beds that range in thickness from a fraction of an inch to 4 inches. These beds are parallel with the top of the deposit, form a clinoform pattern, or diverge and then converge to enclose offshore bars.

Beds composing offshore bars are bounded by wavy, parallel bedding surfaces. These surfaces parallel the sloping seaward top of the bar (fig. 2); they terminate at the flat landward top; but seaward, they extend across the facies change into the offshore-bar transition, although the identity of specific surfaces usually is lost in the area of facies change. Additionally, the beds form a seaward-imbricate pattern as the oldest bed terminates farthest landward and the youngest extends farthest seaward. Also, the beds of offshore-bar sandstone are imbricated parallel with the length of the bar, but the angle of initial dip in this direction is less than it is in the seaward direction. Beds in both the offshore-bar sandstone and its seaward transition range in thickness from 2 inches to 2 feet.

Offshore Siltstone and Mudstone. Siltstone beds in the offshore siltstone and mudstone show truncated wave-ripple or current-ripple laminae, along with wave-ripple-marked surfaces. However, both the siltstone and mudstone beds may consist of even, parallel laminae that are concordant with bedding surfaces, and both may be reworked by burrowing organisms that produce churned and mottled structures. Less important sedimentary structures in offshore siltstone and mudstone include burrows, trails, and minute scour-and-fill structures.

Offshore-Bar Sandstone. Wavy, parallel beds of offshore-bar sandstone are commonly

TABLE 1.—*Sequence of diagnostic trace fossils[1] and foraminifers across the gallup shoreline*

Beach		Shoreface	Offshore-Beach Transition and Offshore Siltstone and Mudstone		
Backshore	Foreshore				
Trace Fossils Associations[1]	Ophimorpha	Ophiomorpha Sigmoidal Teichichnus Hallimondia	Shallow-marine association— Ophiomorpha Sigmoidal Teichichnus Hallimondia Gyrochorte Arthropod tracks	Curvolithus Rhizocorallium Asterosoma Rosselia	Chondrites Phycosiphon Scolicia Taenidium Skolithos
Foraminiferal Assemblage	None	None	Arenaceous[2] Benthonic	Calcareous[2] Benthonic	Planktonic[2]
			Ammobaculites Haplophragmoides Spiroplectammina Reophax Trochammina Anomalina	Dentalina Marginulina Lenticulina Anomalina Gaudryina Frondicularia Heterohelix Globigerina Epistoma Trochammina Haplophragmoides Spiroplectammina Globotruncana	Heterohelix Anomalina Globotruncana Neobulimina Globigerina Dentalina Lenticulina Gyroidina Epistoma Frondicularia Marginulina Gavelinella Cibicides

[1] Trace fossils are described by Walter Hantzschel *in* Moore, R. C., ed., 1962, Treatise on Invertebrate Paleontology. Part W, Miscellanea, New York, Geol. Soc. America, 259 p.
[2] Foraminifers are listed in order of abundance in each assemblage.

churned by burrowing organisms (fig. 4-E), but some show preserved, planar cross laminae. The cross laminae usually dip between 20 and 25 degrees, are convex downward, and become tangential to the base of the bed. Cross laminae in superposed beds dip uniformly in one direction (fig. 5), which is approximately parallel with the length of the offshore bar (fig. 6); their strike is thus nearly perpendicular to the length of the bar. Lamb (1966, p. 851) and McCubbin (1969, fig. 12) also note this pattern of cross laminae. Additional sedimentary structures in offshore-bar sandstone are burrows and trails, wave-ripple marks, truncated wave-ripple laminae, and contorted laminae.

Offshore-Bar Transition. Wavy, parallel beds of offshore-bar transition are characteristically churned by burrowing organisms. Distinct burrows or trails are uncommon. Likewise, original laminae are rarely preserved, although high-angle cross laminae (as in offshore-bar sandstone) or truncated wave-ripple laminae may be present.

Texture and Composition. The offshore siltstone and mudstone facies is an alternation or any gradation between the two lithologies implied by the facies name. Usually mudstone predominates, but siltstone may be more abundant where this facies approaches the transition into beaches or offshore bars. Thin lenses of sand-

stone, sandy skeletal limestone, or skeletal limestone may be present; and carbonaceous grains are locally abundant.

The offshore-bar sandstone is coarser and less well sorted than any of the beach or beach-associated facies. Sorting also decreases into the offshore-bar transition that typically contains 15 or more percent mudstone matrix, even though the sand grains in both the bar and the transition are identical in size. Both of these facies may contain small pebbles in amounts up to 7 percent.

Both the offshore-bar and the transition sandstone consist predominantly of quartz grains, with feldspar grains (although minor) the second most abundant mineral. Disseminated glauconite is especially abundant in the offshore bar, and dolomite and collophane are more abundant in the transition than in the bar. Rock fragments are also present in both facies. Small pebbles are most commonly chert but may be rock fragments. Carbonaceous grains are locally abundant in the transition and are present in the bar sandstone.

Fossils. Megafossils are common in the offshore environment. Molds of ammonites and thin-shelled mollusks appear in the offshore siltstone and mudstone. *Inoceramus* fragments are locally abundant in the offshore-bar sandstone and occur, along with prisms, in the offshore-

Fig. 5.—Aerial view of traces of cross laminae. The overall uniform strike of cross laminae in offshore-bar sandstone persists for over 2 miles (the vertical distance across this photograph). Curving traces are due largely to the topography of the four or more bedding surfaces exposed. Cross laminae dip to the top right of the picture, and subsurface studies show that this dip direction nearly parallels the elongation of the bar.

bar transition and offshore siltstone and mud-stone. Plant-fragment imprints are locally abundant in silty and muddy lithologies and are present in the offshore-bar sandstone.

Foraminiferal assemblages in the offshore siltstone and mudstone change seaward from predominantly arenaceous benthos to calcareous benthos and then to planktonic forms (Table 1). The calcareous benthonic assemblage of fora-minifers characterizes offshore bars and their transitions. The shallow-marine association of trace fossils (Table 1) occurs throughout the offshore siltstone and mudstone as well as in the offshore bars.

COMPARISON OF GALLUP AND RECENT
NEARSHORE FACIES

Both ancient and Recent depositional models must be considered together to satisfactorily understand the specific conditions for formation of particular types of ancient rocks. However, problems arise in making such comparisons. The ancient rocks clearly show bedding patterns and large-scale sedimentary structures that can be observed only in extensive trenches pumped free of water in Recent sediments. Only recently has the approach of trenching Recent sediments been used, and then many of the trenches are too limited in extent to provide good comparisons with ancient rock exposures. [Exceptions are the studies of eolian dunes by Bigarella and others (1969) and McKee (1966).] Consequently, most data about Recent sediments derive from cores—texture, composition, small-scale sedimentary structures, and fossil content—and these are not the most diagnostic characteristics for distinguishing Gallup facies. On the other hand, areas of present-day sedimentation provide the sites for study of depositional processes.

Stratigraphic and Facies Relationships

Curray and Moore (1964) describe a succession of Holocene regressive beaches of the Costa de Nayarit, Mexico, that are an approximate modern counterpart of the lower Gallup beach complex. These author's diagrammatic cross section (their fig. 3, p. 79) is based upon observations of surface morphology, cores, and geophysical records. Except for their basal transgressive sand, their stratigraphic framework is essentially the same as for the Gallup beach complex. However, their data do not allow the refinement of facies possible in Gallup outcrops; and they show only the gross lithologic facies: (1) alluvium, which corresponds to the Gallup coal-swamp facies; (2) littoral sands, which equate with the Gallup beach and shoreface; and (3) shelf muds, which are

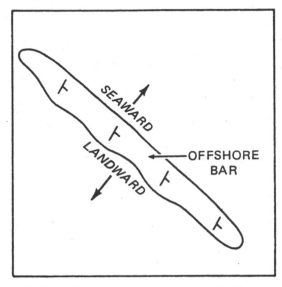

FIG. 6.—Dip of cross laminae compared with elongation of offshore bar. Cross laminae dip parallel with the elongation of the bar, and the bar is elongated parallel with a distant shoreline.

equivalent to the Gallup offshore siltstone and mudstone.

The similarities in stratigraphic framework and facies relationships between the Costa de Nayarit Holocene beaches and the Gallup beaches suggest generally similar origins. In both examples, a number of small rivers seem to have supplied the sand to the shoreline. This sand was worked by waves and longshore currents and formed a strand plain consisting of a seaward-building succession of younger and younger beaches. As the beaches built seaward, alluvial deposits encroached upon the landward margins of the strand plain. In addition, eolian dunes spread over the Gallup strand plain; but dunes are not conspicuous on the Costa de Nayarit.

Dune Sandstone

Of the many illustrations of the interal structures of coastal eolian dunes, most comparable to the Gallup dunes are the dunes diagramed by Bigarella and others (1969) from the coast of Brazil. In faces perpendicular to the wind direction, these authors (for example, their fig. 9, profile IV) show beds bounded by curved, non-parallel bedding surfaces. The cross laminae composing adjacent beds form a crisscross pattern, and the cross laminae dip predominantly northward or at nearly right angles to the illustrated profile. Furthermore, in faces cut parallel with the wind direction, the cross laminae dip parallel with the wind direction between nearly parallel bedding surfaces (see Bigarella and others, fig. 9, profile I). All of these characteristics suggest cross-laminated, trough-shaped beds or the festoon crossbedding of many workers, even though Bigarella and others (1969, p. 19) identify the bed shapes as dominantly tabular planar.

In addition, on many occasions the writer and his colleagues have exposed the cross-laminated, trough-shaped beds characteristic of eolian dunes on Mustang and Padre Islands, Texas. Both Milling and Behrens (1966, fig. 8) and McBride and Hayes (1962, fig. 2) illustrate aspects of the cross-laminated, trough-shaped beds composing eolian dunes on Mustang Island.

Many of the structures observed in Recent eolian dunes are uncommonly preserved in the Gallup dunes, where only the foreset of slipface beds and cross laminae are commonly found. The uncommonly preserved structures include (1) windward dipping beds and laminae (Bigarella and others, 1969, fig. 8; Land, 1964, fig. 2; McBride and Hayes, 1962, fig. 3; McKee and Tibbitts, 1964, fig. 6; and McKee, 1957, Plate 3A-B); (2) various types of slump and contorted structures (Bigarella and others, 1969, figs. 16–18 and 21–28; McKee, 1966, Plates 4B and 5E; and McKee, 1957, fig. 4A); (3) ripple marks shown on most closeup photographs of dune surfaces (for example, McKee, 1966, Plate 2D); and (4) root casts, burrows of rodents and insects, and tracks and trails of animals (common in the writer's experience but seldom illustrated by other workers).

The mechanism of dune formation and preservation explains why only slipface beds and laminae are preserved in ancient dune deposits. As a dune migrates downwind, sand from the windward side is eroded and deposited on the slipface of the dune. Thus a dune that has migrated beyond its initial position will consist almost entirely of slipface deposits in its lower part. Moreover, only the lower part of the dune will usually be preserved, for erosion, both on the windward side of a dune and of a coalescing field of dunes, will be controlled by the height to which the water table rises into the dune sandstone, as suggested by Stokes (1968).

Beach and Shoreface Sandstone

Several workers have illustrated present-day backshore and foreshore beach structures like those in the Gallup beaches. Edges of complexes of laminae filling swales in backshore trenches are illustrated by Milling and Behrens (1966, fig. 5), Soliman (1964, figs. 4 and 5), and McKee (1957, figs. 4 and 8). McKee (1957, figs. 12 and 13) also shows cross sections of a swale fill-

ing. Low-angle cross laminae between parallel bedding surfaces in foreshore trenches were photographed by Andrews and van der Lingen (1969, figs. 2 and 3). Straaten (1959, fig. 19) shows foreshore cross laminae in a core. Between the foreshore and the backshore at the berm crest, both foreshore and backshore layering patterns appear (Andrews and van der Lingen, 1969, fig. 8).

Recent shoreface sedimentary structures are known principally from cores. These cores show the churned structure (for example, Bernard and others, 1962, fig. 10) that characterizes one type of Gallup shoreface, but the cores are too small to show the large-scale truncated wave-ripple laminae found in the other type of Gallup shoreface. However, the writer has excavated this wave-ripple structure in one longshore bar of a low-tide terrace and suspects that this structure may commonly form in the surf zone bordering beaches.

The process of formation of beach and shoreface structures, except for the continuous parallel bedding surfaces through these facies, can be observed on many beaches. Backshore laminasets form in swales on the backshore as storm surges wash sand and water over the berm crest, and the water drains from the swales. The foreshore laminae are deposited by the backrush of the swash from breaking waves. Shoreface large-scale truncated wave-ripple structures presumably form where waves break in their advance across the surf zone, but additional studies of sedimentary structures in the surf zone are needed.

The mechanics of formation of the parallel bedding surfaces that continue through the beach and shoreface have not been observed. Presumably, such bedding surfaces, which dip seaward at a lesser angle than the foreshore cross laminae, form during storms when maximum wave energy erodes the beach to some critical depth. This maximum wave energy would tend to have a constant value because fetch, a critical factor in wave development, is controlled by the constant size and configuration of the body of water bordered by a specified beach. A further control is that the body of water must be subjected to tidal flucation, for the bedding surfaces through beaches bordering nontidal bodies of water are not continuous (Panin, 1967, figs. 5 and 6; Logvinenko and Remizov, 1964, figs. 5–7). Consequently, if the sand supplied to the shores of a tidal sea exceeds the amount removed during storms, a succession of parallel bedding surfaces continuous through the beach and shoreface could be preserved.

Offshore Facies

Gallup offshore facies include (1) siltstone and mudstone that is laterally equivalent to beach and shoreface sandstones, and (2) offshore bars and their transitions that are enclosed or overlain by the offshore siltstone and mudstone.

Offshore siltstone and mudstone—Textures and sedimentary structures like those in the Gallup offshore siltstone and mudstone are illustrated from Recent sediment cores by Curray (1960, fig. 15) as well as by many other workers. Foraminiferal faunas are especially useful in identifying this facies because silts and muds showing similar structures and textures sometimes occur in other environments such as lagoons and bays. Also the gradual change in composition of the foraminiferal population may suggest the approximate distance from the shoreline and the water depth at the time of deposition.

Offshore bars—As no Recent counterparts of the upper Gallup sandstones are known to the writer, interpretation of their origin is an open question. The writer interprets these sandstone bodies to be some type of offshore bars, and this interpretation differs from the strike-valley sandstone interpretation preferred by McCubbin (1965, 1969) and Lamb (1968, p. 827). The concept of strike-valley sandstone follows Busch (1959, p. 2832), who states with such sandstones "... are deposited in low areas between cuestas at the time the land surface is inundated by a transgressive sea." However, the writer's stratigraphic data show only one side to the "valleys"; the "valleys" seem to be seaward-facing terrace edges. As the sandstone bodies described here are generally higher stratigraphically than those described by McCubbin, two origins might be postulated. But both the facies relationships and the internal structures of the upper Gallup sandstones do not seem compatible with a valley-fill origin, unless the sandstone bodies accumulated along one edge of a "valley" many miles wide. Consequently, the writer suggests that the term offshore-terrace bar, rather than strike-valley sandstone, best describes these sandstone bodies.

Irrespective of the name applied to the upper Gallup sandstone bodies, their attributes as previously described suggest that the following combination of conditions prevailed during their formation:

1. Offshore bars form on a shallow shelf that may be as wide as 100 miles perpendicular to the shoreline. The one offshore bar that can be related to contemporaneous shoreline (beach) deposits (fig. 2) lies 12 miles offshore, but other offshore bars lie at several multiples of this distance from shore. In addition calcareous benthonic foraminifers, *Inoceramus,* and other related forms of animal life that occur in offshore bars suggest a shallow-shelf depositional site.

2. A temporary stillstand of sea level is indicated by the absence of landward contemporaneous deposits contiguous to each offshore bar. (If subsidence coincident with deposition occurred, contemporaneous deposits adjacent to the offshore bars should be present. If uplift of as much as 20 feet took place, the bar would have been destroyed.)

3. The source of sand was, presumably, a delta or other shoreline deposit northwest of the depositional area of the offshore bars. This source could have been as far as 200 miles from the depositional site, for the persistent southeastward dip of cross laminae in all of the offshore bars suggests that bars southeastward into the San Juan Basin consist of sand that was swept across the outcrop area studied in detail (fig. 1). Furthermore, the supply of sand must have been limited because the sand does not form a continuous deposit across the shelf. Finally, the grain size of rocks both underlying the bars and forming the closest shoreline deposits is so different from that in the bars as to preclude these materials as possible sources.

4. Bottom currents flowed nearly parallel with both the distant shoreline and the long dimension of the bar. These currents, indicated by the persistent dip direction of cross laminae in the bars, periodically attained sufficient velocity to transport small pebbles, and maintained a velocity great enough to move coarse sand. But the origin of these currents is unkown; to the writer's knowledge, no strong currents have been described that flow approximately parallel with the shoreline on modern shelves.

5. Deposition of offshore bars was localized at a seaward steepening or break in slope of the sea floor (a terrace edge). The changed slope is probably less than 1 degree steeper than the nearly horizontal, landward-extending sea bottom. Further, the trend of the break in slope is nearly parallel with the shoreline and slightly divergent from the bottom current direction.

6. The water depth at the break in slope was at some critical depth. First, stratigraphic reconstruction (fig. 2) suggests that the critical depth of water approximated 200

feet, which is also within the range suggested by the foraminifers. Second, absence of adjacent landward equivalents to the bars, as well as their external geometry and internal bedding pattern, indicates such a critical depth.

Sediment-transporting currents, sweeping across the break in slope at a slight angle to the trend of the break, deposited the offshore-bar sand. The sand fell out of the current immediately seaward of the break and thus started to form a bed. As deposition continued, each bed advanced in a downcurrent direction, along a front that commonly reached 2 miles in length. This front trended approximately perpendicular to the shoreline and to the ultimate long dimension of the growing bar.

In addition, each bed advanced downcurrent as sand spilled over its front and formed cross laminae upon cross laminae. Thus, sand comprising successive cross laminae was swept across previously deposited laminae; this process is a self-truncating one that would maintain a nearly plane scoured surface. Furthermore, as the seaward edge of the bed extended into deeper and deeper water, more and more fine sediment was deposited with the sand. Hence, the bed then grades seaward into offshore-bar transition, which in turn grades seaward into offshore siltstone and mudstone.

Finally, a temporary interruption in the supply of sediment or some other change terminated the formation of each bed, and a period of quiescence commonly followed this change. Then burrowing organisms destroyed most of the original laminae before the process of bed formation again started. Accordingly, each successive bed begins somewhat seaward from the underlying one and extends somewhat farther seaward.

CONCLUSIONS

The cross-sectional depositional model presented here shows the stratigraphic framework of beach and related facies across an ancient shoreline. Within this stratigraphic framework, facies are readily distinguished by bedding patterns and sedimentary structures within beds; but texture, composition, fauna, and flora provide supporting data for a complete environmental analysis.

Both ancient and Recent depositional models should be considered together to satisfactorily understand the specific conditions for formation of particular types of ancient rocks. The ancient rocks clearly show bedding patterns and large-scale sedimentary structures that can be observed only in extensive trenches pumped free of water in Recent sediments. But the interpretation of these large-scale features should then be in accordance with smaller-scale features common to both deposits as well as the knowledge of depositional processes gained from the Recent counterpart.

ACKNOWLEDGEMENTS

The writer expresses his appreciation to the many people who aided this study in various ways. R. N. Van Horn first guided the writer through the study area. D. P. Phillips and B. Biswas served ably as field assistants. Biswas initiated the microfossil study, and J. H. Beard completed identification of the microfossils. Adolf Seilacher visited a part of the area where he identified most of the trace fossils. J. D. Howard made the grain-size analyses. L. J. Eicher and W. E. Murrah accompanied the writer in the field on various occasions and critically discussed the observations and concepts developed in this study. Both J. C. Harms and R. E. Hunter critically reviewed the manuscript. Interpretations and conclusions presented, however, are solely the responsibility of the writer.

The writer thanks the Esso Production Research Company for permission to publish this paper.

REFERENCES

Andrews, P. B., and Van Der Lingen, G. J., 1969, Environmentally significant sedimentologic characteristics of beach sands: New Zealand Jour. Geology and Geophysics, v. 12, p. 119–137.
Bernard, H. A., Leblanc, R. J., and Major, C. F., 1962, Recent and Pleistocene geology of southeast Texas, p. 175–224 in Rainwater, E. H., and Zingula, R. P., eds., Geology of the Gulf Coast and Central Texas, Guidebook of Excursion: Houston Geol. Soc., 391 p.
Bigarella, J. J., Becker, R. D., and Duarte, G. M., 1969, Coastal dune structures from Parana (Brazil): Marine Geology, v. 7, p. 5–55.
Busch, D. A., 1959, Prospecting for stratigraphic traps: Am. Assoc. Petroleum Geologists Bull., v. 43, p. 2829–2843.
Campbell, C. V., 1966, Truncated wave-ripple laminae: Jour. Sed. Petrology, v. 36, p. 825–828.
———, 1967, Lamina, laminaset, bed, and bedset: Sedimentology, v. 8, p. 7–26.
Curray, J. R., 1960, Sediments and history of Holocene transgression, continental shelf, Northwest Gulf of Mexico, p. 221–266 in Shepard, F. P. Phleger, F. B., and van Andel, Tj. H., eds., Recent Sediments, Northwest Gulf of Mexico. Am. Assoc. Petroleum Geologists, 394 p.
———, and Moore, D. G., 1964, Holocene regressive littoral sand, Costa de Nayarit, Mexico, p. 76–82 in Straaten, L. M. J. U. van, ed., Deltaic and shallow marine deposits. Amsterdam, Elsevier Pub. Co, 457 p.

DANE, C. H., 1960, The boundary between rocks of Carlile and Niobrara age in San Juan Basin, New Mexico and Colorado: Am. Jour. Sci. (Bradley volume), v. 258-A, p. 46–56.

LAMB, G. M., 1968, Stratigraphy of the Lower Mancos Shale in the San Juan Basin: Geol. Soc. America Bull., v. 79, p. 827–854.

LAND, L. S., 1964, Eolian cross-bedding in the beach dune environment, Sapelo Island, Georgia: Jour. Sed. Petrology, v. 34, p. 389–394.

LOGVINENKO, N. V., AND REMIZOV, I. N., 1964, Sedimentology of beaches on the north coast of the Sea of Azov, p. 245–252 *in* Straaten, L. M. J. U. van, ed., Deltaic and shallow marine deposits. Amsterdam, Elsevier Pub. Co., 457 p.

McBRIDE, E. F., AND HAYES, M. O., 1962, Dune cross-bedding on Mustang Island, Texas: Am. Assoc. Petroleum Geologists, v. 46, p. 546–551.

McCUBBIN, D. G., 1965, Cretaceous strike-valley sandstones, northwestern New Mexico (abs.): Am. Assoc. Petroleum Geologists Bull., v. 49, p. 349.

———, 1969, Cretaceous strike-valley sandstone reservoirs, northwestern New Mexico: Am. Assoc. Petroleum Geologists Bull., v. 53, p. 2114–2140.

McKEE, E. D., 1957, Primary structures in some Recent sediments: Am. Assoc. Petroleum Geologists Bull., v. 41, p. 1704–1747.

———, 1966, Structures of dunes at White Sands National Monument, New Mexico (and a comparison with structures of dunes from other selected areas): Sedimentology, v. 7, p. 1–69.

———, AND TIBBITTS, G. C., 1964, Primary structures of a seif dune and associated deposits in Libya: Jour. Sed. Petrolegy, v. 34, p. 5–17.

MILLING, M. E., AND BEHRENS, E. W., 1966, Sedimentary structures of beach and dune deposits, Mustang Island, Texas: Inst. Marine Sci. Pub., v. 11, p. 135–148.

PANIN, N., 1967, Structures des depots de plage sur cote de la Mer Noire: Marine Geology, v. 5, p. 207–219.

PENTTILA, W. C., 1964, Evidence for the pre-Niobrara unconformity in the northwestern part of the San Juan Basin: Mtn. Geologist, v. 1, p. 3–14.

SABINS, F. F., JR., 1963, Anatomy of stratigraphic trap, Bisti field, New Mexico: Am. Assoc. Petroleum Geol. Bull., v. 47, p. 193–228.

SHEPARD, F. P., 1963, Submarine Geology, New York, Harper & Rowe, 557 p.

SOLIMAN, S. M., 1964, Primary structures in a part of the Nile delta sand beach, p. 379–387 *in* Straaten, L. M. J. U. van, ed., Deltaic and shallow marine deposits: Amsterdam, Elsevier Pub. Co., 457 p.

STOKES, W. L., 1968, Multiple parallel-truncation bedding planes—a feature of wind-deposited sandstone formations: Jour. Sed. Petrology, v. 38, p. 510–515.

STRAATEN, L. M. J. U. VAN, 1959, Minor structures of some Recent littoral and neritic sediments: Geologie en Mijnbouw, v. 21, p. 197–216.

27

Reprinted from pp. 209, 211–223 of *Carboniferous of the Southeastern United States*, G. Briggs, ed., Geol. Soc. America Spec. Paper 148, 1974, 361 pp.

Beach- and Barrier-Island Facies in the Upper Carboniferous of Northern Alabama

David K. Hobday
Department of Geology
University of Natal
Pietermaritzburg, South Africa

ABSTRACT

A rock model for the upper Carboniferous orthoquartzitic sandstone complexes of northern Alabama is constructed through examination of outcrops which are large enough to show the patterns of both vertical and horizontal variation. This rock model is never present in its entirety in any single exposure, but all outcrops of the orthoquartzitic sandstone can be explained in terms of their relation to the model. The characteristics of the model are consistent with modern shoreline features and establish that the sandstone bodies are representative of beach- and barrier-island situations. Paleogeographic reconstruction, utilizing seaward directions as indicated by the beach-barrier complexes, reveals two directions of sediment influx, one from the northeast and one from the south. These two systems merged in north-central Alabama, causing the sea to withdraw toward the northwest.

[*Editor's Note:* Material, including figure 1, has been omitted at this point.]

FORMULATION OF A ROCK MODEL FOR THE
ORTHOQUARTZITIC SANDSTONE

A preliminary reconnaissance of northern Alabama revealed seven major outcrops of orthoquartzites and associated rocks. These outcrops (Fig. 2) were selected because they are large and continuous, relatively fresh, and aerially distributed so as to allow observation of both vertical and horizontal arrangement of rock types. Other outcrops were examined, but relatively little additional information was obtained.

Of the major exposures, Rockledge exhibits the greatest variation in vertical rock sequences and most clearly displays their lateral facies relations. The rocks in the lower part of this outcrop are assigned to the Parkwood Formation in a sequence consisting of green and gray shale which becomes siltier upward. The sequence is capped by 20 ft. of massive graywacke sandstone. Resting upon the upper surface of this sandstone are thin beds of ferruginous sandstone containing crinoid stems, small brachiopods, and trilobite fragments. The overlying thin, gray shale unit is disconformably overlain by a complex of lenticular orthoquartzitic sandstone beds. This sandstone complex, over 60 ft. thick in parts, pinches out less than half a mile to the northeast where it truncates the underlying dark shale. Vertical repetition of sedimentary structures, minor erosional breaks, and shale partings within the sandstone indicate that this complex consists of several intricately related sequences lying one upon the other (see units I, II, and III of the P1 unit on Fig. 3).

Four distinct varieties of sedimentary structures are present within this orthoquartzite complex. The northeastern ends of I, II, and III are composed mainly of sets of long, nearly parallel laminated beds that dip southwestward at angles of less than 10 degrees. Multiple planes of low-angle truncation are common between these beds. These parallel-bedded units are laterally transitional into trough cross-bedded sand-

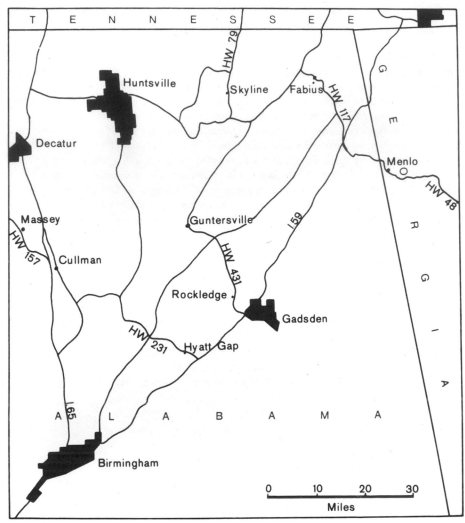

Figure 2. **Map indicating location of seven major exposures at Rockledge, Skyline, Hyatt Gap, Fabius, Cullman, Menlo, and Massey.**

stones, which are inclined in a direction more or less perpendicular to the low-angle parallel beds. Well-defined channels containing massive sandstone are most numerous within the upper portions of the complex (see D at top of P1 unit on Fig. 3). These channels trend both subparallel and normal to the direction of inclination of the planar cross-beds.

At the southwestern end of the outcrop, the II and III sandstone units thin abruptly and interfinger with green-gray siltstones containing discontinuous orthoquartzite beds from 1 to 18 in. thick. The thicker sandstone tongues are commonly cross-bedded, but the thinner units are made up of horizontal layers which have either planar or undulating contacts.

A considerably smaller lenticular orthoquartzite body, designated P2 on Figure 3,

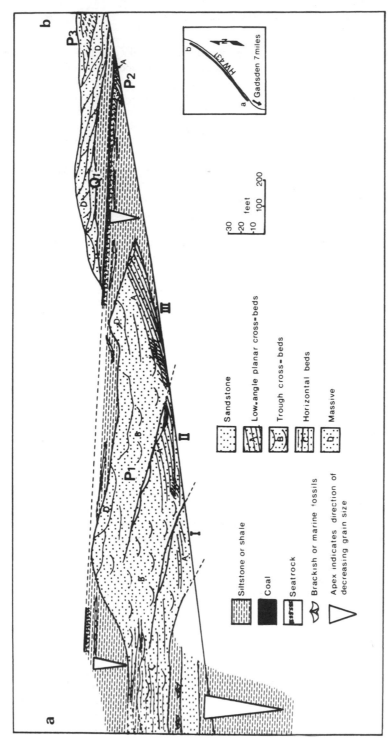

Figure 3. Exposures of P and Q facies at Rockledge, Alabama. Three occurrences of the P facies are shown at P1, P2, and P3. The single occurrence of the Q facies is indicated at Q1.

339

is at approximately the same level as the upper portions of P1. Its northern edge truncates a sequence of dark gray silty shales along a surface which is slightly concave upward and inclined southward at a low angle. At the other extremity, the sandstone interfingers with shale and the uppermost tongues of sandstone extend farthest south. Long, low-angle subparallel beds lie upon the truncated surface and grade southward into trough cross-bedded sandstone. The upper surface of the sandstone is flat, root-marked, and overlain by 2 in. of coal. Units P1 and P2 are thus very similar in terms of their anatomical attributes. Apart from scale, the only marked difference lies in the configuration of the upper surface which, in the case of P1, is not flat but slopes downward toward both interfingering and truncating margins.

The orientation of P1 relative to beds containing marine fossils at the southwestern end of the outcrop and dark shales containing plant fossils at the northeastern end suggests a land-sea transition with sand bodies representing the shoreline zone. If this interpretation is correct, then the coal-capped upward-coarsening sequence of shale and sandstone between P1 and P2 represents an impounded lagoon with a small beach, P2, also facing in a seaward direction.

Resting disconformably upon dark gray shale which overlies P2 is the unit designated Q1 on Figure 3. It consists of en echelon lenses of massive sandstone and laminated beds which are inclined northeastward, directly opposite to the inclination direction of similar structures in units P1 and P2. The contact between Q1 and overlying beds, labelled P3, is an erosional surface. P3 contains sedimentary structures similar in character and orientation to those in P1 and P2 below and is overlain by silty shales and graywacke sandstones which are attributed to a fluviodeltaic environment.

The designation of P and Q units in the Rockledge exposure is based on differences in direction of inclination of the units. Thus, those which are directed predominantly southwestward (in the direction of the marine facies) are termed P, whereas those inclined northeastward are Q. At this outcrop, there is no clear-cut lithic differentiation between P and Q.

At Skyline (Fig. 4), P facies are present, but Q rocks are equally abundant and can be differentiated from P on purely lithic properties. As at Rockledge, the P unit, consisting of low-angle planar cross-beds, high-angle planar and trough cross-beds, and massive sandstones, interfingers southward with green shales and marine limestone. Sparsely fossiliferous, horizontal beds of very fine-grained sandstone dominate the lower parts of the Skyline exposure. A coal bed rests upon the upper surface of the P facies and increases in thickness from 1 to 3 ft. from the southern to the northern part of the outcrop.

The overlying Q unit includes some siltstone and shale but is dominated by very extensive parallel-laminated sandstone beds inclined northward at a low angle. The multiple planes of low-angle truncation, which are typical of the P units, are entirely absent, and the orthoquartzite beds, commonly interlayered with siltstone or carbonaceous shale, are generally conformable. There is considerable admixture of massive sandstone lenses and trough and planar cross-bedded sets, and most of the trough cross-beds are inclined northward in conformity with the extensive planar cross-beds. This is in marked contrast to their most common relations within the P units at Rockledge where trough cross-beds and planar cross-beds tend to be mutually perpendicular. The upper part of the succession is truncated by a series of channels up to 8 ft. deep (D on Fig. 4) which trend westward at right angles to the mean direction of maximum inclination of the other sand beds.

Thus, the Q facies in the Skyline outcrop consist of a series of beds inclined away

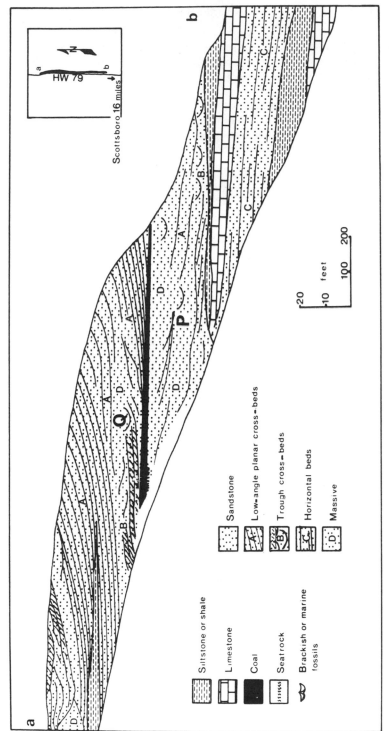

Figure 4. Exposure of P, Q, and shallow marine facies near Skyline, Alabama.

341

from the marine facies and in the direction of thickening of intercalated wedges of coal or carbonaceous siltstone. If the P facies represents a system of seaward-facing beaches, the Q facies appears to be the product of landward overwash consisting of tidal deltas and washover fans. Combining the data from Rockledge and Skyline, a preliminary model can be constructed in which the P facies are similar but the Q facies differ (Fig. 5). At Rockledge, a variable thickness of dark shale intervenes between the backslope of the underlying P1 unit and the base of Q1, but at Skyline only a coal bed separates the two units. The Rockledge Q1 unit is terminated upward by the development of P3, but at Skyline it is vertically more persistent.

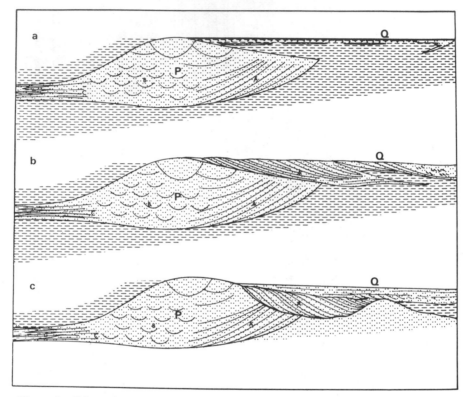

Figure 5. **Schematic model indicating seaward-inclined P facies and variations in Q facies as evidenced at Rockledge, Skyline, and Massey. (a) Backslope of P facies overlapped by upward-coarsening shales, siltstones, and thin sandstones capped by seatrock and coal; for example, Rockledge. (b) Backslope of P overlapped by extensive landward-inclined sandstone beds interfingering with siltstones and coals; for example, Skyline. (c) Backslope of P, markedly channeled and overlain by landward-inclined large-scale cross-beds.**

The character of the shaly marine components is exhibited to a greater degree at the Hyatt Gap exposure (Fig. 2). Silty shales are encountered at five different levels in the lower portions of the Hyatt Gap outcrop, and at each level they become coarser upward by an increase in the average grain size and by the progressive intercalation of massive, cross-bedded sandstones. The upper sequences are more condensed, have thicker sandstones between them, and are 8 to 20 ft. thick as compared with 50 ft. in the lowermost exposed sequence. Occasional marine fossils are encountered within some of the sandstones, and the siltstones are commonly bioturbated within the upper portions of each sequence.

The entire succession is capped by two thick orthoquartzite complexes, both of which are consistent with the P component of the model. The lowermost orthoquartzite is characterized by a very conspicuous development of trough cross-beds, trending perpendicular to, and all but obscuring, the low-angle, seaward-inclined planar cross-beds. Above this lowermost orthoquartzite, the upper portion of the Hyatt Gap exposure differs from previous outcrops in that the horizontally bedded sandstones with associated marine shales (which elsewhere are present in the lower frontal portions of the P component) tend to overlap and interfinger with the trough cross-bedded and low-angle planar-bedded sandstones to a greater degree than previously observed. This produces an additional possibility to the Rockledge-Skyline model (Fig. 5).

Further confirmation and modification of the model comes from the Fabius outcrop (Fig. 2), where the P facies are recognizable in the upper portion of the exposure. Here the low-angle planar cross-beds are cut perpendicular to their mean southward direction of dip by large channels up to 20 ft. deep. Some channels contain massive sandstone, whereas others are filled from the northern side by thick sets of low-angle beds, which resemble the beta and epsilon cross-beds of Allen (1963). Rafted coals are common, and there is a tendency for the larger of these to rest against the steeper channel walls.

In the southern part of the Fabius exposure, the orthoquartzites overlie partially slumped siltstones containing three thin coal beds with seatrock. These siltstones merge southward over a distance of several hundred feet into silty shales with sideritic concretions containing marine fossils. One specific example of this lateral transition is a thin coal with well-developed seatrock that passes southward into a 4-in.-thick band of sideritic clay ironstone overlying a clay devoid of root marks. Thus, the Fabius outcrop provides a modification of the model in terms of the presence of coaly facies underlying the P component.

Near Cullman on Interstate Highway 65 (Fig. 2), two P-type orthoquartzite complexes provide additional information concerning the three-dimensional geometry of the sandstone bodies. The sandstone bodies are viewed approximately in strike section, and their relatively large extent in this direction suggests that the P bodies are essentially linear, with the low-angle planar cross-beds dipping at right angles to the longer dimension. This outcrop further shows that the ends of these linear bodies interfinger laterally with dark shale.

Another exposure which reveals a strike section of the P facies is located at Menlo, Georgia (Fig. 2). The lower 300 ft. of this outcrop consists of upward-coarsening sequences, between 8 and 40 ft. thick, composed of shales and siltstones with minor sandstones. Overlying these is a vertical sequence of eight orthoquartzite bodies averaging 10 ft. in thickness, separated by siltstone averaging 1 ft. in thickness. The sandstones are both massive and low-angle planar cross-bedded and are cut into in their upper part by channels up to 15 ft. deep, which contain bedding types similar to those in the unchanneled portions. This sequence of sandstones and intercalated siltstones seems to represent the landward edges of longitudinally exposed P units. Separated from the top of these sandstones by 50 ft. of silty shale are two superimposed orthoquartzite complexes which clearly illustrate the mutually perpendicular relation between the trough cross-beds and long, low-angle planar cross-beds. The overlying upward-coarsening "bay fill" sequence of siltstones, with horizontally bedded sandstones and a highly carbonaceous shale on top, is capped by low-angle planar cross-bedded sandstones—the characteristics of which are typical of the inner portion of the P component of the model.

The last of the major outcrops is located 5 mi. from Massey in northwestern Alabama (Fig. 2). As at previous outcrops, the P component overlies marine facies but in this case is directed toward the south. The upper surface of this sandstone is deeply channeled, and its landward (or northern) end is occupied by a deep channel filled by a single 20 ft. set of inclined beds of alternating sandstone and shale. These beds dip in an opposite direction to the low-angle planar units in the underlying P sandstone and not only fill the channel but overlap underlying sandstones and shales farther to the north. These inclined beds thus appear to be another form of the Q facies, but unlike those at the Skyline exposure, the landward-directed sandstone deposits lie upon shale and sandstone rather than coal.

ROCK MODEL

All of the attributes of the seven major outcrops can be summarized in a single three-dimensional rock model which consists predominantly of two basic facies, P and Q. The P components are linear bodies, ranging in thickness from 10 to 40 ft. and are commonly superimposed one upon the other, with discontinuous siltstone beds along the contacts. In transverse section, one end of the lenticular P component truncates dark gray, silty shale which contains abundant plant remains. The other side grades into marine facies with considerable variability in the mode of transition. At some localities, the seaward side of the P component slopes abruptly downward, and the thin protruding tongues of sandstone grade outward from planar and trough cross-beds to horizontally bedded varieties and become interbedded with siltstone or shale. Elsewhere, these alternating horizontal beds interfinger with the main thickness of the P component at a higher level, or even overlap it, as is the case at Hyatt Gap. In a lengthwise direction, the P facies sandstones intertongue with dark shale.

The P bodies are cut by channels which tend to be oriented either parallel or perpendicular to the sandstone bodies. These channels, which are most abundant in the upper portions of the complex, range in depth from 2 to more than 20 ft. and are most commonly filled with apparently structureless sandstone. Others are filled from one side by low-angle planar cross-beds or, less commonly, by sigmoidal cross-beds, with or without silty partings.

Figure 6 represents a generalized plan view of the model and indicates how the seven major exposures relate to the model. Internally laminated, low-angle planar cross-beds predominate toward the landward margins of the P component and are inclined in the same direction as the dip of the lower contact of sandstone on shale. The proportion of trough cross-beds increases in the direction of seaward thickening of the P facies. Where the low-angle planar cross-beds interfinger with trough cross-beds, both varieties are inclined seaward, but farther seaward average dip direction of trough cross-beds becomes perpendicular to the dip of the planar cross-beds upon which they are transversely superimposed. The trough cross-beds become progressively more dominant seaward, with planar cross-bed surfaces widely spaced and indistinct. In the frontal portions of the P facies, these cross-beds are succeeded seaward by horizontal beds, with both rippled and planar surfaces. The rippled variety is more common and is present as fine- to very fine-grained sandstone, which is frequently silty and calcareous. These sandstones are bioturbated with varying degrees of intensity, and a ubiquitous characteristic is the abundance of tracks and trails on the undulating bedding planes. Massive sandstone beds, often very poorly sorted and probably heavily bioturbated, are commonly encountered in these frontal portions. The outermost P facies interfinger seaward with red, green, and gray marine shales and, occasionally, with oölitic and bioclastic limestones.

The characteristics of the more variable Q component, which overlaps the backslope of P in a landward direction, are not as precisely known. Some of the observed variations are illustrated in Figure 5. At Rockledge, it consists of a series of landward-inclined planar beds and lenses of massive sandstone, beneath which are dark shales, a coal bed, and an upward-coarsening shaly sequence in downward succession. At Skyline, the shales are absent, and the backslope of the P unit is directly overlain by a coal bed, above which are the landward-inclined beds of the Q facies. The dips on these inclined beds are very gentle, and some individual beds persist laterally for distances of more than 50 ft. Bedding plane partings of clay shale and carbonaceous siltstone thicken down the lower foresets which merge into alternating beds of siltstone and sandstone. Low-angle planar cross-beds and trough cross-beds are both inclined in the same direction, except within channels which trend perpendicularly to the inclination of the planar cross-beds. These channels become progressively larger in a landward direction and obliterate parts of the lower foresets of the planar cross-beds. At Massey, the Q facies are remarkably similar to Skyline, but differ in that the beds occupy a deeply scoured channel and override, in a landward direction, a large mound of trough cross-bedded sandstone. At both Skyline and Massey, an intercalated wedge of dark silty shale thickens markedly landward.

"Bay fill" sequences, which underlie the P facies, or interfinger with their seaward-most portions, generally consist of shales at the base which coarsen upward and contain massive or horizontally bedded sandstone in their upper portions. Certain of these sequences, as at Fabius, Cullman, and Menlo, have a seatrock and a coal bed at the top and therefore are nonmarine, whereas at the remaining four exposures, the upward-coarsening sequences beneath the major orthoquartzites are essentially marine.

INTERPRETATION

The seven major exposures of the upper Carboniferous orthoquartzites in northern Alabama have shown sufficient features in common to formulate a generalized rock model, yet have also displayed enough variation to indicate important modifications. The principal components of the model are designated P and Q and are identified primarily on the seaward or landward directions of inclination of the accretion planes. In addition to these are "bay fill" sequences, which represent finer grained equivalents of the sandstone units.

The landward side of P truncates the underlying sediments, which most frequently are dark silty shales. This suggests that wave scour of an older deposit, generally marsh, was followed by deposition of winnowed quartz sand on the truncated surface. Gould and McFarlan (1959) invoke a similar explanation for the initial development of the chenier beach-ridge deposits of southwestern Louisiana. The low-angle planar cross-

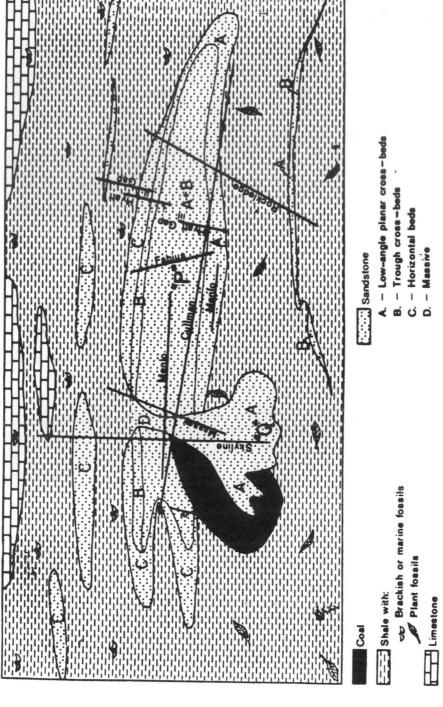

Figure 6. Plan view of schematic rock model indicating positions of major sections as related to the model.

Coal

Shale with:
🐚 Brackish or marine fossils
🌿 Plant fossils

Limestone

Sandstone
A. – Low-angle planar cross–beds
B. – Trough cross–beds
C. – Horizontal beds
D. – Massive

346

beds, which predominate toward the inner truncating margins of the sandstone bodies, can be attributed to deposition on the upper beach foreshore under the delicate balance between swash and backwash (Thompson, 1937; McKee, 1957). The dips, ranging between 3 and 8 degrees, are a measure of the foreshore inclination, and multiple surfaces of low-angle truncation indicate changes in beach profile as documented by Psuty (1966).

Away from the truncating landward margins of the P component are high-angle planar and trough cross-beds which, in the zone of initial interfingering with the low-angle planar cross-beds, are inclined seaward in accordance with the dip of the accretion planes. The sandstones of this transitional zone are generally the coarsest grained of the entire sequence, and contain abundant well-rounded quartz pebbles. These attributes are consistent with observations on certain modern beaches. In the foreshore zone, backwash currents carry particles seaward until they are checked by the uprush of the next wave. The smallest particles are swept up the beach again, but the largest settle and accumulate seaward until they reach a depth where they cannot be moved landward by wave action. "The result is a seaward facing steplike deposit whose upper surface is a continuation of the beach face, and whose outer surface is the angle of repose of the sand" (Bascom, 1964, p. 207). A substantial tidal range would create more than one step, and it is possible that the seaward migration of these steps, concomitant with beach progradation, may have been responsible for the interfingering relations of these two seaward-inclined bedding types.

Farther seaward, trough cross-beds predominate, and the direction of dip of these beds is approximately perpendicular to the seaward-dipping foreshore accretion surfaces into which they are scoured. This arrangement was probably in response to the longshore migration of linguoid or lunate ripples under the impetus of unidirectional currents (C. H. Moore, Jr., 1967, oral commun.). The superpositioning of these cross-beds on parallel, low-angle seaward-inclined surfaces suggests periodic destruction of the foreshore. These foreshore planes become more widely spaced and less distinct toward the frontal portions of the P facies, where the cross-bedded sandstones are succeeded seaward by horizontally bedded varieties.

The rippled, horizontal beds are attributed to offshore deposition where the rate of detrital influx was reduced but where biogenic effects were more pronounced. Burrows, tracks, and trails are particularly abundant at some localities, and marine fossils are occasionally encountered. Certain lenticular sandstone beds have been so effectively bioturbated that no internal structure is visible. The oölitic and bioclastic limestones which sometimes interfinger with the frontal P facies appear to have accumulated as offshore carbonate banks on an open shelf.

One of the major varieties of the Q facies, consisting of shales grading upward into siltstones and capped by thin sandstones, a seatrock and coal, is interpreted as being of lagoonal origin. Graded sandy siltstone laminae suggest settling from suspension in quiet water of poorly sorted sediments. Frequently associated with these are flaser beds, which appear to be the products of starved ripple formation (Coleman and Gagliano, 1965). Abundant coalified plant remains, thought to have been contributed from nearby marshes and swamps, are present throughout, but are more finely divided within the lower shaly portions. The orthoquartzite sandstones within these lagoonal sequences appear to have resulted from two distinct processes. Certain landward-thinning beds appear to have been derived from sands washed over the barrier crest into the lagoon, and some of the thicker units possibly were formed in tidal channels and small tidal deltas. Other orthoquartzite beds originated from the winnowing of sediments of the landward shore of the lagoon (P2 at Rockledge), and the capping seatrock and coal bed at this locality indicate ultimate filling of the lagoon to a level where vegetation could take hold.

Another major variation of the Q facies consists of extensive landward-inclined sandstone beds with finer grained, and sometimes carbonaceous, partings. Intercalated cross-beds are generally inclined in the same direction. These sandstones are thought to have originated as tidal deltas or washover fans. At certain localities they blanket landward-thickening marsh peats and lagoonal shales, but elsewhere they occupy deeply eroded channels backward of the barrier. Channels, trending parallel to the shoreline, are progressively larger landward, probably as a result of the confinement of back-barrier current flow by the encroaching sediments.

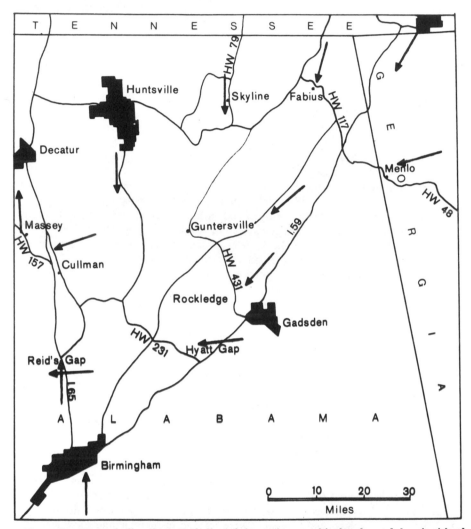

Figure 7. Seaward directions as indicated by orthoquartzitic beach- and barrier-island complexes.

PALEOGEOGRAPHIC RECONSTRUCTION

The rock model, and its genetic interpretation as a barrier-island system, provides a basis for paleogeographic reconstruction in terms of barrier buildup and progradation. The south to southwestward directions of seaward retreat (Fig. 7) are in agreement with several orthoquartzite complexes in eastern Tennessee (R. C. Milici, 1967, oral commun.), but the directions at Hyatt Gap, Cullman, and Menlo are slightly more westward. In marked contrast, the barrier system represented in the Massey outcrop is directed northward, a trend which accords with orthoquartzite complexes farther south in the Birmingham area. These patterns suggest two distinct sources, one in the northeast and the other to the south. The Cullman and Massey exposures apparently reflect these two different directions of sedimentary influx and are only 8 mi. apart. However, at neither locality is the shaly sequence beneath the orthoquartzites very thick nor dominated by sediments related to an earlier developed deltaic or barrier complex. Thus, it would appear that the barriers developed more or less synchronously and that the two sedimentary systems merged in north-central Alabama, expelling the sea toward the northwest.

The character of the sediments between the thick limestones and the base of the orthoquartzites indicates the nature of the platform upon which the barrier complexes developed. Where coals and graywacke sandstones intervene, the initial development of a delta system, followed by partial reworking and the superimposition of beach barriers, is suggested. On the other hand, orthoquartzites resting directly upon marine facies indicate the development of barriers in the absence of any prior major deltaic influx. Thus, the Rockledge, Fabius, and Menlo exposures, together with the lower half of Hyatt Gap, represent the former situation, that is, barriers associated with delta abandonment and decay. Elsewhere, barriers built directly out over a shallow marine platform.

ACKNOWLEDGMENTS

This paper is based on a dissertation, supervised by John C. Ferm, at Louisiana State University. Others who helped during the study were James M. Coleman, Sherwood M. Gagliano, George F. Hart, Clyde H. Moore, and James P. Morgan. The field work was made possible by Louisiana State University Geology Department summer grants. The Alabama Geological Survey provided financial assistance and allowed free access to certain data. Alabama state geologists Charles Copeland, Jim Drahovzal, and Thornton Neathery were a source of considerable help and encouragement. Helpful suggestions in preparation of this manuscript were provided by Garrett Briggs, John Ferm, Richard Inden, and Robert Milici.

David K. Hobday

REFERENCES

Allen, J. R. L. 1963. The classification of cross-stratified units, with notes on their origins. *Sedimentology* **2**:93–114.

Bascom, W. 1964. *Waves and Beaches; the Dynamics of the Ocean Surface*. Garden City, N.Y.: Anchor Books, 260pp.

Coleman, J. M., and S. M. Gagliano. 1965. Sedimentary structures—Mississippi River deltaic plain. In G. V. Middleton, ed., *Primary Sedimentary Structures and Their Hydrodynamic Interpretation*. Soc. Econ. Paleontologists and Mineralogists Spec. Pub. 12, pp. 133–148.

Gould, H. R., and E. McFarlan, Jr. 1959. Geologic history of the chenier plain, southwestern Louisiana. *Gulf Coast Assoc. Geol. Soc. Trans.* **9**:261–270.

McKee, E. D. 1957. Primary structures in some Recent sediments (U.S. and Mexico). *Am. Assoc. Petroleum Geologists Bull.* **41**:1704–1747.

Psuty, N. P. 1966. *The Geology of Beach Ridges in Tabasco, Mexico*. Louisiana State Univ. Coastal Studies Inst. Tech. Rept. 30.

AUTHOR CITATION INDEX

SUBJECT INDEX

Abandonment phase, 166, 214, 254, 257, 262, 267. *See also* Destructive deltas; Transgressive phase

Accretion
  lateral, 36–37, 39, 66, 111, 113–116, 126, 134
  vertical, 9, 10, 35–38

Aerial photographs, 111, 127, 133–134, 330

Aggradational phase, 213–214, 254–255. *See also* Constructive deltas

Alabama, northern, 288, 336–350

Alaska, 49

Alberta, 213

Algodones dunes, 187, 189

Allegheny Formation, 260–268

Alluvial environments, 24–143
  channel dimension, 38–39, 44, 50, 65–67, 70–71, 73–75, 111, 114–115, 119, 120–121, 124, 129–130
  dependent variables, 50, 61, 64–78, 110
  derived functions, 61, 67, 110, 113
  discharge, 38–39, 73, 114–115, 121, 137
  flow strength, 50, 73–78
  friction factor, 44, 50, 71–73
  grain size, 50, 65, 120
  length, 111, 116, 123–124
  recognition
    outcrop, 218–225
    subsurface, 228–233
  slope, 44, 50, 67, 74–75, 78, 111, 115, 121–122, 124
  velocity, 38, 111, 114–115, 122

Alluvial-fan facies, 8–23, 41–47, 169. *See also* Bajada

Alluvial fans
  ages
    Archean, 8
    Cenozoic, 9
    Cretaceous, 9
    Devonian, 8–9, 12–23
    Paleozoic, late, 9

    Precambrian, late, 9
    Triassic, 8, 10, 41–47
  formations
    Molasse, 8
    Mount Toby, 10, 41–47
    New Red Sandstone, 9
    Peel Sound, 9, 12–23

Alluvial-plain deposits
  ages
    Carboniferous, 49, 110–111, 113–116
    Cenozoic, 48–49, 110
    Cretaceous, late 111, 117–125
    Devonian, 48–50, 53–82, 83–97, 98–109
    Jurassic, 126–131
    Miocene, 111, 132–141
    Paleozoic, late, 49
    Permian, 49
  formations
    Ferron, 117–125
    Molasse, 48–49
    Old Red Sandstone, 48–49
    Peel Sound, 98–109
    Red Marl, 83–97
    Scalby, 126–131
    Wood Bay, 50, 53–82

Alluvial-plain facies, 48–167, 220–225, 227–233. *See also* Delta-plain facies; Finning-upward cycles; Lacustrine facies; Meandering-river facies

Analcime, 144, 150

Antidunes, 10, 41–47, 72

Appalachians, central, 213–214, 260–268, 287

Archean, 8, 287

Arctic Canada, 9, 12–23, 50, 98–109, 213

Arid Climate, 15–16, 168, 202, 208

Avalanche structures, 185, 188

Avulsion, 96. *See also* Crevasse-splay deposits

Axial cycles, 255

About the Editor

FRANKLYN B. Van HOUTEN is Professor of Geology at Princeton University, where he has taught since 1946. He completed his undergraduate work at Rutgers University in 1936, and received the Ph.D. degree from Princeton University in 1941.

Professor Van Houten's research has focused largely on sediments that accumulated in continental realms, especially in mountain belts and rift systems. He has also investigated red beds and the nature of the iron oxides minerals in them, as well as the occurrence of zeolites and clay minerals in sedimentary rocks.

He is currently serving as Chairman of the Awards Committee of Sigma Xi, and is an honorary member of the Society of Economic Paleontologists and Mineralogists and of the Colombian Geological Society.